Urban Agriculture

Series editors
Christine Aubry, AgroParisTech, INRA UMR SADAPT, Paris, France
Éric Duchemin, Université du Québec à Montréal Institut des Science de Environment, Montreal, Québec, Canada
Joe Nasr, Centre for Studies in Food Security, Ryerson University, Toronto, Ontario, Canada

The *Urban Agriculture* Book Series at Springer is for researchers, professionals, policy-makers and practitioners working on agriculture in and near urban areas. Urban agriculture (UA) can serve as a multifunctional resource for resilient food systems and socio-culturally, economically and ecologically sustainable cities.

For the Book Series Editors, the main objective of this series is to mobilize and enhance capacities to share UA experiences and research results, compare methodologies and tools, identify technological obstacles, and adapt solutions. By diffusing this knowledge, the aim is to contribute to building the capacity of policy-makers, professionals and practitioners in governments, international agencies, civil society, the private sector as well as academia, to effectively incorporate UA in their field of interests. It is also to constitute a global research community to debate the lessons from UA initiatives, to compare approaches, and to supply tools for aiding in the conception and evaluation of various strategies of UA development.

The concerned scientific field of this series is large because UA combines agricultural issues with those related to city management and development. Thus this interdisciplinary Book Series brings together environmental sciences, agronomy, urban and regional planning, architecture, landscape design, economics, social sciences, soil sciences, public health and nutrition, recognizing UA's contribution to meeting society's basic needs, feeding people, structuring the cities while shaping their development. All these scientificfields are of interest for this Book Series. Books in this Series will analyze UA research and actions; program implementation, urban policies, technological innovations, social and economic development, management of resources (soil/land, water, wastes…) for or by urban agriculture, are all pertinent here.

This Book Series includes a mix of edited,coauthored, and single-authored books. These books could be based on researchprograms, conference papers, or other collective efforts, as well as completedtheses or entirely new manuscripts.

More information about this series at http://www.springer.com/series/11815

Francesco Orsini • Marielle Dubbeling
Henk de Zeeuw • Giorgio Gianquinto
Editors

Rooftop Urban Agriculture

RUAF Foundation
Resource Centres on Urban Agriculture & Food Security

ResCUE-AB

Alma Mater Studiorum
Università di Bologna
A.D. 1088

Springer

Editors
Francesco Orsini
Department of Agricultural Sciences
University of Bologna
Bologna, Italy

Henk de Zeeuw
RUAF Foundation
Leusden, The Netherlands

Marielle Dubbeling
RUAF Foundation
Leusden, The Netherlands

Giorgio Gianquinto
Department of Agricultural Sciences
University of Bologna
Bologna, Italy

ISSN 2197-1730 ISSN 2197-1749 (electronic)
Urban Agriculture
ISBN 978-3-319-57719-7 ISBN 978-3-319-57720-3 (eBook)
https://doi.org/10.1007/978-3-319-57720-3

Library of Congress Control Number: 2017954386

Printed on acid-free paper

This Springer imprint is published by Springer Nature
The registered company is Springer International Publishing AG
The registered company address is: Gewerbestrasse 11, 6330 Cham, Switzerland

Preface

> ...We are faced today with a grave threat, not one solely based on the fact that we don't have answers to burning problems in society, but even more to the point that we don't possess a clear apprehension of what the main problems are and clear understanding of their real dimensions... [Slavoj Zizek, Slovenian Philosopher, 13 May, 14:35, HRT "Nedeljom u Dva" ("Sunday at 14:00PM," Croatian Television, HRT 2012)]

Food Urbanism and Beyond: Toward a Sustainable Urban Agriculture

The complexities of contemporary global urban, political, economic, and environmental issues are evident. It is not hyperbole to say that we human beings are now confronted with the greatest challenge that we have ever faced; in fact, it is a matter of life and death. The planet has recently been experiencing a convergence of natural and man-made crises that are unprecedented in our lifetime. We are also facing the consequences of accelerating and rapid urbanization, the scarcity of natural resources and their mismanagement, the impact of major errors in our responses to disasters, and the increasing demand for and complexity of greatly expanding transportation flows. Our societies have also undergone rapid and radical shifts in terms of age and class, increasing the inequities between the rich and poor and intense demand for affordable and high-quality housing. All of these major challenges require immediate solutions from architects, urban planners, urban designers, landscape architects, horticulturists, and urbanists; actually, we need the combined efforts of all good people who are concerned with the physical condition and future of our cities. We need these professionals and experts to contribute their most imaginative, pragmatic, resilient, innovative, and just solutions.

As we have in many ways entered both the age of the "triumph of the city," where cities are at their peak performance in innovation, growth, culture, technology, urban expansion, opportunity, as well as competition, so have we also entered a "beyond the urban age" where cities also find themselves confronted by issues of justice and the equitable distribution of wealth, opportunity, and power to all people in society. We therefore need to start rethinking what "the good city" and "cities for all" should be. Thinking that slums or poor neighborhoods are appropriate (dense or splintered) urban forms that make a contribution to ecological footprint and sustainability through resource allocation and recycling is a dangerous path to take. Implying that being poor is ecologically sound is simply wrong, because it is a matter of pure necessity—not of ecological awareness or choice—to live in a decent and healthy urban environment.

According to new UN DESA report, "World Population Prospects: The 2015 Revision," the current world population of 7.3 billion is expected to reach 8.5 billion in 2030, 9.7 billion in 2050, and 11.2 billion in 2100, whereas China and India remain the two largest countries in the world, each with more than 1 billion people, each with growing megacities and representing 19 and 18 % of the world's population, respectively. But by 2022, the population of India is expected to surpass that of China. A "perfect storm" of food shortages, scarce water, and insufficient energy resources threatens to unleash public unrest, cross-border conflicts, and mass migration as people flee from the worst-affected regions, the UK former government's chief scientist, Professor Sir John Beddington, has warned recently. So in the midst of converging crises of climate change and beyond, epidemic diseases, decaying infrastructures, international terrorism and regional wars, economic collapses, rifts in society, uncontrolled migrations, and other calamities, *water scarcity and food security* truly remain the two major "perfect storms" that will hit us in the decades to come in the path to "long emergency." With such a surge in population, human agriculture exerts a tremendous toll on the planet, from water draws to pollution and from energy use to habitat loss.

We must now recognize that we have to, aside from innovative solutions in inner cities (farming the city at eye level, vertical farming, rooftop agriculture, etc.), start to reorganize the landscape for local food production, as industrial agriculture will be one of the prime victims of our oil predicament. The successful places in the future will be places that have a meaningful relationship with growing food close to home. In relation to that, clearing out the terminology when it comes to "growing food" is extremely necessary in order to understand the crux of the problem. The dichotomy between the idea of the dense cities and high-rises for growing food vs. the yard and countryside of growing food in smaller scales will remain. Both are needed but both need to be carefully reconsidered and thought about. How people will live in the countryside under the condition where their lives will be centered on growing and producing the food—the fundamental of the new agrarian village—remains to be seen; moving beyond intentional communities will be linked with issues of economy, demographics, ecology, societal structurations, and spatial transformations. As architects, urban designers, landscape architects, horticulturists, environmental engineers, urban ecologists, and urban planners of city landscapes, these professions hold a vital tool in the growth of communities centered on food production. Food is both a local and global issue. The lack of productive urban land, food insecurity, uncontrolled urban growth, the lack of stable local food markets, land use conflicts in the periurban areas, and a general lack of societal knowledge of food growing and preparation fuel these discussions from all sides.

Andrés Duany, the co-founder of the New Urbanism movement and the proponent of Traditional Neighborhood Development, has cleared up the terminology of "food and urbanism." Agrarian urbanism, as he explains, is different from both "agricultural retention" (deploys an array of techniques to save existing farms, including farmland trusts, greenbelts, and transfer of development rights), "urban agriculture" ("cities that are retrofitted to grow food"—the food produced is supported by distribution and processing systems such as farmers' markets, community kitchens, food cooperatives, and contracted restaurants), and "agricultural urbanism" ("when an intentional community is built that is associated with a farm—land is cultivated within existing cities and suburbs, sometimes using parcels in depopulated sectors"). Duany thinks bigger: "Agrarian urbanism is a society involved with the growing of food." Agrarian urbanism refers to settlements where the society is involved with food in all its aspects: *organizing, growing, processing, distributing, cooking, and eating it*. The concept is based on the English Garden City, Israeli kibbutz, 1960s commune, and US master-planned golf course community. Promoting the growth and vitality of these (current and future) urban agricultural spaces through coordinated policy, planning, and action across scales—from individual decision making to municipal planning to national and global policy—remains the grand task ahead.

The inner cities and the idea of "farming the city—urban agriculture" (as mentioned above in the "agrarian" terminology) through different approaches bring another set of complexities and also beg for clearance of terminology and approaches. A number of unresolved issues need to be addressed before we can consider "farming the city—urban agriculture" as a permanent solution to our (future) food needs. Although community gardens, kitchen gardens, organic micro-farming, and rooftop farming are very positive for a city's sustainability and for the well-being of its inhabitants, they are not without problems. Issues of lack of space and the rationale of farming in the middle of urban areas remain; the open question of high water requirements for agricultural activities is still not solved (nor is it for green lawns in suburbia); possible soil and water pollution that can lead to waterborne diseases and issues of inner city air pollution that are related to contaminated food and serious health problems are yet to be resolved fully; last but not least, the aesthetic issues linger but far more the "dark" rise and justification of the high-rise development. Some would say that "urban agriculture" was a new justifiable label used to "sugar-coat the pill to maintain conventional farming in the city" or to develop mega urban projects—like high-rise buildings—that otherwise would have been taken very badly by the people living in or on prospective sites. "Farming the city—urban agriculture" needs to be seen both as food (a tool for today's urbanization) and a resource (a tool for tomorrow's resilient post-urban age).

Whatever the case *(organic) urban agriculture*—also known as urban farming, pop-up food cultivating, guerilla farming, foodscaping, organic city repair, DIY guerilla gardening, and many other terms relating to agricultural practices in the middle of the city—is becoming a major activity in societies all over the world; also let us not forget the predecessors in the shape of allotment—community gardens. Urban agriculture provides many benefits, including food security for people in the city, a reduction of energy used in conventional agricultural practices and food service, a reduction of carbon footprints, and environmental services for cities in terms of providing open green space. All over the world, people are turning unused lots, backyards, and even rooftops into gardens. Imagine if this movement could grow so massive that cities would no longer have to depend on rural and suburban agriculture to produce food for their own citizens. Great projects such as HK Farm (Hong Kong), Brooklyn Grange (New York City), Dakakker (Rotterdam), City Farm (Tokyo), and Lufa Farms (Montreal) are some of the examples testimony to this. Testing the grounds for social change—citizen-generated alterations of the built environment that are intended to improve the public realm or put underutilized space in service to the community seem to be the calling of the day. When trying to determine if urban agriculture may contribute to a sustainable future, the primary question to ask is will this agriculture be at the service of the inhabitants? Brian Clark Howard of National Geographic sees rightly benefits of urban farming in that it can add greenery to cities, reduce harmful runoff, increase shading, diminish "food miles" associated with long-distance transportation, get the freshest produce, and counter the unpleasant heat island effect. Garden plots can help people reconnect with the Earth and gain a greater appreciation for where our food comes from.

After all, the primary reason for designing wonderful built environments is to improve the lives of people; thus, incorporating elements of psychology and sociology into our designs is powerfully beneficial. By engaging in experimentation, research, innovation, and intellectual synergy between urban design and applied social science, we will truly achieve an integrated and holistic approach to analyzing, understanding, planning, and designing our built environments. The quality and the livability of the urban environment in our cities, towns, districts, and neighborhoods are the deciding factors in the social, cultural, economic, and environmental performance of societies and the quality of life of all its citizens. We now stand on the threshold of the greatest challenge for our professions. This challenge is no less than improving people's lives through optimally designing their urban environments and sustaining life on our planet. This wonderful anthology *Rooftop Urban Agriculture* is certainly a step in the right direction.

Urban Planning and Urban Design, Tigran Haas
Centre for the Future of Places (CFP),
School of Architecture and the Built Environment
KTH—Royal Institute of Technology
Stockholm, Sweden

Contents

The original version of this book revised.
An erratum to this book be found at https://doi.org/10.1007/978-3-319-57720-3_21

Part I
The Status and Challenges of Rooftop Agriculture

Marielle Dubbeling

Introduction

Marielle Dubbeling, Francesco Orsini, and Giorgio Gianquinto

Abstract In a world characterised by growing urbanisation, urban agriculture is gaining relevance due to its potential for increasing resource efficiency, contributing to city food security and enhancing associated ecosystem and social services. In cities, however, spaces available for cultivation are limited, thus leading to the need to explore innovative growing solutions, such as cultivating building rooftops. Rooftop agriculture can also contribute to addressing specific city challenges such as climate change. Experiences are sprouting all over the world, scientific evidence on most suitable growing solutions, policies and potential benefits is growing. The present review will address the main features of rooftop agriculture, providing an interdisciplinary assessment of different approaches for development and the multi-faceted forms that rooftop agriculture may assume in different contexts, bringing together existing experiences as well as suggestions for planning of future sustainable cities.

Introduction

With the urban population having surpassed the rural one (Batty 2015), the relevance of urban food production is today commonly recognised among national and international bodies (Orsini et al. 2013; De Zeeuw and Drechsel 2016). Given the scarce land availability and high land costs in cities, different agricultural production and value chain intensification strategies are explored in a number of cities and towns across the world. These include: (1) optimising land/space rent of agricultural production by intensifying soil-based cropping and animal husbandry, developing non-soil based production systems (hydroponics, containers) and/or switch to above ground, building-borne systems (like rooftop gardening); (2) optimising

The original version of this chapter was revised.
An erratum to this chapter can be found at https://doi.org/10.1007/978-3-319-57720-3_21

M. Dubbeling (✉)
RUAF Foundation, Leusden, The Netherlands
e-mail: m.dubbeling@ruaf.org

F. Orsini • G. Gianquinto
Department of Agricultural Sciences, University of Bologna, Bologna, Italy
e-mail: f.orsini@unibo.it; giorgio.gianquinto@unibo.it

F. Orsini et al. (eds.), *Rooftop Urban Agriculture*, Urban Agriculture,
https://doi.org/10.1007/978-3-319-57720-3_1

income-adding value to agricultural production (including processing and direct producer-consumer relations); (3) optimising multiple urban functions of agricultural value chains (including recreation, landscape management and other functions) and (4) optimising resource utilisation, improving the spatial connectivity of agricultural activities (promoting waste-water re-use in agricultural production; better linking waste management, production, processing and marketing-promoting food hubs) (Mougeot 2010).

Urban agriculture is understood as the production of food, and related activities such as waste recycling, processing and marketing, in urban, suburban and peri-urban areas close to a city. Today's urban agriculture is about food grown on open spaces or water bodies (aquaculture) in and around the city, on farmlands in peri-urban areas, on backyards, roofs, windowsills, balconies and patios. It can take the form of household gardening, community gardening, social or commercial production. Resulting from its close interaction with the city environment, urban agriculture is inherently multifunctional and produces more than food. It also plays a role in waste recycling, community cohesion, education and learning, landscaping and climate resilience for example (De Zeeuw and Drechsel 2016).

In terms of urban land use planning, different types of urban agriculture suit different planning objectives (for example social development, urban food security, local economic development, climate change adaptation) and different types of land uses as described above. Rooftop agriculture is one specific form of urban agriculture. As described in the next chapter, rooftop agriculture is undertaken by a large range of stakeholders, applying a variety of production systems and technologies for a diversity of reasons and aims. Rooftop agriculture contributes to specific city planning and development objectives and is specifically suitable in certain circumstances. Most notably this includes optimisation of the use of urban space and addressing of specific climate change impacts.

Optimising Use of Urban Space

One of the greatest unused resources or capacities of cities is flat roofs, especially in denser and inner-city areas where other growing spaces may be lacking –or polluted- and city space is generally quite expensive.

As in many other cities, uncontrolled and rapid urbanisation in Kathmandu (Nepal) (Fig. 1) has resulted in a rapid decrease in agricultural land. Loss of these production areas, that traditionally provided Kathmandu city with rice, grains, vegetables, poultry and dairy, made it more vulnerable to disruptions in food supply. The city now has to depend on the produce of either rural areas or imports from India or China. The only major access road is often blocked due to floods or landslides, while the changing climate is likely to increase the frequency of such natural disasters. Protection and preservation of remaining peri-urban agricultural lands is deemed highly necessary to enhance city resilience. Next to this, the potential of using built-up spaces, and specifically rooftops, could provide an interesting

Fig. 1 Kathmandu: a built-up city with no space for growing? (Photo: P.S Joshi, UN Habitat-Nepal)

opportunity to grow food in inner-city areas, in addition to productive use of existing –but often limited- open land areas in cities for food production.

Use of rooftops may specifically provide an alternative to use of open inner-city areas that are contaminated because of former industrial use, waste dumps or other forms of pollution, provided that clean soil or other substrates are used for rooftop growing.

(Informal) research results indicate available rooftop areas of 880 ha within Melbourne city council boundaries for greening and sustainable projects. Respectively 236 ha could be used for intensive green vegetation and 328 ha of rooftop space for lighter vegetation (http://www.theage.com.au/victoria/new-maps-show-melbournes-unused-rooftops-are-ripe-for-greening-20151109-gku4yq.html). Moreover, according to Amsterdam Rooftop Solutions, in the Dutch capital alone the available rooftop space is 12 km^2 (http://www.smart-circle.org/smartcity/blog/smart-city-event-2015-smart-sightseeing/).[1] MacRae et al. (2010) identified the need for 1243 ha of rooftop growing space to meet the target of providing 10% of

[1] How much of this potentially available space is already used for other purposes (storage; laundry or recreational uses; water tanks or electricity units) or how much of this space is fully accessible taking safety concerns and load requirements into account is not indicated.

Toronto's fresh vegetable supply, or about 25% of the rooftop space theoretically identified earlier by researchers (Nasr et al. 2010).

New buildings can be built with agriculture roofs, or rooftop greenhouses. Old buildings can be rehabilitated using growing containers, soil-based or hydroponic systems (for more examples and specific cases see chapters "A Panorama of Rooftop Agriculture Types", "Soil Based and Simplified Hydroponics Rooftop Gardens" and Part V). Although technological solutions are available (chapter "Technology for Rooftop Greenhouses"), altogether with details on technological requirements for rooftop greenhouses (see part 3 for a discussion on rooftop agriculture management), policy regulations supporting rooftop agriculture is still remarkably absent.

Various rooftop policy tools have been used in cities like Toronto, Montreal (Canada), Melbourne (Australia) and several European cities. These include bylaws, density bonuses, incentive programs, grants, fees, and levies (usually related to stormwater runoff from buildings). The challenge is how to modify existing instruments to promote food production on rooftops. This will include design elements: food production usually requires deeper soil than required for green roofs; access to the roof: growers need daily access to the roof during the growing season and the capacity to readily move material up and down; insurance: wide applicability (looking at both retrofitting existing roofs as well as to new constructions and zoning). If modifying existing policies and programs proves too difficult, a specific food-production bylaw might have to be introduced (Nasr et al. 2010). See for a further discussion on policies chapter "Rooftop Farming Policy".

Addressing Climate Change

Rooftop agriculture is also specifically suitable for addressing climate change in densely built up human settlements (see also further chapters "Resource Efficiency and Waste Avoidance", "Community Building and Social Justice Aspects of Rooftop Agriculture" in this publication). By covering and protecting the roof from direct solar radiation (directly shading the building surface which would otherwise absorb heat), rooftop gardens can reduce heat flux into the building, thus increasing – in periods of high temperature – thermal comfort for rooms located directly under the rooftop. Green and agricultural roofs thus reduce heat transfer through the roof and also ambient temperatures on the roof surface, as concrete building mass also radiates the stored heat again to the environment.

Temperature effects on thermal comfort on the rooftop and in apartments below the rooftop are dependent on the percentage of rooftop area covered and of coverage throughout the year. As a general rule, 80% of the total rooftop surface needs to be covered throughout the year (but at least during periods with highest temperature) to have significant temperature effects.

Apart from having a direct impact on building temperature comfort and on ambient temperatures above the rooftop, rooftop gardens may also contribute to cooling the city. By covering a roof with a layer of vegetation, evapotranspiration provides

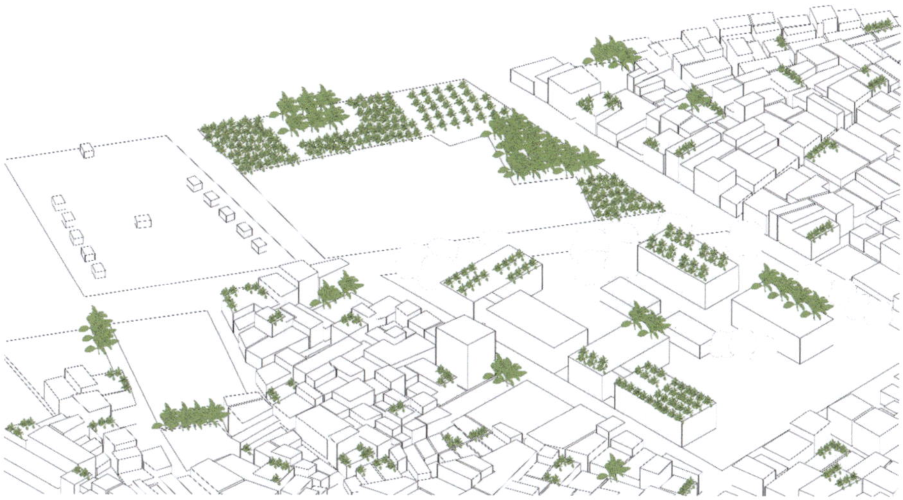

Fig. 2 Combined area approach, combing private and institutional rooftop farms (and possibly in future also open-space farming) (Source: PDP RUAF Foundation 2014)

cooling effects. This cooling effect may be most notable at night, when night heat is released from green roofs, while heat is "trapped" in building mass and densely built-up spaces. Ambient cooling effects on city (or neighbourhood level) can however only be expected if larger areas of (preferably geographically concentrated) rooftops – and other open spaces – are covered with vegetation.

Of course, in wintertime or colder climates rooftops can also have a positive effect on building insulation. In addition, rooftops can contribute to carbon sequestration and storm water management (RUAF Foundation 2012) and biodiversity (see chapter "Biodiversity of Flora and Fauna").

International research shows that to reach minimum climate/temperature threshold impacts with regards to temperature effects, an area approach is needed to concentrate a larger number of interventions (covering a larger area of rooftops and possibly additional ground space) in specific settlements and parts of settlements (Fig. 2). Probably minimum 50% of rooftop space should be covered to reach such effects.[2] Notwithstanding this, individual rooftops will still have positive

[2]A scenario study implemented in Melbourne (Australia), indicated that for a 2009 scenario, Average Summer Daily Maximum (ASDM) temperatures would be reduced by 0.3 °C by doubling the density of vegetation in the Melbourne central business district, or by 0.4 °C with green roofs (Green roof vegetation was 0.5 m high and covered 50% of building rooftops completely). Increasing vegetation density both at ground level and with green roofs reduced ASDM temperatures by 0.7 °C. The same relative effect of vegetation on ASDM temperatures was predicted for 2050 and 2090 scenarios following expected climate change trends. A 2005 study in Toronto, Canada modelled the effect of implementing green roofs on low-rise buildings with low slope and flat roofs of areas greater than 350 m^2, and concluded that green roofs, implemented as a city-wide strategy, could mitigate the heat island effect by reducing local ambient temperatures by 0.5–2 °C (www.growinggreenguide.org/).

temperature effects on thermal rooftop comfort as well as thermal comfort in apartments below rooftops (RUAF Foundation 2014). Specific elements for rooftop garden design, to for example enhance their environmental performance, are described in chapter "Elements of Rooftop Agriculture Design" of this publication.

Conclusion

Based on the large amount of open rooftop space available, the specific climate change impacts as well as the many other economic, social, environmental and ecological benefits of rooftop gardening as described further in this publication, the conversion of paved rooftops into urban green productive infrastructures seems a suitable strategy for many cities. As outlined in the following chapters, further technological and policy development is required to design efficient rooftop agriculture systems that optimise space and their different benefits, development that can built on the many examples and innovations provided in this book.

References

Batty M (2015) Cities in a completely urbanised world. Environ Plan B abstract 42:381–383

De Zeeuw H, Drechsel P (2016) Cities and agriculture; developing resilient urban food systems, RUAF Foundation and IWMI. Earthscan (Routhledge), London/New York

MacRae R, Gallant E, Patel S, Michalak M, Bunch M, Schaffner S (2010) Could Toronto provide 10% of its fresh vegetable requirements from within its own boundaries? Matching consumption requirements with growing spaces. J Agric Food Syst Community Dev 1(2)

Mougeot LJ (2010) Agropolis: the social, political and environmental dimensions of urban agriculture. Routledge, London

Nasr J, McRae R, Kuhns J (2010) Scaling up urban agriculture in Toronto-building the infrastructure. Metcalf Food Solutions, Toronto

Orsini F, Kahane R, Nono-Womdim R, Gianquinto G (2013) Urban agriculture in the developing world. A Rev Agron Sustain Dev 33:695–720

RUAF Foundation (2012) Rooftop gardening- a climate change perspective. RUAF Foundation, Leusden. Accessed on 25 Oct 2016 at: http://www.ruaf.org/publications/practitioner_briefs

RUAF Foundation (2014) Feasibility study on urban agriculture models and up-scaling for informal settlements in greater Cairo region. RUAF Foundation, Leusden. Accessed on 25 Oct 2016 at: http://www.ruaf.org/projects/evaluation-rooftop-farming-pilot-project-and-feasibility-study-urban-agriculture-approaches\

A Panorama of Rooftop Agriculture Types

Joe Nasr, June Komisar, and Henk de Zeeuw

Abstract Recent projects demonstrate that rooftop urban agriculture is not only possible but beneficial to the communities, organizations or businesses that maintain them. These undertakings vary considerably in type depending on the main objectives for cultivation (commercial, social, educational, therapeutical, ecological, or other), the technologies applied (from simple to very advanced), the type of building where these gardens or farms are located (apartment building, school, hospital, industrial building, etc.), climate considerations and other factors. The strategies for farming on a variety of building roofs, as well as the numerous types of stakeholders and their farming needs, will be discussed here. Different ways of categorizing rooftop growing can be used, taking one of the above-mentioned factors as the starting point. This section will present a panorama of different types of rooftop agriculture according to the main purpose intended for rooftop gardening or farming projects, combined with the context in which this takes place.

Introduction

Rooftop agriculture is undertaken by a large range of stakeholders for various reasons, applying a variety of production systems and technologies. In this section we will provide a panorama of the different types of rooftop agriculture that have been developed over the last decade in cities around the world, showing the large variety

---

The original version of this chapter was revised.
An erratum to this chapter can be found at https://doi.org/10.1007/978-3-319-57720-3_21

J. Nasr (✉)
Centre for Studies in Food Security, Ryerson University, Toronto, ON, Canada
e-mail: jnasr@ryerson.ca

J. Komisar
Department of Architectural Science, Centre for Studies in Food Security, Ryerson University, Toronto, ON, Canada
e-mail: jkomisar@ryerson.ca

H. de Zeeuw
RUAF Foundation, Leusden, The Netherlands
e-mail: h.dezeeuw@ruaf.org

F. Orsini et al. (eds.), *Rooftop Urban Agriculture*, Urban Agriculture,
https://doi.org/10.1007/978-3-319-57720-3_2

in growing spaces used (including rooftops of schools, residences, businesses, offices, restaurants and hotels), the different types of stakeholders involved (including individual households, community groups, small entrepreneurs, non-governmental organisations, educational institutes, larger businesses, real estate companies) and the different agendas that may lie behind the installation of a rooftop garden or farm, ranging from bringing a neighbourhood together through a productive social space, to increasing food security for a financially challenged family or community, to generating a financial return that can be sustained over time.

Rooftop agriculture ranges from growing vegetables and herbs in bins or containers on a terrace, to more farm-like expanses that use an engineered lightweight soil applied directly on top of a soil-ready roofing surface, to using simple or more advanced hydroponic systems in the open air or in greenhouses. In addition, some roofs are fairly closed off to visitors while others are purposely designed to be accessible to the public (as volunteers, as students learning about healthy nutrition and local food growing, or as customers of the restaurant or vegetable shop at the rooftop garden). Figure 1 may be helpful to provide an overview of the diversity among the types of rooftop agriculture. It presents a number of dimensions, however it is not meant to offer exhaustive coverage.

Understanding the variety of rooftop food production systems, the context set by the building type (and related issues such as tenancy, rooftop structure and access), the stakeholders involved and their main objectives can provide insight into the suitability, potential impacts and most frequent problems of each rooftop production system.

We will present below several examples of different types of rooftop agriculture using their primary objective as the main structuring principle for this chapter. However, it is quite common for a rooftop garden to develop from one type into another over time: e.g. gardens where production becomes increasingly important in order to survive economically, or where a small productive garden coffee shop expands into a restaurant cum event/workshop space resulting in the productive

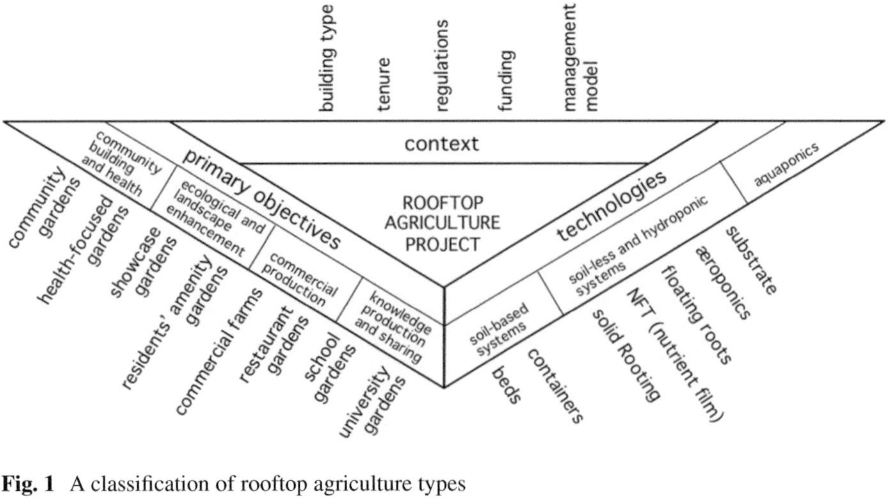

Fig. 1 A classification of rooftop agriculture types

garden itself evolving into merely an attraction and green environment for the restaurant/events centre.

There are also various rooftop initiatives that are not easily categorized: a farm might act as a social development initiative and stress its social programmes while functioning as a private commercial enterprise. This obfuscation can be a result of the need for volunteer help and community support for many urban farming initiatives to survive. Some projects involve many intersecting initiatives, for example, a restaurant that puts a greenhouse on its rooftop to make productive use of food scraps and the waste heat produced in the kitchen, can be seen as a commercial farm or a farm focused on ecological and landscape enhancement.

The technologies applied in rooftop agriculture range from self-fabricated growing containers from recycled materials to fully automated large-scale rooftop greenhouses with multi-layer hydroponic systems. When presenting the cases, the technology used and type of building on which the farm or garden is located will be briefly indicated. These technologies will be further explained and discussed in **part 2** of this handbook.

The cases marked with ** (at the end of the paragraph concerned) are presented in more detail in **part 5**. A number of the cases presented here are described more extensively in the Carrot City initiative's book (Gorgolewski et al. 2011) or website (Carrot City).

Community Building and Health

Some rooftop agriculture projects are developed to fulfil social functions as their main objective. These projects champion programmes such as food literacy, eco-education and community building initiatives, providing jobs and meaningful work for the disadvantaged and enhancing their access to nutritious food, and more. In such projects, food production is mainly a means to a social end and productivity or profitability is not a primary aim. Produce may be distributed among the participants or sold to enable the sustainability of the garden but the main goal is to create a stronger, healthier community.

Roof and Terrace Home Gardens

Especially in several countries of Asia and the Middle East/North Africa, many households grow food and herbs on rooftops and larger balconies of individual houses to enrich the diets of the household with fresh vegetables and herbs, save on food expenditure, and earn a small additional income from selling eventual surpluses. In most cases simple production systems are implemented like clay pots, bins, bags, boxes, racks and tables filled with soil, mixtures of soil and compost or other growing mediums but also residents use low or higher plastic tunnels and shade nets and sometimes simple hydroponics systems.

In many cities in the South, local governments and local NGOs, sometimes supported by international organisations, are actively promoting such terrace gardening to enhance urban food security and urban resilience. In India, the Department of Environment and Climate Change of Kerala State, in cooperation with local NGOs, is actively promoting terrace gardening by providing training, 25 grow bags and a micro irrigation system with automatic timer to each interested household at highly subsidized rates. The Department claims that thousands of terrace gardens have been established already (Suchitra 2015).

In 2014, the German Corporation for International Cooperation (GIZ) in cooperation with Schaduf, a private company that develops urban agriculture systems, implemented a project in Cairo's informal settlements to assist poor and vulnerable households to install a simple hydroponics system and train the participating households throughout the whole crop cycle. Schaduf purchases the surplus produce to sell in the city's upscale markets (Sarant 2015).**

In Nepal, that same year, the Kathmandu Environment Management Department in cooperation with the local NGO ENPHO began assisting households to establish rooftop gardens and providing training to residents in a variety of techniques that are adapted to the context of the urban roof including vermicompost preparation, nursery management, rain-water harvesting, and plant pruning and protection (Bogaty 2015).

Between 2010 and 2013, in the Gaza strip, the Food and Agriculture Organization (FAO) implemented a project with local female-headed poor and food insecure households installing simple aquaponic production units (fish and vegetable growing) on the rooftops of their homes (UN FAO).**

In 2010, the Municipality of Amman (Jordan) initiated a rooftop gardening programme, providing inputs (seeds and fertilizers) and training/technical advice (on crop production and protection, grey water recycling, etc.) to households interested in creating a rooftop garden (Greater Amman Municipality 2010).**

Community Rooftop Gardens

The roof is traditionally seen as a quintessentially private, secluded, inaccessible space. Increasingly though, the rooftop is starting to be seen as a potential community space, serving as a platform for many community-building activities, including outreach, education, socializing, and other purposes.

In Toronto along Danforth Avenue, one of the city's main arteries, the Carrot Green Roof serves the community. Designed for public access it includes multipurpose spaces among the garden spaces on the roof. It sits atop Carrot Common, a retail strip anchored by a well-known cooperative supermarket, the Big Carrot. When the roof had to be replaced, the board that governs the not-for-profit cooperative decided to invest into transforming the roof into a hybrid community and growing space, including a demonstration garden that shows many growing techniques for edible, medicinal and herbal plants, an outdoor kitchen and a gathering space for a variety of activities (including meetings and even summer theatre). An open stair from ground level to the roof provides full public access – on nice days, shoppers can grab lunch from the supermarket and eat while enjoying and learning from the garden (Carrot Green Roof).

The spacious (2200 m^2) "park in the sky" on top of Gare Perrache, one of the two main train stations in Lyon, France, was rescued after years of gradual neglect. A local association (*Les Jardins suspendus de Perrache*) transformed part of the park into a productive garden. It combines a relatively large communally gardened area and two dozen smaller individual plots (Le passe jardins; Les jardins suspendus de Perrache).

Another emerging trend is the inclusion of a rooftop garden as an amenity for residential structures, enabling residents to grow vegetables, herbs and ornamental plants and to use the garden as a meeting place and outdoor recreational space. Inclusion of opportunities for gardening on terraces and rooftops in apartment buildings for higher income groups is already quite common in a number of cities. In Toronto, Daniels Corporation, the development firm that is erecting most buildings in the revitalized Regent Park neighbourhood, is starting to integrate such rooftop gardens routinely in their market-rate condominium apartment complexes in that neighbourhood and elsewhere across Toronto (Daniels Grows).

Productive rooftops for apartment dwellers are starting to be included in social housing projects as well. In the South Bronx, New York City, Via Verde is an example of a large affordable housing complex designed as a showcase for the possibilities new forms of social housing may provide, including many terraces that serve multiple functions for the residents – including cultivation (Via Verde).

In Bologna, Italy, residents of three apartment buildings along Via Gandusio, have collaborated since 2010 with Biodivercity, a non-profit association with ties to the University of Bologna, in the creation of rooftop gardens applying simplified hydroponic systems. The main goal, beyond food production and physical activity, is to generate social inclusion and strengthen relationships among the residents in this lower income neighbourhood with a culturally very diverse population (Orsini et al. 2014).**

The builders of new apartment complexes for elderly people are starting to value gardening for physical activity, nutritional enhancement, and more generally improved quality of life. In Toronto's Regent Park, where new market-rate apartment buildings have productive garden spaces, the roof of One Oak, a 10-story apartment building built by Toronto Community Housing Corporation for seniors and families, was designed with raised beds that are tall enough for residents to access comfortably without bending over (Kearns Mancini Architects).

In Asia, a new model has emerged: the rooftop allotment garden. The East Japan Railway Company has created "Sorado farms" on top of five of its railway stations where several hundreds of small plots (3 m^2 each, using containers or raised beds) are rented out to commuters or nearby residents. The gardens are managed by a professional farmer who also acts as advisor for the gardeners, many of whom are beginners. Renters not only receive tools, equipment and advice, but they can have someone weed their plot, check for bugs or even harvest their produce, at an extra cost (Meinhold 2014).

Rooftop Gardens for Social Inclusion

Some social housing projects feature rooftop gardens that connect to particular populations with special needs or vulnerabilities. For example, New York City's Bronxscape targets youth transitioning out of foster care, providing them a space to learn gardening and cooking skills on top of the Louis Nine House, a new building that was created to cater to their special needs (Neighborhood Coalition for Shelter 2008).

In Chicago, the rooftop garden atop the Gary Comer Youth Center is a high-design sheltered space laid (with 60 cm of soil) atop a new building in the disadvantaged South Side of Chicago. In this case, gardening is seen as part of the extracurricular education that would enable students living in a difficult context to be prepared for life beyond high school (Gary Comer Youth Center Green Roof).

Another example is Peachtree-Pine Rooftop Garden in downtown Atlanta, Georgia. The garden is located on the large shelter managed by the Metro Atlanta Task Force for the Homeless. Since 2009, the shelter's residents have tended 80 low raised beds using organic methods as well as kept a number of beehives. The rooftop thus serves as a training facility, as the homeless volunteers learn to garden on the roof, then move on for further training in entrepreneurial farming and marketing at a larger garden elsewhere in the city; they can obtain a certificate after completing a 6-month internship, in partnership with Truly Living Well, an urban farming training centre (Metro Atlanta Task Force for the Homeless; Tatum 2015).

Therapeutic Rooftop Gardens

In some cases health is the primary focus for rooftop gardens, especially for gardens that are placed on buildings used specifically for health-related purposes. The connection between gardening and health has a long history – in fact, the medicinal garden is deeply anchored in the origin of the hospital, the hospice, the asylum and other places for care or isolation of individuals with any health issues. Commonly, such a garden would have been located in the ground as such facilities tended to be placed at the edge of town or in the countryside, where there was little pressure on space. However, increasingly, urban expansion has meant that health-focused gardens started to be placed on the rooftop of clinics and hospitals in dense urban areas. The objective nowadays is more connected to horticultural therapy than to the production of medicinal herbs – especially as studies have increasingly shown the strong relation between healing and greenery.

Therapeutic rooftop gardens exist atop hospitals in Seoul (Korea), Toledo, Ohio, USA, and on the island of Oahu, USA. There are many examples of therapeutic gardens on other health facilities as well. A health clinic run by Access Alliance, a not-for-profit health organization in Toronto now has a rooftop garden for their facility at AccessPoint on Danforth. Access Alliance created this garden when the hub was established to address a variety of their clients' health needs through hands-on activities. The garden includes growing spaces, small trees, a trellised shade structure that supports grapevines, a storm water collection system, and a sitting area (Access Alliance Multicultural Health & Community Services).

The Municipal Institute for People with Disabilities (IMDP) in Bologna is preparing the establishment of a multi-functional garden ("social orchard") where orthotherapy can be performed by people with disabilities through gardening activities. The garden will consist of growing boxes with substrates (Personal communication).**

Rooftop Gardens as an Amenity for Employees

Gardening is starting to be seen as an amenity in work environments. Providing workers with areas to grow food to supplement their income has an older history and is still applied, especially in developing countries. However, this practice has largely disappeared in the global North, where the current revival of the staff garden is taking a different form: it is now approached as an amenity for employees just like an office gym, promoting mental and physical health. In countries like Japan, where high density and increasing environmental consciousness combine with long work hours, rooftop growing projects are becoming increasingly popular and are being introduced by corporations in the largest cities there (Ozawa 2008).

The Toronto high-rise building that serves as headquarters of Telus, one of the largest Canadian telecommunication companies, contains several planted terraces that serve as storm water catchment areas but also function as gathering spaces for employees. Under the impetus of a small group of committed employees, one of the planted terraces was transformed in 2013 into an attractive productive garden space with easy-to-grow herbs and vegetables adapted for the local climate. Volunteers learn how to grow and what to grow in their own gardens. Others spend their break time in this pleasant space, demonstrating how edible landscaping can be both productive and attractive (Bayens 2013; Wong 2014).

Commercial Production

Forms of rooftop agriculture that produce for the market to generate income as their main objective are highlighted in this section. The scale of the farms as well as the technologies applied may vary widely. Farms may have additional secondary functions (e.g. education, events, green building certification, etc.). What connects all these examples is that, no matter what other functions may be present, the profitability of the activity is an overarching objective with these cases.

Small- and Medium-Scale Commercial Farms

Urban Canopy is a medium scale urban farm in Chicago that was established in 2011 on top of "The Plant," a former pork-packing plant in one of the food desert neighbourhoods of Chicago. The farm is commercially viable because it uses the community supported agriculture (CSA) model to enhance its sustainability. The creation of the farm transformed this building into a showcase of waste diversion, food production and recycling. Among its innovations, Urban Canopy has a small vertical hydroponic system on the roof. They also collect kitchen scraps from members of their "compost club" to use as fertilizer. Their outreach to the community extends to inviting the public to volunteer as farmers, tour their rooftop facility and/or to buy "shares" of the future yield. By engaging the public in these ways, and by producing in several growing spaces (next to the rooftop garden they produce microgreens in an indoor space and also work an in-ground farm), they are thriving (Plant Chicago).

Since 2011, EnerGaia has operated a facility for the commercial production of the microalgae Spirulina on the rooftop of the Novotel in the Siam area of Bangkok. EnerGaia leases the rooftop from Novotel with a two-year (renewable) contract. The closed system uses 100 food-grade 250-l semi-transparent polypropylene tanks placed in two circles, with an air blower, an aquarium pump and a harvester with bag located at the end of each circle. The tanks function as photo bioreactors: the spirulina biomass develops inside the tanks thanks to sun radiations that reach the plant biomass through the transparent walls and circulation of the nutritive media in the tanks. The farm produces about 4 tons of spirulina per year which is sold in various forms (dry powder, fresh paste, frozen paste, etc.) to restaurants and local producers of pastas, chocolate truffles, ice creams, and gluten-free rice noodles fortified with Spirulina (EnerGaia).**

Fed Square Pop Up Patch garden was created in 2012 on the rooftop of the Federation Square car park in the Central Business District of Melbourne. The garden extends approximately 1000 m^2 and is managed by the start-up company Little Veggie Patch Co (LVPC) that rents this roof from the owners of Federation Square, who also contributed to the initial investment. LVPC rents out growing boxes of 1.5 m^2 to local residents and restaurants to grow food for their own use. The rent includes seeds and seedlings, pest and disease control and technical advice/education by the farm manager (The Little Veggie Patch Co).**

Rooftop Republic is a small business that set up and manages rooftop gardens on top of the Bank of America and the Confucius School buildings in Hong Kong. Fringe Club, a contemporary arts space, requested Rooftop Republic to manage the rooftop garden (90 m^2 with raised beds made from lightweight materials) that was earlier established by another firm. Rooftop Republic gets free use of the rooftop garden plus supply of water and electricity in exchange for delivering the produce to the Fringe Club restaurant. Hence, the income earned by Rooftop Republic is not from selling produce but from the many paid workshops, cooking classes, tours and even yoga classes they organise in the rooftop garden (some 32 workshops in 2015 (Cam 2014).**

In Singapore, Comcrop, an aquaponic farm (280 m^2) is located at the rooftop of the Scape Mall. Optimizing the rooftop space has been achieved by adopting twelve 4.5 m high racks of plastic tubes fitted with holes for vegetable crops. The tubes are connected to tanks for raising tilapia. A tent-like shading structure cools the fish tanks and the plants in the racks (Fresh Fruit Portal 2014; Weise 2015).**

Large-Scale Commercial Farms

UrbanFarmers AG is a Swiss company that is also operating an aquaponic rooftop farm in Basel. In 2016 it opened its second aquaponics farm in The Hague, the largest urban rooftop farm in Europe, occupying the rooftop and the floor below of the iconic "De Schilde" building (a former office building). The rooftop is covered with a 1200 m^2 hydroponics greenhouse while the 6th floor below is designed for fish farming, processing and packaging (Challen 2015).**

In Bad Ragaz, Switzerland, ECCO SA set up the ECCO-Jaeger rooftop farm in 2014 at the rooftop of the headquarters of the company. The farm combines a rooftop greenhouse and an aquaculture system for both vegetable and fish production. While the greenhouse of 1000 m^2 occupies the roof of the warehouse, the aquaculture system is placed within the building (2000 m^2). The waste heat generated by the refrigeration units in the warehouse is used to heat the aquaponic farm (Ecco-Jaeger).

Gotham Greens is a pioneer in North America, with four hydroponic greenhouse farms and more farms in the planning stages. Their newest site, on top of a soap manufacturing plant located on the south side of Chicago, is the largest urban rooftop farm in the world. The 7000 m^2 greenhouse has a high yield year round due to control systems that monitor and adjust the environment. Gotham Greens seeks to offset the high electricity demands with solar panels, passive ventilation design and thermal curtains to contain heat and recycling of the irrigation water (Gotham Greens).**

Lufa Farms, a Montreal-based farming company also uses hydroponic techniques in their rooftop greenhouses. Both of their farms are constructed on top of industrial buildings for a total of almost 7000 m^2 of growing space that they optimize by growing vine crops to extreme heights and recycling their irrigation water (Lufa Farms).

In contrast to these more controlled greenhouse environments, Brooklyn Grange (Brooklyn Grange Rooftop Farms) and Eagle Street (Eagle Street Rooftop Farm) in New York are two more traditional open-air farms on industrial buildings. For both of these farms, it was crucial to find a rooftop that could sustain the weight of at least 15 cm of engineered, lightweight soil on a waterproof membrane. Brooklyn Grange now covers two rooftops with vegetables grown rather traditionally, directly on the rooftop, and has expanded its scope to include an apiary with 30 beehives. Education outreach is integral to the farm through a youth/young adult program, "City Growers," that enables many workshops from nutrition to earth science.

The nearly 1000 m^2 rooftop farm established by UpTopAcres in 2015 in Bethesda, a suburb of Washington, DC, in partnership with Federal Realty Investment Trust, is another open air farm. The existing green roof was retrofitted into a rooftop farm to grow yearly 50,000 kg of a wide range of crops including salad mixes, herbs, beets, carrots, radishes and microgreens. Produce is distributed through a 35-member CSA scheme and to nearby restaurants and retail (UpTop Acres).

Rooftop Gardens Serving a Restaurant, Institution or Shop

A special type of commercial rooftop gardens is the farm-to-table model. In this case, food grown on the roof of a restaurant, hotel, hospital, supermarket or large enterprise with a staff canteen uses the roof of its own building to produce fresh vegetables and herbs for its own kitchen. Other motives for the creation of a productive garden on top of a restaurant, hotel or enterprise are often the wish to make their business more environmentally sustainable, make productive use of wastes (organic materials, excess heat, and cooling water) by such a business and/or create an attractive green environment for their clients.

Restaurants have emerged as leaders in this movement, with projects often led by the chefs themselves. Established restaurants like Parts & Labour, Vertical, and Beast in Toronto have involved rooftop growing, either on the roof of the restaurant itself or nearby. In some cases, restaurant gardens use simple growing systems: raised beds built out of timber, or simple plastic containers. In other cases, the restaurant invests in more sophisticated systems like the aeroponic vertical tower growing system that is used to maximize productivity on the small roof of the Bell, Book and Candle restaurant in Greenwich Village (Bell, Book and Candle; Moran 2013) and the 900 m^2 of greenhouses established by Zabar's Vinegar Factory café in New York on its rooftop to grow vegetables (mainly different varieties of tomatoes) and herbs from fall to spring, using waste heat from the bakery below (Eli Zabar).

Sometimes the restaurants contract a specialized firm to manage the rooftop garden professionally. The Seattle Urban Farm Co. (SUFCo) is providing such services for several restaurants in Seattle, working in close cooperation with the chefs to select appropriate crops to produce in each season according to the chef's needs, employing convertible cold-frame greenhouses that can be fitted with plastic covers in winter or shade cloth in the summer months (Seattle Urban Farm Company).

In Toronto, for more than a decade, the venerable Fairmont Royal York Hotel, built in 1929, has been a pioneer in rooftop gardening to supply its restaurant. Using a roof on the 18th floor to produce greens, herbs and arctic berries, high platforms for the growing containers help the chefs to harvest easily. Composters use kitchen waste, and beehives provide honey for the restaurant. The entire hotel chain, from Montreal to Singapore, is now committed to providing fresh produce in this way. To date they have 28 gardens, many on rooftop terraces that guests can visit (Fairmont).

In Paris, the Frame Brasserie at the Pullman Eiffel Tower hotel benefits from strawberries and salad greens grown on its ground-level roof above its garage. This was set up and continues to be managed by Topager, an edible landscaping enterprise growing quality organic vegetables in a rich growing substrate developed in cooperation with INRA-AgroParisTech (Topager [a]). Similar developments are taking place in other hotel chains, such as the Waldorf Astoria in New York City (Waldorf New York 2014).

McCormick Place, the convention centre in Chicago, had already set up a green roof over part of its structure when it decided to transform this space into a rooftop farm to supply fresh vegetables and herbs to its restaurants. This soil-based farm is managed by the Chicago Botanic Garden and is directly providing the convention centre's users with more than 10,000 servings per year through its food service company. Initially covering over 1800 m^2, it aims to expand to over a hectare of the centre's roof, which would make it the biggest rooftop farm in the US (Barclay 2013).

Fenway Farms, located on top of the administrative building of the legendary Fenway Park baseball stadium in Boston, is a 650 m^2 garden managed by Green City Growers. It is using standardized plastic crates to supply 1800 kg of produce in one season for the club kitchen (Fenway Farms; Green City Growers 2015).**

Some hospitals have created rooftop gardens that do not function as a therapeutic garden, but rather as farms supplying fresh nutritious food and herbs for the hospital kitchen, often including specific crops needed for the specific dietary needs of certain patients (The Lempert Report and the Center for Food Integrity 2012). Examples include the rooftop garden of the Stony Brook University Hospital on Long Island, New York that is managed by staff nutritionists and dietetic interns (Stony Brook University 2012), and the rooftop farm of Weiss Hospital in Chicago that also includes an apiary and a chicken coop (Hartocollis 2012; Odway 2015).

Various supermarket chains are also creating rooftop farms on top of their buildings, including the open-air farm on the Whole Foods Market in Lynnfield, Massachusetts. The purpose-built roof holds 25 cm of engineered soil. The farm is maintained by Green City Growers who also manage Fenway Farms in Boston. The green roof provides an estimated 5000 kg of produce per year, which is then sold inside the store (Whole Foods Market – Lynnfield, MA).

Ecological and Landscape Enhancement

A number of rooftop agriculture projects are created as part of the eco-design of the building to contribute to its sustainability (some obtained certification for it) or as an attractive green context for a hotel, restaurant, or other public space. The productive side of the garden is secondary to its ecological or landscaping function, although products may be used in a hotel restaurant or enterprise canteen, or used by its staff who take home some fresh food. Across these cases, environmental quality – in terms of ecology and/or aesthetics – is fundamental.

Ecology-Focused Gardens

The Changi General Hospital in Singapore created (already in 1998) a rooftop farm on its roof mainly to reduce the heat and the glare from the bare concrete roof of its atrium. It uses a flat, low-cost technique of modified hydroponics to produce tomatoes and herbs for use in the hospital (Wilson 2005; Changi General Hospital).

Shopping Eldorado, a mall in Sao Paulo, Brazil, created a rooftop garden as part of its sustainability plan mainly to enable the recycling of organic wastes generated in its food court. A composting machine was installed in the mall's basement and the compost produced is used as growing medium in the hundreds of growing containers on the roof, reducing waste. The rooftop garden is tended by a specialized landscaping company and the produce is distributed to the mall's employees as a benefit (Ecoeficientes).**

The NTT company in Tokyo created a green roof by growing sweet potatoes (which spread out into a thick cover) in order to reduce heat on the roof, enhance insulation and lower the energy use in the building. The potatoes are harvested annually and distributed among the participating employees (Ozawa 2008).

In London, the Bloomsbury Street Hotel uses a bag-like container garden system called "pocket habitats" to reduce water runoff and increase local produce as part of a push by the local business improvement district for a more sustainable neighbourhood (Mavrogordato 2013).

In Montreal, the *Palais des congrès* sought to define itself as a model green convention centre by opening up its roof to a number of local groups for cultivation, with a design that allows observation from indoors. The result has been a proliferation of different experiences since 2010, from researchers experimenting with innovative techniques, to a group working with homeless women, to a beekeeper cooperative, to 450 containers producing approximately 1000 kg of vegetables each year for the main caterer of the Palais. In 2016, the project was expanded in collaboration with the Urban Agriculture Lab, starting with a 600 m^2 pilot that would be developed into a vertical urban farm of 4000 m^2 (Montreal 2013)**

The HERO MotoCorp created a major green house with hydroponic systems on top of its new Factory and Global Parts Center in Neemrana, India as part of its sustainability plan, earning HERO the LEED-Platinum certificate from the Indian Green Building Council, with the highest scores for any factory. The greenhouse plays a role in temperature and water management of the building and reusing excess heat and CO^2 generated by the factory (Hero MotoCorp 2014).

Roosevelt University in Chicago built a 32-story LEED-certified building in 2012, the Wabash building, that includes five green roofs, two of which are now used to grow vegetables for the university dining halls. Organic wastes from the dining halls are composted and used in the rooftop gardens. These two rooftop gardens are also an educational opportunity, as the student growers see what thrives in each garden (Hustad 2015).

Landscape-Focused Gardens

In recent years some rooftop gardens have been established above all for aesthetic purposes. Not coincidentally, many of these gardens, while on a rooftop, are on a relatively low surface that is overlooked by a taller building from which users (whether hotel guests, employees or residents) can gaze at the layout, pattern, colour and other design dimensions of the garden.

In London, the substantial renovations of the Southbank cultural complex along the river Thames included the creation of a series of green terraces that flow up the outside of Queen Elizabeth Hall, including the roof. The latter was designed by the Eden Project in 2014 as a lush garden that includes a café/bar, a wildflower meadow, a small forest and an allotment area. Here, gardening takes place in the midst of masses of Londoners and tourists milling around, sitting or lying in the grass. Edibles are showcased by their simple inclusion in a public rooftop space, with little effort to communicate or educate (Southbank Centre).

In Amsterdam the office building Zuidpark created a rooftop garden that is visible from higher buildings surrounding it. The 1000 staff working in the building use the rooftop garden during lunch times and breaks and may take home some produce if they want, while the small restaurant in the building also uses some fresh greens regularly (Levenston 2012).

In other cases, productive gardens are used to create an attractive edible landscape on top of parking garages. The *Jardin de la Duche* in the Swiss town of Nyon is a carefully designed park on top of a large municipal garage that was implanted in a steeply sloping site in the heart of the town just below the historic castle. Using a series of terraces, plants that celebrate the traditional crops of the region are used ornamentally. Rows of grapevines, fruit trees, bushes and other plantings that would be common in local kitchen gardens are interspersed (Jardin de la Duche). This park is thus a large showcase of edible landscaping for residents of the city and its environs.

Knowledge Production and Sharing

In today's cities, given the limited access to space, the use of rooftop spaces for generating and sharing knowhow about food growing is becoming increasingly common. Whether a school lacking a yard to teach children where food comes from, or a university seeking to provide a laboratory for research on growing conditions in cities, or a non-profit organization wishing to provide outreach to improve cultivation practices, the roof is often used as a place for knowledge production as well as dissemination, as the examples in this section illustrate.

Research-Oriented Rooftop Farms

In the last few years, several universities and enterprises and have installed rooftop gardens and greenhouses to undertake experimentation related to food growing in the city.

The University of Toronto's "sky garden" began in 2009 as a small project on the Faculty of Engineering roof, using lightweight semi-hydroponic containers with a drip irrigation system in which the roots grow in a granular medium. Participating students have been experimenting with a range of crops and other variables to develop moveable, lightweight rooftop gardening systems (Irving 2015).

In Paris, a rooftop garden was installed in 2012 atop the historic building of AgroParisTech, the country's leading agriculture school. This setting, which is quite common across the city, was used to undertake research on different substrates, variability in growing conditions, pollution levels, and productivity. While scientific studies for scholarly publications were conducted here, involvement of young innovators served to incubate Topager, a successful enterprise providing edible landscaping services (Topager [b]).

Similarly, in Bangkok, Kasetsart University's Urban Agriculture programme, centred on the top floor of a campus parking garage, has combined research and education since 2014. Spearheaded in this case by a team of landscape design researchers based at the Faculty of Architecture, this initiative has installed container-based gardens on several unused roofs across campus, including the main library, the Office of Agricultural Extension and Training, and the Architecture Faculty itself. The success of this project has depended on active involvement by staff based in each of these buildings (KU Urban Agriculture).

The Institute of Environmental Science and Technology (ICTA) of the Autonomous University of Barcelona has operated two experimental innovative building-integrated rooftop greenhouses since September 2014. Each greenhouse is 125 m^2 and placed on top of the ICTA-ICP building. The greenhouses are innovative in two ways. First, the water, energy and CO_2 flows of the greenhouses are integrated with the ICP building systems, reducing greatly the building's total energy and water consumption along with the combined greenhouse gas emissions of the building and the greenhouse. Second, the greenhouses are used to experiment with different hydroponic systems and crops (Sanyé-Mengual 2015).

An example of an experimental rooftop garden established by a technology provider is the small hydroponic greenhouse on top of a shipping container, with fish tanks and the cleaning/recirculation system that was established by ECF in Berlin in 2012. This aquaponics unit acted as the experimental stage and showcase of innovation for the larger scale aquaponics farms constructed by ECF in later years (Vidinopoulos 2013).

Educational Rooftop Gardens

Realistically, many schools in larger cities will not have growing space on ground level, so the creation of new school gardens must include realizing the existing potential of school roofs. The rooftop school garden (sometimes in the form of a greenhouse) is thus becoming increasingly common. In many cases, this has taken very simple forms, using basic growing techniques on underused roof or balcony spaces. Less often, showcase projects have been created on school roofs.

In Toronto, five non-profit organizations have allied to champion the creation of a productive garden in every school. One of these organisations, FoodShare, runs schoolyard farming projects that seek to teach students skills related to running market gardens – from production to marketing. In partnership with Toronto's Eastdale Collegiate School, FoodShare operates a large container garden called School Grown RoofTop. The school building has a 1000 m^2 roof space originally built as an outdoor tennis court, which is very fortunate because the engineering and life-safety requirements for this initial use were ideal for a heavy load of soil-filled containers and for the safety of the student gardeners. Most of Eastdale's roof is now covered with dozens of large modular growing containers while a portion of the rooftop has been set aside for events and gatherings within the garden. Here, the students learn growing, harvesting, cooking and marketing the food (Brown 2013).

In parallel, a campaign by the New York non-profit organization NY Sun Works is aiming to install 100 school greenhouse labs by 2020. New York's first school greenhouse was built in 2011 for PS (Public School) 333 (NY Sun Works). Since the construction of this pioneering project, a number of other school rooftop greenhouses have emerged across New York City.

In Cincinnati, the Rothenberg Preparatory Academy is a historic school located in the inner-city neighbourhood of Over the Rhine where a roof top garden was created as part of the renovation effort. The garden's 25 beds provide hands-on experience for around 450 students annually. It offers service learning and contributes to science, technology, engineering and math education as part of the curriculum. The garden is independently managed by a "Garden Guild" that is also a fundraising entity for financing the gardening activities. This project shows how gardens atop schools can be an integral part of the school and its teaching while managed professionally and independently (Over the Rhine Foundation 2014).**

In Bangkok, 50 raised garden beds were constructed from 100 donated wooden pallets and recycled plastic sheeting on a concrete roof by the staff and students of the NIST International School. More recently, NIST is also experimenting with aquaponics. The organic produce is sold to the parents and staff, which provides income to buy more seeds and supplies (Johnson 2014a; Johnson 2014b).

Also, numerous universities now have rooftop growing projects, though the nature of these projects varies greatly.

Toronto's Ryerson University has set up an urban rooftop garden run by a student-focused garden group called Ryerson Urban Farm (formerly Rye's HomeGrown). The farm-like productive garden on the university's engineering building was created through the conversion of an existing green roof. Raised beds were created with a mixture of compost and soil, reducing soil depth where paths were needed. The garden, run by a professional urban farmer, provides hands-on learning to volunteers and paid interns. Produce is sold in the on-campus farmers' market and served in the school cafeteria (Ryerson Urban Farm).**

In downtown Montreal, at University of Quebec at Montréal, a student association started rooftop gardens in 2008. Three sites for shared rooftop gardening now allow students and employees to experiment with a variety of cultivation techniques to produce vegetables. Hives located on a campus rooftop produce honey while pollinating the gardens on and off campus. Run by the Urban Agriculture Lab, a research group, a rooftop greenhouse is also used for winter vegetable production and for indoor projects such as aquaponics, hydroponics, and aeroponics (CRAPAUD – UQAM).

In the same city, Concordia University built a greenhouse on the 13th floor of a science building in 1966 to serve as teaching laboratories for the Biology Department. After this department moved to a new science building, different compartments of the sizable complex of interconnected glasshouses were rented out to student groups for independent projects. The greenhouse also hosts many food-oriented workshops and has become a catalyst for a range of urban agriculture activities on and off campus. Most unusual perhaps is the large indoor composting operation, where five million worms compost up to 24 million tons per year of organic waste generated by food facilities in the building (City Farm School; Concordia Greenhouse).

School and university buildings are not the only locations for rooftop educational gardens. The roofs of other buildings are sometimes used for outreach to a broader public. In 2011, a rooftop garden was established on *top of the building in Durban where the Urban Management Zone Department of e-Thekwini Municipality has its offices.* The rooftop garden has an important ecological and food educational function: regularly guided tours are organised for school groups who are informed about various aspects of organic nutrition, growing methods and how to maintain sustainable family food security. The rooftop garden also plays a role in providing practical information and examples to individual citizens, students, garden associations and others about urban gardening, sustainable food security, recycling, green buildings and repurposing (McNulty 2012).

The Rotterdam Environment Centre (RMC), an NGO for environment and nature in the Netherlands created a 1000 m^2 soil based rooftop garden in 2012. The project, named "DakAkker" is located on top of the Schieblock building in Rotterdam. The garden produces vegetables, herbs, and fruits. They are served to the customers of the small rooftop restaurant and sold in a weekly vegetable market on the roof and to other restaurants in the neighbourhood. However, the main goal of the rooftop garden is to provide a hands-on educational programme for elementary schools about urban agriculture, healthy food, climate change, and all that grows on the DakAkker (Het Schieblock).

The Chulalongkorn University in Bangkok has recently initiated a project called Siam Green Sky that is located at the 7th floor of a large shopping mall, Siam Square One, close to the university in the heart of an intensely busy commercial and transportation hub. The main purpose of this project is educational but its outreach (as a demonstration garden and space for workshops and other events) is intended to the urban population at large rather than university students. The garden is thus maintained by a professional staff, as well as by the public during training events (Barrow 2015).

Conclusion

The examples presented above show the great variety of types of rooftop agriculture that have been created until now, with different stakeholders and different objectives in mind. They make use of many varied opportunities to create spaces for urban rooftop agriculture.

Although we have been presenting these cases according to their main objective, many rooftop gardens and farms have one or more secondary objectives in addition to their main one such as production plus education, building sustainability or edible landscape plus food for the customers, production for the market plus social aims, engineering plus education. Multi-functionality is thus the norm in rooftop agriculture.

To realize a rooftop that includes food production in one form or another, there are a number of challenges, as will be dealt with in subsequent chapters of this handbook. Some are infrastructural and technical in nature, others require addressing regulatory issues (such as bylaws and building codes). Moreover, many common cultivation challenges and practices have to be figured out to achieve any success in the specific growing conditions on a roof: season extension, appropriate crops, pollination, and so on. Despite these numerous challenges, quite a few productive rooftops have been implemented in recent years, in very diverse contexts, showing creative solutions to these challenges.

This has helped urban agriculture become one of the most innovative and dynamic sectors in cities today. Innovations in rooftop agriculture are likely to continue in the future.

Bullet Points

- The proliferation of rooftop greenhouses, particularly for commercial production, will probably expand much more, building on the experiences of the pioneers that are expanding their practices and being replicated by other companies.

In particular, the use of hydroponic and aquaponics production will probably become increasingly common as it moves from a novelty to an established practice. This is illustrated by the above-mentioned case of UrbanFarmers, which developed from experimentation in a container in Switzerland to a substantial project in The Hague only to be replicated by another firm in Bad Ragaz.

- The use of rooftops as a common component of educational space for urban agriculture is spreading fast and can be expected to expand further as a space for educational programming as well as physical exercise in tight urban contexts – in both schools and universities.
- The use of rooftops as an amenity for users (residents, employees, shoppers…) is rapidly becoming a common feature of many residences, offices, shopping centers and other buildings in increasingly dense cities, addressing several purposes simultaneously including leisure, aesthetic, status and environment. This is enabling specialized landscape design and maintenance firms to develop a professional niche.
- The proliferation of "terrace gardens" will continue: container gardens and small-scale hydroponics on balconies and flat roofs of houses and apartments are increasing, especially in warmer climates and in developing countries.

These are among the trends that can be expected to continue to develop in future years. What is clear is that rooftop agriculture, in its many forms and for its many purposes, is likely to become more common, with a variety of innovations continuing to emerge in decades to come.

Literature

Access Alliance Multicultural Health & Community Services. The green roof at AccessPoint on Danforth. Available at: http://accessalliance.ca/wp-content/uploads/2015/03/AAMHCS-Green-Roof-overview.pdf. Accessed 30 Dec 2016

Barclay E (2013) Rooftop farming is getting off the ground. The Salt, NPR, 25 Sep. Available at: http://www.npr.org/sections/thesalt/2013/09/24/225745012/why-aren-t-there-more-rooftop-farms. Accessed 30 Dec 2016

Barrow R (2015) Siam green sky is the largest urban rooftop garden in Bangkok. Thai Travel News, 25 May. Available at: http://www.thaitravelblogs.com/2015/05/siam-green-sky-is-the-largest-urban-rooftop-garden-in-bangkok/. Accessed 30 Dec 2016

Bayens A (2013) Team members get growing at the new rooftop community garden of TELUS House Toronto. Available at: http://blog.telus.com/community/team-members-get-growing-at-the-new-rooftop-community-garden-of-telus-house-toronto/. Accessed 30 Dec 2016

Bell Book & Candle. Available at www.bbandcnyc.com. Accessed 30 Dec 2016

Bogaty T (2015) Kathmandu, Nepal: Rooftop farming training to cover 150 households. The Himalayan Times, 23 Sep. Available at: http://www.cityfarmer.info/2015/10/01/kathmandu-nepal-rooftop-farming-training-to-cover-150-households/#more-308807. Accessed 30 Dec 2016

Brooklyn Grange Rooftop Farms. Available at: http://www.brooklyngrangefarm.com/. Accessed 30 Dec 2016

Brown L (2013) Eastdale collegiate opens its roof to to urban agriculture. Toronto Star, 25 Dec. Available at: http://www.thestar.com/yourtoronto/education/2013/12/25/eastdale_collegiate_opens_its_roof_to_urban_agriculture.html. Accessed 30 Dec 2016

Cam L (2014) Hong Kong's best rooftop farms. TimeOut Hong Kong, 4 Nov. Available at https://www.timeout.com/hong-kong/attractions/hong-kongs-best-rooftop-farms. Accessed 30 Dec 2016
Carrot City. Available at www.carrotcity.org. Accessed 30 Dec 2016
Carrot Green Roof. Carrot city website. Available at: http://www.ryerson.ca/carrotcity/board_pages/rooftops/carrot_green_roof.html. Accessed 30 Dec 2016
Challen J (2015) UF002 De Schilde: Construction commences for world's biggest rooftop farm. FutureAg, 10 Sep. Available at: http://futureag.info/tag/uf002-de-schilde/. Accessed 30 Dec 2016
Changi General Hospital. The greenroofs & greenwalls projects database. Available at: http://www.greenroofs.com/projects/pview.php?id=565. Accessed 30 Dec 2016
City Farm School. Our history. Available at: http://www.cityfarmschool.com/our_history/
Concordia Greenhouse. History of the greenhouse. Available at: http://concordiagreenhouse.com/about/history-of-the-greenhouse/. Accessed 30 Dec 2016
CRAPAUD – UQAM. Carrot city website. Available at: http://www.ryerson.ca/carrotcity/board_pages/community/crapaud.html. Accessed 30 Dec 2016
Daniels Grows. Available at: http://danielshomes.ca/innovative-programs/daniels-grows. Accessed 30 Dec 2016
Eagle Street Rooftop Farm. Available at: http://rooftopfarms.org/. Accessed 30 Dec 2016
Ecco-Jaeger. Available at: http://www.ecco-jaeger.ch/. Accessed 30 Dec 2016
Ecoeficientes. Veja como funciona essa Horta Urbana no coração de SP. Available at: http://www.ecoeficientes.com.br/horta-urbana-sao-paulo/. Accessed 30 Dec 2016
Eli Zabar. The vinegar factory. Available at: http://www.elizabar.com/The-Vinegar-Factory.aspx. Accessed 30 Dec 2016
EnerGaia. Available at: http://energaia.com/. Accessed 30 Dec 2016
Fairmont. Fairmont herb gardens. Available at: http://www.fairmont.com/promotions/fairmontherbgardens/. Accessed 30 Dec 2016
Fenway Farms. The greenroofs & greenwalls projects database. Available at: http://www.greenroofs.com/projects/pview.php?id=1729. Accessed 30 Dec 2016
Fresh Fruit Portal (2014) Singapore: Comcrop pioneers urban rooftop farming. 14 Mar. Available at: http://www.freshfruitportal.com/news/2014/03/14/singapore-comcrop-pioneers-urban-rooftop-farming/. Accessed 30 Dec 2016
Gary Comer Youth Center Green Roof. The greenroofs & greenwalls projects database. Available at: http://www.greenroofs.com/projects/pview.php?id=998. Accessed 30 Dec 2016
Gorgolewski M, Komisar J, Nasr J (2011) Carrot City: making places for urban agriculture. Monacelli, New York
Gotham Greens. Our farms. Available at: http://gothamgreens.com/our-farms/. Accessed 30 Dec 2016
Greater Amman Municipality (2010) Greater Amman Municipality (GAM) Cultivates rooftops of 100 houses in Citadel. 4 Dec. Available at: http://www.amman.jo/en/resource/snews.asp?id=278DE745-CEFC-4B45-8272-024715257FD2. Accessed 30 Dec 2016
Green City Growers (2015) Fenway Farms. Available at: http://greencitygrowers.com/fenway-farms/. Accessed 30 Dec 2016
Hartocollis A (2012) Challenging the reputation of hospital food on a rooftop farm. The New York Times, 18 Oct. Available at: http://www.nytimes.com/2012/10/19/nyregion/challenging-the-reputation-of-bland-hospital-food-on-a-rooftop-farm.html. Accessed 30 Dec 2016
Hero MotoCorp (2014) Hero garden factory. 21 Oct. Available at: https://www.youtube.com/watch?v=r9v4BiVQjO0. Accessed 30 Dec 2016
Het Schieblock. 701 – Schieblock DakAkker. Available at: http://www.schieblock.com/index.php?pageID=8&studioID=70. Accessed 30 Dec 2016
Hustad K (2015) Want to check out Roosevelt's sutainability innovation? Look up. Chicago Inno Beat, 20 July. Available at: http://chicagoinno.streetwise.co/2015/07/20/urban-farming-roosevelt-university-utilizes-vertical-space/. Accessed 30 Dec 2016

Irving T (2015) A bumper crop atop UofT engineering's sky garden. UofT Engineering News, 15 Oct. Available at: http://news.engineering.utoronto.ca/sky-garden-a-bumper-crop-from-this-u-of-t-engineering-rooftop/. Accessed 30 Dec 2016

Jardin de la Duche. Carrot City website. Available at: http://www.ryerson.ca/carrotcity/board_pages/rooftops/jardin_duche.html. Accessed 30 Dec 2016

Johnson B (2014a) Rooftop school garden in Bangkok. Mother Earth News, April/May. Available at: http://www.motherearthnews.com/organic-gardening/school-garden-zm0z14amzkin.aspx. Accessed 30 Dec 2016

Johnson B (2014b) Life on the roof: NIST international school. A Day Bulletin Life, no. 38. 27 Nov. Available at: http://issuu.com/adb_life/docs/adb_life38/4. Accessed 30 Dec 2016

Kearns Mancini Architects. Regent Park, Block 11, One Oak Street. Available at: http://kmai.com/projects/regent-park-block-11-one-oak-street. Accessed 30 Dec 2016

KU Urban Agriculture. Available at: https://www.facebook.com/kuurbanagriculture. Accessed 30 Dec 2016

Le passe jardins. Available at: http://lepassejardins.fr/spip.php?article124. Accessed 30 Dec 2016

Lempert Report and the Center for Food Integrity (The) (2012) Rooftop hospital farms. The Food Journal and Food, Nutrition & Science, 28 Oct. Available at: http://www.foodnutritionscience.com/articles/rooftop-hospital-farms/. Accessed 30 Dec 2016

Les jardins suspendus de Perrache. Available at: http://jardins.suspendus.over-blog.com/. Accessed 30 Dec 2016

Levenston M (2012) Zuidpark – Europe's largest rooftop farm opens in Amsterdam. City Farmer News, 28 June. Available at: http://www.cityfarmer.info/2012/06/28/zuidpark-europes-largest-rooftop-farm-opens-in-amsterdam/. Accessed 30 Dec 2016

Lufa Farms. Available at: https://lufa.com/en/. Accessed 30 Dec 2016

Mavrogordato T (2013) Offices are turning their roofs into edible gardens and bee sanctuaries. The Guardian, 13 Aug. Available at: http://www.theguardian.com/sustainable-business/offices-roofs-edible-garden-sustainable-cities. Accessed 30 Dec 2016

McNulty B (2012) Priority zone. TravelStart Blog, 15 May. Available at: http://www.travelstart.co.za/blog/priority-zone/. Accessed 30 Dec 2016

Meinhold B (2014) Rooftop farms on Japanese train stations serve as community gardens. Inhabitat, 25.3. Available at: http://inhabitat.com/rooftop-farms-on-japanese-train-stations-serve-as-community-gardens/. Accessed 30 Dec 2016

Metro Atlanta Task Force for the Homeless. Our rooftop garden! Peachtree + Pine Works. Available at: http://atlantataskforceforhomeless.org/roof-top-garden-watch-us-grow/. Accessed 30 Dec 2016

Montreal (2013) Rooftop gardens of the Palais des Congrès. Blog, 25 July Available at: http://www.tourisme-montreal.org/meetings/2013/07/25/palais-des-congres-2/. Accessed 30 Dec 2016

Moran T (2013) Future growing rooftop tower garden farm at the bell book and candle restaurant in Manhattan. ABC News Nightline, 19 Sep. Available at: https://www.youtube.com/watch?v=cwvIXc-mtZE. Accessed 30 Dec 2016

Neighborhood Coalition for Shelter (2008) Bronxscape: an exciting new project. 15:2 (Fall). Available at: http://www.slideshare.net/Farrah85p/e2t380. Accessed 30 Dec 2016

NY Sun Works. The sun works center at PS333. Available at: http://nysunworks.org/projects/the-greenhouse-project-at-ps333. Accessed 30 Dec 2016

Odway K (2015) Weiss memorial hospital's rooftop farm & farmers market take root in uptown for 6th season. Lincoln Park Patch, 25 June. Available at: http://patch.com/illinois/lincoln-park/weiss-memorial-hospitals-rooftop-farm-farmers-market-take-root-uptown-6th-season. Accessed 30 Dec 2016

Orsini F, Gasperi D, Marchetti L, Draghetti S, Bazzocchi G, Gianquinto G (2014) Exploring the production capacity of rooftop gardens (RTGs) in urban agriculture: the potential impact on food and nutrition security, biodiversity and other ecosystem services in the city of Bologna. Food Sec 6:781–792

Over the Rhine Foundation (2014) Rothenberg preparatory academy rooftop garden. Available at: http://www.otrfoundation.org/Docs/Background%20Information%20-%20Rothenberg%20Rooftop%20Garden.pdf. Accessed 30 Dec 2016

Ozawa H (2008) Tokyo – Rooftop and underground urban farming lures young Japanese office workers. Agence France-Presse, 5 Nov. Available at: http://www.cityfarmer.info/2008/11/10/tokyo-rooftop-and-underground-urban-farming-lures-young-japanese-office-workers/#more-595. Accessed 30 Dec 2016

Plant Chicago. Tour the plant. Available at http://plantchicago.org/tour/. Accessed 30 Dec 2016

Ryerson Urban Farm. Available at: https://ryersonurbanfarm.wordpress.com/. Accessed 30 Dec 2016

Sanyé-Mengual E (2015) The ICTA-ICP rooftop greenhouse lab: coupling industrial ecology and life cycle thinking to assess innovative urban agriculture. Poster for CILCA 2015 conference. Available at: http://icta.uab.cat/ecotech/jornada/CILCA2015/CILCA_poster_RTG_20150408.pdf. Accessed 30 Dec 2016

Sarant L (2015) Food needs drive Cairo residents to the rooftops. Middle East Institute, 30 Nov. Available at: http://www.mei.edu/content/at/food-needs-drive-cairo-residents-rooftops. Accessed 30 Dec 2016

Seattle Urban Farm Company. Projects. Available at: https://seattleurbanfarmco.squarespace.com/projects. Accessed 30 Dec 2016

Southbank Centre. Queen Elizabeth hall roof garden. Available at: http://www.southbankcentre.co.uk/venues/Queen-Elizabeth-Hall-Roof-Garden. Accessed 30 Dec 2016

Stony Brook University (2012) Stony brook university hospital rooftop Farm helps supplement nutrition for patients. Available at: Stony Brook News, 19 Sep. http://commcgi.cc.stonybrook.edu/am2/publish/General_University_News_2/rooftop_farm.shtml. Accessed 30 Dec 2016

Suchitra M (2015) Rooftop revolution. Down to earth, 30 June. Available at: http://www.downtoearth.org.in/coverage/rooftop-revolution-5016. Accessed 30 Dec 2016

Tatum G (2015) Task Force shelter producing pounds of veggieson rooftop garden. San Francisco Bay View, 25 July. Available at: http://sfbayview.com/2015/07/task-force-shelter-producing-pounds-of-veggies-on-rooftop-garden/. Accessed 30 Dec 2016

The Little Veggie Patch Co. Pop Up Patch. Available at: http://littleveggiepatchco.com.au/pages/pop-up-patch. Accessed 30 Dec 2016

Topager [a]. Available at: www.topager.com. Accessed 30 Dec 2016

Topager [b]. Potager sur le toit d'AgroParisTech. Available at: http://topager.com/portfolio-item/potager-sur-le-toit-d-agroparistech/. Accessed 30 Dec 2016

UN FAO. Urban agriculture in the Gaza Strip through vertical gardens and aquaponics. Available at: http://www.fao.org/fileadmin/templates/FCIT/PDF/Fact_sheet_on_aquaponics_Final.pdf. Accessed 30 Dec 2016

UpTop Acres. The farm at elm and woodmont. Available at: http://uptopacres.com/the-farm-at-elm-and-woodmont.html. Accessed 30 Dec 2016

Via Verde. Carrot city website. Available at: http://www.ryerson.ca/carrotcity/board_pages/housing/via_verde.html. Accessed 30 Dec 2016

Vidinopoulos T (2013) Efficient city farming Berlin. Open Journal. Available at: http://www.openjournal.com.au/ecf-berlin/. Accessed 30 Dec 2016

Waldorf New York (2014) The Waldorf Astoria rooftop garden. Waldorf Blog, 27 June. Available at: http://www.waldorfnewyork.com/blog/the-waldorf-astoria-rooftop-garden/. Accessed 30 Dec 2016

Weise E (2015) Crops in a concrete jungle: Singapore's rooftop farms. USA Today, 4 June Available at: http://www.usatoday.com/story/tech/2015/06/04/singapore-rooftop-farm/28357581/. Accessed 30 Dec 2016

Whole Foods Market – Lynnfield, MA. The greenroofs & greenwalls projects database. Available at: http://www.greenroofs.com/projects/pview.php?id=1612. Accessed 30 Dec 2016

Wilson G (2005) Singapore's new business opportunity: Food from the roof. Talk to the Singapore society for soilless culture, July 13. Available at: http://www.cityfarmer.org/singaporeroof.html. Accessed 30 Dec 2016

Wong D (2014) Telus launch green rooftop with community garden. Carbon49, 24 June. Available at: http://www.carbon49.com/2014/06/telus-green-rooftop-garden/. Accessed 30 Dec 2016

Rooftop Farming Policy

Tim Delshammar, Sofie Brincker, Kristian Skaarup, and Livia Urban Swart Haaland

Abstract This chapter describes how rooftop farming is affected by different policies. A policy is a set of rules or guiding principles that are the basis for actions within governmental as well as commercial and non-govermental organisations. For an understanding of how rooftop farming might be affected by policies, one has to regard explicit as well as implicit policies. A case study from Denmark shows how a rooftop farm may have to comply with a set of supportive and restrictive policies of a variety of governmental and private actors. Since none of the policies that came into action were actually directed at, or explicitly mentioning, rooftop farming, to understand how rooftop farming relies on policies it is important to identify the mechanisms that may facilitate or hinder it.

Introduction

A policy is a set of rules or guiding principles that are the basis for actions within an organisation. Sometimes also decisions or actions are regarded as policy. Policies may be an important driver of or obstacle for rooftop gardening (Freisinger et al. 2015). Dubbeling et al. (2010) mentions four types of policy instruments that may support urban agriculture development: legal instruments like laws, by-laws and ordinances; economic instruments like tax incentives or subsidies; communicative/educative instruments and; urban design instruments (Dubbeling et al. 2010). Provision of space available for roof top farming is an obvious example of how urban design may support roof top farming.

Policy may concern actions in general, like ethical policies, or it may concern actions within a specific domain, like food production or building construction. Policies are set up within governmental organisations on national, regional or local

T. Delshammar (✉)
Street and Parks Department, City of Malmö, Malmö, Sweden
e-mail: tim.delshammar@malmo.se

S. Brincker
ØsterGRO Rooftop Farm and Copenhagen University, Copenhagen, Denmark

K. Skaarup • L. Urban Swart Haaland
ØsterGRO Rooftop Farm, Copenhagen, Denmark

F. Orsini et al. (eds.), *Rooftop Urban Agriculture*, Urban Agriculture,
https://doi.org/10.1007/978-3-319-57720-3_3

levels. There are a number of cities that have decided on food policies. These are examples of policies within governmental organisations on local level. But governments are not the only type of actor that sets up policies or that may support rooftop farming. Also corporations and non-governmental organisations use policies.

However, there seems to be very few, if any, examples of policies specifically geared at promoting or regulating rooftop farming. Policies for green roofs have been developed for a couple of years. Rooftop farming policies still have to follow, with for example the Kathmandu Metropolitan City having developed a draft rooftop farming policy in 2014 (Nepal Forum for Environmental Journalism 2014). For an understanding of how rooftop farming might be affected by policies, one has to regard different types of policies. There are for instance, both explicit and implicit policies. An implicit policy can guide decisions without explicitly mention the subject in question. Thus, an urban food policy (or a building or environmental policy) is likely to affect decisions on rooftop farming without explicitly mentioning rooftop farming. A review of policies bearing on rooftop farming was done in Nepal in 2013, leading to the drafting of the rooftop farming policy mentioned above (Nepal Forum for Environmental Journalism 2013). In the absence of good examples of explicit rooftop farming policies, this chapter will rather deal with implicit policies. Furthermore, it will deal with policies that are supportive of roof top farming as well as policies that restrict it.

Though laws and spatial plans are not usually regarded as policies, they do in some cases have the same properties as policies: e.g. to guide action within specific organisations. There are for example laws that regulate or guide the actions within governmental organisations. Sometimes spatial plans are decided by a municipality to guide the planning decision within the same municipality. For that reason it makes sense to bring up laws and spatial plans in the same context as formal policies.

Policies, planning and regulation, ownership, access and use rights are aspects of the institutional framework for governance. An understanding of policies must regard the entire institutional framework as well as other aspects of governance like scale, the urban context, actors and coalitions, resources, and the current discourses (Lawrence et al. 2013).

The following paragraphs will discuss how rooftop farming is treated in some existing city food policies, thereafter some examples of supportive and restrictive other policies, laws and plans will be discussed. Lastly, a case of roof top farming will be presented and analysed from the perspective of policies.

Urban Food Policies

The Healthy and Sustainable Food for London, UK (London Development Agency 2006) deals with all stages of the food chain, from primary production to eating and disposal. These are discussed in relation to five key policy themes: health, environment, economy, social & cultural and food security. The follow-up implementation plans also support local production, mainly by means of suggestions for public sector food procurement (London Development Agency 2007; Greater London Authority 2011). Although the suggested actions to increase food production within London supports rooftop farming, they however do not necessarily ensure more rooftop farms.

In addition to promoting local food production, the Vancouver (Canada) Food Strategy does specifically highlight the importance of rooftop farming. It is suggested that beekeeping on roofs should be reconsidered to make it possible to have beehives on institutions and higher density locations. Further, it calls for the education and awareness about growing food on rooftops in neighborhoods to be promoted. (City of Vancouver 2013).

Other Regulations, Plans and Policies

Besides food strategies, urban spatial planning is probably the most important policy domain for rooftop farming. Taking the Swedish planning system as a starting point, it is an interlaced system of laws on national level, policies on regional and local levels and plans on local level. The Swedish Planning and Building Act sets the generic goals, like sustainable use of land. It also frames the planning process in terms of the local planning authority's responsibilities and limitations. The municipalities produce comprehensive plans with planning guidelines for the entire municipality. These are elaborated in detail in the detailed plans that determine the land use and e.g. building heights uses and design. (SFS 2010:9009).

The planning system obviously differs from country to country. However, a unifying trait is that this system might be supporting as well as restricting rooftop farming. It is unlikely that planning legislation will demand rooftop farming, but there are possibilities to use the local detailed plans to promote or even prescribe rooftop farming (Freisinger et al. 2015). On the other hand, rooftop farming might be prohibited intentionally or unintentionally in detailed plans. Restrictions for building heights or access (including safety regulations) may for instance hinder greenhouses on rooftops. Other municipal planning issues – spatial or not – might also relate to rooftop farming, e.g. green structure planning, climate planning, planning for ecosystem services, leisure and social cohesion. Rooftop farms that become a part of the green structure, for leisure or business may be supported by governments that seek new ways of achieving multiple goals.

Most urban development is affected not only by governmental planning policies, but also policies guiding other actors. Many large building companies adhere to quality assessment schemes like BREEAM (Great Britain), LEED (USA), CASBEE (Japan) or Green Star (Australia). These schemes are intended for guiding the planning and building process. The inclusion of goals like green roofs or local food production in these systems may support rooftop farming.

Other important actors are the financers of building and development projects. Not only banks, but also international development agencies and residents societies finance building and development projects. Rooftop farming might be explicitly or implicitly in line with the policy goals of these organisations as a way to achieve sustainability. On the other hand, rooftop farming might be out of scope or even perceived as being in conflict with the policies. For example, there might be issues of security or risks of damages to buildings that prohibit rooftop farming. Under all circumstances, financing is a key issue and most organised financers are bound to be guided by some kind of policy.

Case Study: The Danish Rooftop Farm "ØsterGRO"

ØsterGRO is the first rooftop farm of its kind in Scandinavia and also the first and only market driven urban farming project in Denmark. The farm was established in the spring of 2014 as a bottom up project, initiated by the three young architects with a dream of creating an actual agricultural farm in the center of the dense, urban city of Copenhagen. The farm is located in the northern part of Copenhagen in the district of Østerbro. The name ØsterGRO referring accordingly to the local neighborhood and the Danish word for "grow".

On the 600 m^2 rooftop there is a farming area of 350 m^2. The planting beds on the concrete surface consist of 30 cm deep and 40 cm wide layer of substrate. In addition there are four beehives, 14 chickens and a 28 m^2 greenhouse. The greenhouse houses in the evenings the restaurant Stedsans that had about 6000 guests during 2016. The garden is placed on top of a 15 m high building, with three storeys of companies underneath. The vision for the project is to provide Copenhageners with a new recreation platform for leisure, interaction and learning e.g. knowledge about organic farming. The farm is managed through a community supported agriculture (CSA) model, with two fulltime farmers (two of the founders) that cultivate the farm and distribute the vegetables to the 40 CSA members every week. One key actor in the process of establishing the farm was the city renewal office in the district. The office is a decentralized part of the municipality of Copenhagen with the task to regenerate the area by creating social, economic and physical sustainability in the neighborhood. The strategy is to network with actors in the neighborhood, with the aim to renew public spaces and public accessible spaces. A local rooftop farm was in line with many of the goals that the office had. It would supply citizens with a publicly accessible green space, provide a local platform for further networking and help moderating storm water runoff. The network that the office had established helped the initiators of the farm to find a property owner that would support the idea of a rooftop garden. The initiators got access to the rooftop of a privately owned building. An important aspect of the network was that trust between local actors had been built up. This made it easier to overcome barriers that would otherwise be likely to hinder unusual solutions. Besides the contact between the initiators and the property owner, the office also helped with contacts at the municipality and with other local networks. Last, but not least, the office financed parts of the projects development costs (Fig. 1). The roof top farm was clearly consistent with the policy of the local municipality development office. However, it was not likewise compatible with the existing planning regulations. It was necessary to apply for a building

Fig. 1 The city renewal office has been an important link between the central parts of the municipality, the local property owner and an extensive local network of residents and businesses

permit at the municipality before constructing the farm. The permit was to ensure that the payload and safety measures were met. During the application process, it was found that the municipality had nine parking lots registered on top of the building. This was due to the fact that the building was an old car auction house and the roof had earlier been used to store cars. The farming project collided with a municipal policy of not abolishing any parking lots. Despite the fact that the parking lots were not for public use and situated on a privately owned roof the permit took more than 4 months to get and ended up costing almost 10.000 Danish kroner.

The property owner did not only give access to the roof, but was also supportive in the process of getting a building permit. This made the contacts with the municipal planning office smooth.

ØsterGRO is currently an association that in 2016 counts about 40 members. All members get a bag of vegetables weekly during the harvesting season, June-November. About 50% of the vegetables are produced on the rooftop and the other part comes from a local farm. In addition to the supply of vegetables to members, the garden also supply the restaurant. Being an association instead of a company means that there is no mandatory permit for selling food and that the farm does not have to comply with the regulations of the Danish Veterinary and Food Administration. On the other hand, an association cannot get the organic certification that is highly esteemed and regarded as a safety for the consumer. However, with a small number of consumers and a close relation between the producer and consumer trust is not an issue.

Bullet Points

- The case of ØsterGRO shows how a rooftop farm may have to comply with a set of supportive and restrictive policies of a variety of governmental and private actors. It is notable that the local government had policies that were both supportive and restrictive. None of the policies that came into action were actually directed at, or explicitly mentioning, rooftop farming. Though ØsterGRO as a western European case does not give a worldwide comprehensive picture of how policies interact with the creation of rooftop gardens, it still gives a hint of the complexity.
- To understand how rooftop farming relies on policies it is important to identify explicit as well as implicit policies; how policies affect different actors and stakeholder and different types of polices (laws, economic instruments, communication and design strategies).

References

City of Vancouver (2013) What feeds us: vancouver food strategy. January, 2013. http://vancouver.ca/files/cov/vancouver-food-strategy-final.PDF. Accessed 24 Oct 2015

Dubbeling M, van Veenhuizen R, de Zeeuw H (2010) Cities, poverty and food: multi-stakeholder policy and planning in urban agriculture. Practical Action Pub, Rugby

Freisinger UB, Specht K, Sawicka M, Busse M, Siebert R, Werner A, Thomaier S, Henckel D, Galda A, Dierich A, Wurbs S, Grosse-Heitmeyer J, Schon S, Walk H (2015) There's something growing on the roof. Rooftop greenhouses. Idea, planning, implementation. Leibniz Centre for Agricultural Landscape Research (ZALF), Muncheberg

Greater London Authority (2011) The Mayor's food strategy: healthy and sustainable food for London. An implementation plan 2011–2013. https://www.london.gov.uk/sites/default/files/London%20Food%20Strategy%20Implementation%20Plan%202011-13.pdf. Accessed 24 Oct 2015

Lawrence A, De Vreese R, Johnston M, Konijnendijk van den Bosch C, Sanesi G (2013) Urban forest governance: towards a framework for comparing approaches. Urban For Urban Green 12(2013):464–473

London Development Agency (2006) Healthy and sustainable food for London. The Mayor's food strategy. May 2006. https://www.london.gov.uk/sites/default/files/FoodStrategy2006.pdf. Accessed 24 Oct 2015

London Development Agency (2007) Healthy and sustainable food for London. The Mayor's food strategy implementation plan. September 2007. https://www.london.gov.uk/sites/default/files/London%20Food%20Strategy%20Implementation%20Plan%202007.pdf. Accessed 24 Oct 2015

Nepal Forum for Environmental Journalism (2013) Rooftop gardening and management in Nepal: requirements for specific policy, acts and regulations. NEFEJ, RUAF Foundation and UN Habitat. http://www.ruaf.org/sites/default/files/Rooftop%20gardening%20in%20Nepal-%20requirements%20for%20specific%20policy%20and%20regulation_2.pdf. Accessed 3 Sept 2016

Nepal Forum for Environmental Journalism (2014) Policy for Rooftop gardening in Kathmandu Metropolitan City. NEFEJ, RUAF Foundation and UN Habitat. http://www.ruaf.org/sites//default/files/Rooftop%20garden%20policy%20Kathmandu%20Metropolitan%20City%20-final_1.pdf. Accessed 3 Sept 2016

SFS 2010:900. Plan och bygglag (Swedish Planning and Building Act)

Part II
Design of Rooftop Agriculture Systems

Francesco Orsini

Elements of Rooftop Agriculture Design

Silvio Caputo, Pedro Iglesias, and Heather Rumble

Abstract This chapter focuses on the elements that must be considered when designing rooftop gardens and integrating them within buildings. Different types of rooftop gardens and how they can be integrated within existing and new buildings in order to enhance their environmental performance, better connect with their users and contribute to the amelioration of the urban environment are presented together with a description of necessary factors for implementation. These include: techniques and technologies for cultivation (i.e. simple planters, green roofs and hydroponics), necessary structural loadbearing capacity of the host building and protection from wind. The chapter also gives an overview of existing innovative and experimental projects of rooftop gardens, ranging from those that require little to high investment.

Introduction

The last decade has seen a surge in rooftop farming internationally, with numerous small and large scale projects arising on almost every continent. This is a result of a renewed interest in urban agriculture as a practice for urban resilience, which can also generate new economic opportunities. The availability of new building products and technologies developed specifically for green roofs has facilitated the implementation of urban agriculture projects on rooftops, enabling efficient use of space in increasingly dense urban centres. In many forms and for many centuries, green roofs have been designed as an integral component of buildings providing thermal insulation (Getter et al. 2009; Berardi et al. 2014). However, modern green roof technologies started being developed only over the last decades e.g. in Germany,

S. Caputo (✉)
School of Architecture, University of Portsmouth, Portsmouth, UK
e-mail: silvio.caputo@port.ac.uk

P. Iglesias
Faculty of Architecture, Universidad Politécnica de Madrid, Madrid, Spain

H. Rumble
Department of Geography, University of Portsmouth, Portsmouth, UK
e-mail: heather.rumble@port.ac.uk

F. Orsini et al. (eds.), *Rooftop Urban Agriculture*, Urban Agriculture,
https://doi.org/10.1007/978-3-319-57720-3_4

Switzerland and Scandinavia (Dvorak and Volder 2010). Since then, green roofs have diffused quickly, mainly (but not exclusively) as a way to increase the environmental efficiency of buildings, decrease water run-off, increase urban biodiversity and counter the urban heat island effect (Getter et al 2009; van Woert et al. 2005; Kadas 2006; Berardi et al. 2014; Feng et al. 2010). In the US alone, the green roof industry has grown by more than 35% between 2007 and 2008 (Greenroofs 2008), with cities such as Chicago estimated to have over 600 public and private green roofs either built or in planning. Legislation encouraging the installation of green roofs has also supported their implementation in cities such as Berlin, Germany, Copenhagen, Denmark or Toronto, Canada (Litichevskaya 2010).

The design of green roofs requires particular attention to detailing, dimensioning of structural elements, connection with building systems and landscaping. Landscaping can become a predominant design feature depending on the use of the roof. Extensive green roofs, i.e. roofs that are shallow in depth, planted with hardy vegetation and low in maintenance (Grant 2006), have been used to attract local species of plants, insects and animals, to varying levels of success (Williams et al. 2014). An example of this type of roof includes the 32,000 m^2 Rolls Royce green roof, the largest green roof in the UK. Intensive green roofs have deeper substrate and often more traditional landscaping features. They are typically designed for the enjoyment of buildings' occupants, adding visual and also economic value to the building, but may also be designed to mimic a particular natural habitat. The Vancouver Convention Centre, Canada, is a good example of this, integrating over 400,000 indigenous plants into its design (Vancouver Convention Centre 2015).

Technologies used in all types of green roofs are similar and may be suitable to rooftop farming depending on the depth of soil, which varies according to the type of plant. Activities for horticulture, however, unlike some green roofs that do not require frequent maintenance, necessitate constant access to the roof, especially in the growing seasons. In turn, this requires different and sometimes more complex design approaches from green roofs.

Existing projects vary substantially. Many of these projects (herein called *informal*) have been self-built and designed by users, thus resulting in implementation of highly affordable, sometimes makeshift solutions for cultivation. Others (herein called *formal*) are the result of the implementation of green roofs that can support the cultivation of edible plants. Finally, a third type of rooftop farms (herein called *technological*) uses advanced technologies, sometimes with environmental control systems (e.g. greenhouses) or soil-less systems. Each one of these types requires particular design considerations, which will be presented and discussed in the following paragraphs of this chapter. It must be noted that formal aspects of the rooftop farming design and successful integration within the building structure and envelope should not be underestimated. Researches in urban agriculture (Gorgolewski et al. 2011) highlight the importance of new formal solutions for food cultivation in cities that can visually communicate new practices and generate consent amongst the general public. Rooftop farming will probably necessitate attention on this aspect too. This chapter will first present some cross cutting issues that must be taken into account when designing any rooftop farming project and will subsequently focus on each of the three types (informal, formal and technological) in depth. Finally, it will discuss the different approaches to the implementation of rooftop farming projects, both on new and existing buildings, considering design decisions and construction costs, and comparing them to other solutions.

Design aspects discussed here have been observed on a number of projects mainly across Europe, North America and China. They are therefore based on empirical evidence and sometimes supported by existing literature. All of these projects required the upgrading of existing buildings, which, from a design perspective, is more challenging than the design of new build which can be designed from the onset with a roof capable of supporting horticultural activities.

Common Design Elements

Each project's objectives will inevitably influence the selection of techniques and technologies as well as the approach to design of the rooftop farm generally. Sanyé-Mengual et al. (2015a), in their study on an experimental project on the rooftop of a social housing building in Bologna, Italy, maintain that in terms of environmental impact, conventional cultivation on soil in recycled timber containers appears to be more advantageous than conventional hydroponics (e.g. floating systems or nutrient film technique) in winter months. The perspective changes, however, if the objective shifts to maximized production, with hydroponic systems demonstrating a higher yield than conventional agriculture techniques with equal areas of cultivation (Grewal and Grewal 2012). Another influencing factor can be the availability of buildings as well as the order of economic investment available. The former refers to the need for the building to withstand higher loads due to the weight of soil and other equipment, as well as the availability of water and energy on the rooftop. The latter refers to the implementation cost which will inevitably determine the technologies used. Finally, the approach to design will be influenced by socio-cultural parameters. Regardless of the economic and environmental factors, community projects can privilege spatial and practical arrangements (rather than maximised production) enabling, for example, the gathering of a small to medium number of volunteers and visitors for demonstration purposes. Polyculture gardens can be more water intensive than monoculture cultivations, in which water usage is calibrated on the crop (Sanyé-Mengual et al. 2015a). Therefore, in order to be environmentally efficient, it is important that the design of rooftop gardens takes into account the real use of resources.

As mentioned in the introduction, the chapter clusters existing projects under three different types, each one with its own design approach informed by environmental, economic and social considerations. Before proceeding to the description of these design approaches, some common design elements with which all projects will need to contend with are discussed.

Access to the Rooftop

If not independent, access to both non-residential and residential buildings' rooftops can be problematic in different ways (e.g. privacy of occupants and closure after working hours). Agreement with building occupants is therefore necessary if vertical circulation must be shared. Alternatively, it may be necessary to build external stairs, which in turn can impact the appearance of the building, its external circulation and access, and the adjacent open space generally. A case in point is a rooftop farming project in Crouch End, North London (which terminated in 2014) located on the top of a supermarket. In order to access the roof without disrupting commercial activities and out of working hours, an external staircase was built with prefabricated components for scaffolding at the back of the building, on an area for parking, near the entrance for suppliers. Because of the particular location, the stair was not visible from the main road, thus not affecting the appearance of the main façade of the building and the public circulation. Le Jardin Perché, which is hosted on a sports center in Paris, has a similar arrangement. In both cases, however, the design of the access is a crucial factor to be considered at a planning stage. Clearly, flat roofs are better placed to be used for farming, although it may be possible exceptionally to transform inclined roofs. A case in point is the underground car park in Geneva with a sloping roof that has been landscaped as a vineyard.

Health and Safety

In existing buildings, rooftops are often not designed for public use. In order to upgrade them, compliance with local regulatory frameworks for Health and Safety is necessary. For example, adequate railings, whenever these are missing, must be built. Perimeter railings must be sufficiently tall and provide screening, such as those built on the rooftop of RISC in Reading (a charity with the aim of raising awareness about development issues). These railings are made out of timber strips interwoven, providing privacy to roofs that are overlooked by neighboring buildings, as well as providing privacy for neighbours. Other rooftops such as Jardin sur le Toit in Paris, use metal grid panels which are transparent. Multi-functional metal grid panels were used on the rooftop of Food from the Sky, North London, both as railings and for support for climbing plants (i.e beans). It must not be underestimated that railings, especially if tall, may need to be particularly robust in construction and connected to structural elements in locations with strong winds as they can act as wind breakers. Rooftops can be exposed to higher average wind speeds than at ground level (Hanna et al. 2006), therefore plants must be protected from wind to avoid excessive disruption (Oberndorfer et al. 2007). For example, the authors have witnessed pinned out fruit trees on green roofs at risk from wind damage. Whilst fruit trees at ground level may be pinned out, on a green roof this effectively creates a wind sail and so is not appropriate.

Attention must also be given to existing air conditioning units on the roof of many commercial buildings, which will need to be fenced off to prevent access from the public and still be reachable for maintenance. Besides, air vents can damage plants due to heat, so their deployment must be consequently planned.

Structural Loads

A common classification of green roof cover considers three main categories according to the substrate depth: extensive (25–100 mm), semi-intensive (120–250 mm) and intensive (150–400 mm). Growing media for each category has different weights and implementation costs (see Table 1). However, this classification is not universally used and different organizations utilize different categories. For example, the Green Roof Centre at the University of Sheffield differentiates green roofs in extensive, semi-extensive and intensive, corresponding to slightly different depth and weight ranges (see www.thegreenroofcentre.co.uk). Other scholars distinguish only between intensive and extensive green roofs, the former characterized by soil depth below 150 mm and the latter above 150 mm (Getter et al. 2009). ZinCo, one of the largest German producers of green roofs, recommends for rooftop farming intensive green roofs with a substrate between 200 and 400 mm, depending on the plants that will be grown, with a weight of saturated substrate of approximately 300 kg m^{-2} for a build-up depth of 250 mm. According to the company, plants that can be grown with a 200 mm substrate include: lettuce, onions, herbs, courgettes, aubergines, pumpkins, cabbage, melons and strawberries.

In general, the deeper the substrate, the better for food production. The depth of the substrate, as well as the ratio of organic matter (such as compost) to mineral (such as expanded shale), will dictate the strength of natural buffers within the soil. Soil temperature varies greatest at the soil surface and lessens with deepening soil (Hillel 1982), thus deeper soils may protect plant roots and soil organisms from the extreme temperature fluctuations often found on rooftops. Deeper soils have also been shown to hold more water (Chang and Hong 2012), lessening the need for irrigation. Finally, higher levels of organic matter can also retain water better, although then risks of both waterlogging and excessive weight shall be considered. Thus, the interplay between depth, substrate type and organic matter content affects the success of plant growth on agricultural roofs, in addition to design features such as number/size of

Table 1 Main substrate features of extensive, semi-extensive and intensive green roofs

	Extensive	Semi-extensive	Intensive
Depth of substrate	60–200 mm	120–250 mm	150–400 mm
Weights	60–150 kg/m^2	120–200 kg/m^2	180–500 kg/m^2
Cost	Low	Periodic	High

Adapted from www.livingroofs.org

drainage points and amount of bare, unvegetated substrate (Carson et al. 2013). There are a few previous studies on the "ideal" combination (e.g. Papafotiou et al. 2013), but more work by the scientific community is needed in this area.

There are a number of lightweight "off the shelf" substrates specifically designed for rooftop agriculture, such as that manufactured by ZinCo Green Roof Systems Ltd. These substrates, as well as being lightweight, are also free draining to reduce load on the building. Lightweight substrates typically consist of a proportion of mineral, such as expanded shale or crushed brick, with organic matter added in proportions higher than found in common extensive rooftop agriculture.

Loads associated with the minimum depth for rooftop farming may not constitute a major problem if the farm is integrated in a new building and loads are taken into account from the onset of the design process. In existing buildings, such a load may not be compatible with the carrying capacity of the roof. Castleton et al. (2010) suggests that typically, buildings over 30 years old have higher bearing capacity, whereas newer buildings have been designed with higher structural efficiency (thus with little spare bearing capacity) (see also Knepper, 2000). Existing buildings with steel or timber structural frames have roofs with even lower bearing capacity that may not be suitable for rooftop farming without an upgrading of these frames. A typical bearing capacity for a roof is 150 kg m^{-2} (Whittinghill et al. 2013; Castleton et al. 2010), although this will greatly vary depending on the age of the building, its use (commercial, residential, etc.) and the regulatory framework of the country. According to the UK Building regulations, new buildings must be designed and constructed with a minimum load bearing capacity of 150 kg m^{-2} of distributed load or 180 kg m^{-2} of concentrated load (BS 1988). Given the great diversity of buildings, roofs and regulatory frameworks, a survey from a structural engineer is therefore necessary prior to any design of a rooftop farm.

This is valid also for farms with greenhouses, which may not necessarily have a great weight in terms of structure and external cladding, due to the construction consisting of aluminum elements with an envelope of polycarbonate panels. However, depending on the equipment used for method of cultivation, the total load may critically increase. A key consideration for greenhouses is the load of winds, which becomes bigger with the height of the building, thus requiring particular attention to the robustness of the connection of the greenhouse to the existing building structure and of the greenhouse itself. Finally, great consideration must be given to water tanks representing a major load on parts of the roof.

Implementation and Costs

Rooftop farming can be implemented both in new and existing buildings. New buildings offer a wider range of opportunities for the installation of different solutions and the rational use of roof surfaces since access and structural loads can be designed from the onset of the design process. Existing buildings still offer a great number of possibilities for farming activities as can be seen in the portfolio of projects presented by "The Living Greens" (http://thelivinggreens.com).

Green roofs make use of a special substrate which differs from substrates used in traditional farming in its physic-chemical properties, in its weight and in the way it retains water. It must be specifically formulated for the particular crops of the installation in order to enhance soil properties and prevent the spread of pathogens, and survive in extreme conditions, with high winds, sudden changes in temperature, and total solar exposure. It can be installed in almost every type of supporting structure on top of different roof solutions that include pitched roofs, in which a supplementary retention grid must be installed in order to prevent slippage of the cover. For further information on substrate properties and selection procedure, please refer to chapters "Soil Based and Simplified Hydroponics Rooftop Gardens" and "Integrating Rooftop Agriculture into Urban Infrastructure".

Other installation systems, as hydroponic cultivation, do not require a supporting surface but instead make use of a supporting system that can be placed in a great variety of new and existing roofs and grow species on trays. Among most technological approaches to this technique, there are the solutions offered by VertiCrop (http://grow.verticrop.com/vertical-farming/) with yields up to 20 times higher than field agriculture production requiring only a small amount of water, grown on a fully automated tray system.

The cost of the implementation of green roof in new buildings has been proved to have a return period of 7–11 years, just taking into account the wide range of environmental benefits and the extended life cycle of the waterproof materials (Carter and Keeler 2008; Clark et al. 2006; Lee 2004), being the additional roof unit construction costs the water retention films, the growing media, the plants and the installation costs, which may have variations in different regions. Benefits derived from the farming activities may prove that the increase of building costs is negligible.

Installation costs of a rooftop farm on an existing building have to be studied more thoroughly. As mentioned above accesses and health and security measures can take an important part of the budget and require consultation with an expert. On the other hand retrofitting an old roof structure can be an excellent opportunity to replace it with a green surface in which to develop rooftop farming at an affordable cost with the associated benefits of lower maintenance expenses. An example of costs for green roof agriculture projects can be found in Table 2.

Water and Power – Minimising Inputs

Buildings can generate waste or collect resources which in turn can become nutrient for horticulture. Waste can include greywater, heat recovered from indoor air and organic waste which can be composted. Rainwater can also be collected from parts of the rooftop not used for farming and power produced through on-site microgeneration. The utilization of all these resources can substantially reduce the

Table 2 Examples of green roof costs in Europe

	Lower range (€/M²)	Upper range (€/M²)	Lower range (€)	Upper range (€)
New buildings[a]				
Insulation, waterproofing and improved topsoil	55	70		
Plant species	15	50		
Total	**70**	**120**		
Existing buildings[a]				
Implementation cost for existing buildings, including demolition of old roof and preparation of the base	80	185		
Plant species	15	50		
Total	**95**	**235**		
Comparison with a standard roof[a]				
Standard roof, including insulation, waterproofing and coating	25	60		
Greenhouse				
Steel and polycarbonate	15	35		
Aluminium and polycarbonate	40	75		
Facilities[b]				
Climatic control			2500	5000
Irrigation control			1200	1800
Drip irrigation system	1.5	5.5		

Data collected from: [a]CYPE Ingenieros; iTeC; DANOSA; TEXSA; [b]Andalucia Regional Government

emissions associated with farming practices. However, adapting building systems in a way that enables by-products to be used for cultivation may require not only additional time and economic investments but also careful evaluation of what is available in terms of resources from the building and considerable design efforts. Cerón-Palma et al. (2012) conceptualise the 'rooftop eco greenhouse' as a closed loop system deeply interconnected with the building systems, producing crops and improving the performance of the host building. Beyond the potential that technologies offer to use efficiently resources utilised to operate buildings, any type of rooftop farm can improve the environmental performance of buildings (e.g. thermal insulation and water run-off). It must be therefore designed with this in mind.

Different Design Approaches to Rooftop Agriculture

Layouts, use of materials and components as well as spatial and technological solutions for rooftop farming are influenced by the objectives of each farming project, which can be, for example, commercially rather than community-oriented. What follows is a presentation of the design solutions that have been used in some existing farms, divided into informal, formal and technological farms.

Informal Rooftop Agriculture

This category includes a wide variety of projects, all of which use containers (i.e. planters) positioned directly, or on some intermediate surface, onto the roof. In this category, planters used are typically raised beds made of treated timber (which is often recycled), built specifically for each farm and shaped according to the needs of the farmers or the particular conditions of the site. Longitudinal beds can be made out of recycled timber crates, with a size appropriate to the configuration of the roof available for cultivation and a layout designed to maximise production. Typically, 'informal' rooftop farms are the ideal solution when financial resources are limited. Community groups and organizations that have access to a rooftop and wish to start a rooftop farm, but cannot afford the installation of a green roof, often rely on volunteers to collect material and assemble it with sometimes ingenious arrangements. In these cases, production is often a means to community engagement, and other educational and social purposes. The roof is used as an interface between farmers and the public (e.g. students from schools, local communities, apprentices, etc.). Some projects rely on the sales of produce and the organization of gardening courses to support their broader activities (see Food from the Sky, Table 3), aimed at engaging with local communities through food growing practices.

Many roofs in existing buildings have not been originally designed for public use and their exposed surface is merely a bitumen, waterproof layer (rather than hard paving), which may not be suitable for heavy transit and heavy planters. In a few projects like the experimental temporary farm Via Gandusio in Bologna, Italy, the Very MK Rooftop Farm in Hong Kong and Food from the Sky, London (Table 3), the planters were installed on the existing bitumen layer. In this case, it is advisable for planters to be raised from the floor in order to avoid stagnation of water that could damage materials such as timber (and the bitumen layer) in the long-term. In general, bitumen layers do not possess sufficient mechanical strength and damage can lead to water leaks. These surfaces can be utilized for short-term projects, but it is advisable to install stronger surfaces for longer-term use. For example, RoofKrete offers on the market a 'waterproof' membrane that is recyclable, non-toxic, with high tear resistance and a lifecycle of 100 years (http://www.roofcrete.co.uk/html/green_roof.html). Other recommended wearing courses could include concrete tiles, or other hard tiles (Deplazes 2005). Another possible design solution is that adopted in le Jardin Suspendus, Paris (Table 3), in which raised bed have been positioned on the existing surface of the roof, with walkable platforms in between them. In this way, the soil is at the same level of the platforms.

Table 3 Location, size and dedicated website of projects reviewed in this chapter

Project	Location	Size	Website
INFORMAL			
Wayside Chapel	Sidney, Australia	200	www.thewaysidechapel.com/communal-garden.php
Food from the Sky	London, UK	400	www.foodfromthesky.org
HK Farm	Hong Kong, China	N/A	http://www.hkfarm.org
Via Gandusio	Bologna, Italy	N/A	http://rescue-ab.unibo.it/website/en/project
Kowloon City and Causeway Bay Schools	Hong Kong, China	N/A	http://www.oxy.edu/sites informal
Bank of America Tower	Hong Kong, China	N/A	https://gogreenhk.wordpress.com/2014/08/30/rooftop-farming-in-hong-kong/
University of Hong Kong Rooftop Farm	Hong Kong, China	400	http://www4.hku.hk/photos/index.php/gallery/event/detail/331
Very MK Rooftop Farm	Hong Kong, China	N/A	http://www.oxy.edu/sites
Priority Zone Rooftop Garden	Durban, South Africa	N/A	http://urbanearth.coza/articles/roof-top-gardening-inner-city-durban's-priority-zone-rooftop-garden
Tlhago RooftopGardes	Johannesburg, South Africa	N/A	http://blogs.worldwatch.org/nourishingtheplanet/an-urban-gardening-initiative-greens-johannesburg-rooftops-in-a-bid-to-tackle-climate-change/#more-16536
AFHCO Rooftop Garden	Johannesburg, South Africa	N/A	http://www.cityfarmer.info/2011/11/20/johannesburg%E2%80%99s-first-rooftop-food-garden/
Flyover Farm	Mumbai, India	N/A	http://foodtank.com/news/2015/06/greening-mumbai-bringing-agriculture-to-the-rooftops-of-indias-largest-city
The Bachelor Farmer	Minneapolis, USA	N/A	http://thebachelorfarmer.com/menu/food/

Linden Tree Place	Vancouver, Canada	N/A	http://www.kitshouse.org/programs/housing/rental-process-for-8th-and-vine/
FORMAL			
RISC	Reading, UK	200	www.risc.org.uk/gardens
Museum of London[a]	London, UK	3500	http://www.museumoflondon.org.uk/files/3913/7327/5939/GreenRoofsatMuseumofLondon.pdf
University of Greenwich, Stockwell St Building	London, UK	3400	http://www.buildingcentre.co.uk/case_study/blackdown-green-roof-provides-unique-space-at-greenwich-university
Jardine Matisse	Paris, France		www.veniverdi.fr
Jardin Perche'	Paris, France	600	www.jardinons-ensemble.org/spip.php?article235
Jardin Suspendus	Vincennes, France	1500	http://www.jardinssuspendus.org/
Technoseum	Mannheim, Germany	500	http://www.zinco-greenroof.com/EN/references/rooftop_farming.php
Chongqing Green Roof	Chingqing, China	10,000	http://myvegua.com/10000-square-meter-rooftop-farm-in-chongqing/
Pujia Primary School	Hangzhou, China	2000	http://inhabitat.com/organic-rooftop-farm-grows-atop-an-elementary-school-in-china/
Green Potato Project	Tokyo, Japan	N/A	http://www.cityfarmer.info/2008/11/10/tokyo-rooftop-and-underground-urban-farming-lures-young-japanese-office-workers/
Eagle Street	Brooklyn, USA	560	http://rooftopfarms.org/
Gary Comer Youth Centre	Chicago, USA	760	http://www.greenroofs.com/projects
TECHNOLOGICAL			

(continued)

Table 3 (continued)

Project	Location	Size	Website
Greenpoint	New York City, USA	1400	http://gothamgreens.com
Gowanus	New York City, USA	1860	http://gothamgreens.com
Vinegar Factory	Eli Zabar, NYC, USA	N/A	www.marthastewart.com/270946/eli-zabars-rooftop-garden
Whole Foods	New York City, USA	1900	http://wahaso.com/projects
Arbor House	New York City, USA	930	http://wahaso.com/wahaso-provides-a-sustainable-water-system-for-the-rooftop-greenhouse-in-bronx-ny/
Roof Water-farm	Berlin, Germany	N/A	www.roofwaterfarm.com/en/news/
UF001 LokDepot	Basel, Switzerland	300	http://urbanfarmers.com/projects
ICTA-ICP Lab	Barcelona, Spain	800	http://www.fertilecity.com/en/index.php?p=1_13_goals
LUFA Farms Ahuntsic	Montreal, Canada	2970	http://lufa.com/en/our-farms.html
LUFA Farms Laval	Montreal, Canada	4000	http://lufa.com/en/our-farms.html
Rooftop Greenhouse Project	London, UK	N/A	http://www.rooftopgreenhouse.co.uk/index.html
EPSE Rooftop Greenhouse	Jerusalem, Israel	4500	http://www.epse.org/rooftop-greenhouse

[a]NB this is not a rooftop agriculture project but has been included to demonstrate technological approaches

The height of planters, whether tall and free-standing, or integrated to give a "ground-level" feel should be informed by the accessibility of end-users. Whilst low planters replicate regular ground-level farming and can facilitate a wide range of users (e.g. adults and children with no mobility restrictions), installing higher planters could improve accessibility for gardeners, especially those with specific mobility requirements. Planters designed for standing adult gardeners without mobility restrictions are recommended at around 1 m (Thrive 2008). Wide surrounds may be installed to allow gardeners to sit, in which case a height of approximately 0.7 m is appropriate (Thrive 2008). There are examples of rooftop agriculture projects suitable for wheelchair users, such as Linden Tree Place, Vancouver (Table 3). Relf (1995) suggests that a planter height of 0.8 m is suitable for wheelchair users. Whilst Linden Tree Place is a rare example, there is a growing body of evidence to suggest that gardening is a valuable therapeutic tool for those with mental health conditions (Clatworthy et al. 2013) and that older adults in particular value gardening as an essential activity for their psychological and physical wellbeing (Scott et al. 2015) and so access to rooftop gardens could be a serious consideration in future years, with limited availability of urban space.

Informal rooftop gardens utilize a wide variety of recycled materials to construct planters. Whilst recycled timber, particularly recycled pallets, are very common, other solutions to planter systems have been employed. The Wayside Chapel, Sydney (Table 3), combines traditional plastic garden pots, with large plastic tubs in which are grown over 50 varieties of fruits and vegetables and includes small trees. Food from the Sky rooftop garden in London, UK (Table 3), utilized municipal recycling boxes donated by the local council in which it plant their crops (Barnett 2011). The only requirement for rooftop planters is the ability to store soil and allow water drainage (i.e. include drainage holes). The Bachelor Farmer rooftop garden, Minneapolis, USA (Table 3), even utilizes buckets and children's paddling pools as rooftop planters to supply their restaurant. Less rigid structures may also be used as planters, such as the tough plastic bags typically used to deliver substrates or to collect municipal compost. There are numerous commercial adaptations of these bags available (e.g. potato planter bags, see: http://www.jbaseedpotatoes.co.uk/potato-planter-bags), as are a range of "off the shelf" rigid containers, such as treated marine ply planters (See: http://www.deepstreamdesigns.com).

Rooftops where water supply is already available on or in the proximity of the roof can reduce general costs of installation. Depending on the configuration of the roof, it may be possible to harvest rainwater. For example, if a green roof is surrounded by higher roofs, it may be possible to channel water from these surfaces, though this is a considerable undertaking so is more common in formal arrangements (e.g. The Museum of London, Table 3). The larger, and thus heavier, the tank, the less suitable for locating on the roof itself, though it should be noted that the perimeter of a roof is usually more rigid and thus stronger than central roof areas, so can support more weight. Further tanks, or one large tank, may be installed within the basement of the building, accompanied by a pump back up to the roof.

Whilst rainwater can be harvested, nutrients can also be supplied at little cost. For farms located on residential buildings, the collection of organic waste from households can generate substantial quantities of compost. A case in point is the Food Loop project (www.foodloop.ork.uk) in which waste from a social housing development of more than 450 households is collected and used for a communal garden, from which agricultural products are harvested for sale. Food from the Sky (see Table 3) utilized waste from the supermarket occupying the building below to produce compost for their rooftop farm. Municipal compost is also available at a low cost in many countries and is suitable for enriching existing soil mixes or to make potting compost (see: http://www.verticalveg.org.uk/municipal-compost-is-it-a-good-thing/). In the UK and USA, this compost conforms to standards to ensure levels of, for example, trace metals are below allowable limits. However, there is some evidence to suggest that this is not the case in all countries (see Mandal et al. 2014), so this should be verified in each individual country. If no information is available, chemical analysis by a commercial lab could be required. Generating compost from the buildings below a roof or obtaining local municipal compost not only reduces travel distances to transport commercial composts, but also reduces the volume of peat used in green roof systems, reducing their carbon footprint. Thus, these nutrient input systems are a definite consideration if building a rooftop agricultural system concerned with environmental impact.

It is important to note that the visual character itself of these rooftops can be informal; specialists or designers are rarely involved in their design and construction process. By necessity, they may lack the level of detail and care for spatial solutions that can be expected in projects developed by professionals, even with limited budgets. This may exacerbate neighbors' perception that farming may interfere with their privacy or even affecting negatively views from their windows. This is demonstrated in a recent court case where Brighton residents were overlooked by a neighboring green roof (see: http://egiewcms-test-auth.elasticbeanstalk.com/legal/r-on-the-application-of-barker-v-brighton-and-hove-city-council/). Screening off parts of farm at the perimeter of the space used may be necessary. On the green roof of the University of Greenwich's Stockwell Street Library (Table 3), native beech (*Fagus sylvatica*) hedges were constructed around the roof perimeter for the same purpose. Finally, particular care must also be taken to ensure neighbouring buildings do not block out sunlight.

In response to a resurgence of urban gardening not only on soil but also on urban unused hard-paved open spaces both in Europe and in other countries (see Caputo et al. 2016) new products are on the market that have been designed to grow edible crops on terraces and roofs. The Indian company Greentechlife (www.greentechlife.in), for example, launched a number of containers and other accessories such as a composting bin that are relatively small in size therefore apt to be used in small spaces. Farm:Shop (http://farmlondon.weebly.com/) in Hackney, London, is an enterprise that designs products for urban farming such as micro fish farming tanks. On the rooftop of their shop-showroom they have installed a chicken coop, showing that animal farming is also possible on top of buildings.

Formal Rooftop Agriculture

The 'formal' rooftop farming relies on the use of green roof technologies that allow greater freedom in the landscaping of the entire roof surface, which can be used both for intensive and/or ornamental planting. This type of intervention usually requires higher financial resources compared to 'informal' farms as well as the aid of specialist suppliers and consultants. Although some of the existing green roofs used for farming have been installed by volunteers (e.g. RISC in Reading, Eagle Street Farm in New York; Brooklyn Grange in New York; Table 3), supervision by green roof experts and consultation with structural and system engineers is fundamental. As mentioned above, it may be possible to grow some plants with a limited depth of soil with light-weight components. However, the weight of even this substrate may exceed the typical loadbearing capacity of a roof with a reinforced concrete frame, thus requiring necessary structural reinforcement. Moreover, a systems' engineer may be necessary for a correct execution of the water drainage, collection and discharge. Finally, some of these projects, being commercial for-profit or non-profit enterprises, tend to occupy medium-to-large rooftops allowing levels of yield that can give sufficient financial returns. It is therefore more likely that architects are involved in their coordination and design. In fact, some of the existing 'formal' rooftop farms show particular attention to, for example, deployment of beds and circulation, thus offering a more rational image of the farm.

Most existing projects are either commercial or educational. Educational roofs include projects such as the Gary Comer Youth Center green roof in Chicago, USA (Table 3). This roof not only provides lessons in agriculture for local school children but is an aid to teaching botany and cooking skills. In fact, particularly in the UK, many formal green roofs used to grow food are included on school buildings (for a guide to installing green roofs on schools, see: Garden Organic n.d.). Depending on the nature of the activity, the design of the farm changes substantially, with the commercial privileging the space for production and the educational designed to accommodate several areas of use, including the gathering of visitors for workshops. Educational rooftop farms are also landscaped with a higher attention to the visual quality of the project. Buildings selected for this type of farm are usually non-residential since they offer several advantages: they usually have larger roof spaces than residential buildings; their use does not necessarily conflict with farming activities (as opposed to residential buildings in which some occupants may deem such activities – i.e. passage of farmers, material and tools for farming and visitors – incompatible with their privacy); the access to the rooftop – or the construction of an independent access – may be less problematic than for residential buildings for the same reason outlined above; and so would be the eventual upgrading of the roof structure or insulation.

Given the scale of intervention and the related financial investments it is possible for some of these 'formal' rooftops to include large scale drip irrigation systems, which, coupled with rainwater harvesting can offer a closed loop system. In particularly large buildings, the collection of rainwater and greywater can yield substantial quantities. The green roof of the Museum of London (Table 3), for example, uses water collected from a surface of the 850 m^2 on adjacent buildings in a 25,000 l tank positioned in the basement, reducing their need to draw from mains water supplies and reducing runoff to local sewer systems. Whilst runoff on informal rooftop gardens with planters can be considerable due to areas of empty, flat roof, in formal systems where the entire roof is planted, collecting water from adjacent roofs should be considered. Even shallow, extensive green roofs have been shown to retain up to 80% of the rainwater that falls upon them, with few studies reporting less than 45% (Berndtsson, 2010). This is, of course, why green roofs are installed as part of SuDS projects; they are superb at reducing runoff (for an anecdotal example of this, see: http://www.sustainablemerton.org/living-roofs-the-benefits/). However, this will, of course, limit the amount of water than can be collected and recycled from the green roofs themselves. It may also have implications for the load bearing capacity of the structure of the roof, which will need to be designed accordingly.

In addition to recycling water to produce a more closed-loop system, large roof surfaces allow space for on-site energy production. In terms of supplying the roof itself with energy, formal and informal gardens need little power, so this type of feature (and investment) seems more in line with the technological type of farm. However, in terms of supplying the building below with energy, planting and solar panels can be (contrary to popular belief) complementary. PV output is increased when combined with green roofs (compared to PV on gravel surfaces) primarily due to the fact that plants undertake evapotranspiration, cooling their local environment, including nearby PV panels (Lamnatou and Chemisana 2015). Solar PV panels become less efficient as they warm, so cooling panels by this process enhances their output (Lamnatou and Chemisana 2015). Whilst these systems have, to date, only been tested with extensive planting regimes (e.g. Sedums), the potential for combining solar PV with rooftop agriculture projects is promising.

Technological Rooftop Agriculture

Rooftop farms in this category utilize technologies for indoor environmental control, aquaponics and hydroponics, which require the construction of greenhouses or other dedicated indoor spaces on the roof. The design of these rooftops is determined by functionality and commercial logic, although, given the investment and the nature of the enterprise, attention is paid to the visual integration with the host building generally.

Due to the large investments necessary for these projects, farms are run by commercial ventures or designed and installed for research into new modalities (i.e. urban) of agricultural production. UF001 Lok Depot in Basel, Switzerland (Table 3) combines both of these purposes, as a pilot aquaponic/hydroponic farm with a view to commercializing production. It has been operating since 2013 and yearly produces around 5 tons of vegetables and 850 kg of fish in a 250 m^2 greenhouse. Its success has enabled plans for UF002 to be developed, with funding secured in 2015 to build this in The Hague, Netherlands. In this system, vegetables such as lettuce and tomatoes are grown in hydroponic greenhouses, with the organic waste of fish from the on-site aquaponic farm utilised as nutrient for plants.

At the more experimental end of the scale, a rooftop greenhouse constructed by the ICTA-ICP in Barcelona, Spain (Table 3) has been implemented to investigate the potential for rooftop and vertical farming in urban environments (see Table 3). This greenhouse system was used to grow tomatoes and determine if rooftop growing could be more economical than growing crops in polytunnels elsewhere. Sanyé-Mengual et al. (2015a) found that in this system the sustainability of growing tomatoes was improved and the cost lowered for buyers at the point of consumption (i.e. the general public), but that this was likely to vary depending on crop and season.

In terms of commercial projects, there are several large rooftop greenhouses in operation, the majority of which grow leafy greens in hydroponic systems. The most well-known of these are probably the Gotham Greens glasshouses in New York City and Chicago, USA (Table 3). These glasshouses employ high tech monitoring and control systems to ensure a stable greenhouse environment, as well as having complementary solar PV systems and combined heat and power plants. Together, these establishments produce over 700 tons of greens and tomatoes each year and have won several innovation and design awards.

The lightweight construction of greenhouses may not require higher bearing capacity than the 'formal' type of farming. Furthermore, many of these farms rely on soil-less systems of production or containers with different types of growing media, which can be lighter than the substrate used for green roofs. However, water tanks as well as fish tanks used for aquaponic farms can be of considerable weight, depending on their size and form, thus requiring particularly interventions to strengthen the roof structure. These types of cultivation require a paved surface for the installation of planters, shelving systems or any other equipment. Another important element, as mentioned above, is the lateral load of the wind and the way greenhouses are connected to the existing structures to resist such loads. A rooftop greenhouse in Jerusalem, Israel (Table 3), addresses this problem by using clear polycarbonate sheets that overlap at joins to provide strength in high winds. For more traditional glasshouses, steel frames are bolted together and to the roof surface, with particular attention given to sealing the base to prevent uplift (Mandel 2013).

Similarly to the other types of rooftop farms, the majority of existing technological projects have been implemented on existing buildings. The addition of greenhouses or other constructed add-ons to the top of existing buildings necessitate planning consent, which, in turn, imply not only compliance to building regulations and any other relevant code but also harmonization of the new construction with adjacent buildings and the local built environment at large (documentation from Leibniz Centre for Agricultural Landscape Research available online at http://www.zalf.de/htmlsites/zfarm/Documents/leitfaden/Rooftop%20greenhouses.pdf). Some of the largest existing projects are in technological or commercial parks (e.g. Lufa Farms and ICTA-ICP laboratories; Table 3) where this last issue is less problematic. Other projects are located on the rooftop of buildings located in central areas of cities such as New York (see for example the Rooftop Greenhouse Project on the roof of a public school – http://nysunworks.org/thegreenhouseproject/quotes). New York planning framework permits such constructions and imposes control over factors such as height, transparency and distance from the roof edges (www.nyc.gov/html/dcp/pdf/ap/zr_75_01_guidelines.pdf). Height, proportions, positioning within the rooftop surface area and combination with other built volumes (e.g. containers for fish tanks similar to those used for the LokDepot in Basel; Table 3) must all be carefully evaluated. Ultimately, technological rooftop farms can have strong iconic value, promoting new forms of urban agriculture and alternative food chain systems.

Technological urban farms can be particularly energy and resource intensive and so must be considered carefully in terms of geographical location. Studies conducted in the ICTA-ICP farm (Table 3) using LCA methodology found that, even with reduced distances between rooftop growers and consumers, production at certain times of the year was more energy intensive depending on the level of yield attained and the length of the supply chain (Sanyé-Mengual et al. 2015b). Experimentation with closed loop systems is, therefore, vital to projects wishing to reduce their carbon footprint. Many farms include renewable energy production based on PV panels (e.g. Gotham Greens; Table 3) as well as rainwater harvesting (e.g. the LEED Platinum awarded Arbor House, New York, USA; Table 3) and water efficient irrigation systems. However, hydroponic cultivation can be water intensive, thus higher integration with the building is necessary in order to utilise its resources, for example by integrating with greywater systems, as will be utilised on the Whole Foods rooftop greenhouse, New York, USA (Table 3). In countries where heating systems are required for year-round production, integrated heating systems can also be applied. A promising pilot utilizing this technology is the Rooftop Greenhouse Project in London (Table 3), in which a greenhouse has been installed on the rooftop of a two-storey building. Heat produced in the building is used by the greenhouse above to attain optimal indoor temperatures in winter. In summer, parts of the greenhouse open to increase the potential for natural ventilation of the entire building. One important aspect is that this is a retrofit project, thus demonstrating that sophisticated levels of integration, in which the greenhouse enables higher levels of performance for the entire existing building, are possible.

Conclusions

With urban agricultural practices growing worldwide, rooftop farming becomes an attractive option. Rooftops offer space for cultivation in dense urban environments, where land values and pressure for development make it difficult for new land to be allocated for gardening on soil. Drawing on a number of existing projects, this chapter offers a spectrum of design approaches as well as techniques and technologies that can be used to implement rooftop farming, thus showing that projects can be started also with low investments. Alternatively, at the other end of the spectrum, projects can utilise state-of-the-art technologies with substantial upfront investments and high yields.

Three design approaches have been identified (informal, formal and technological), each presenting particular design challenges that include access to the rooftop, adequate structural loadbearing capacity and harmonization with adjacent buildings. Despite such design challenges, benefits that this particular type of farming can offer are many. Perhaps, one of the most important points the chapter suggests is that rooftop farms can be viewed not only as a way to rationalize the use of empty spaces in cities whilst making buildings more productive but also as an opportunity to design new and renovate existing buildings as integrated and productive systems that use and reuse resources, thus augmenting their efficiency.

Bullet Points

- Some of the common challenges that can be encountered when designing a rooftop farm include: adequate loadbearing capacity of the roof; independent access to the roof; health and safety issues; and water and energy supply. Whilst these can be reasonably addressed when designing new build, they can be problematic for existing buildings. Clarity is therefore essential from the onset of each project regarding what to grow (with consequent depth and weight of soil), how to organize activities and how planning consent can be achieved;
- Because of the above challenges, consultants' advice must be sought in order to identify upgrading works, the necessary planning procedures and the entity of the investment;
- Three types of rooftop farming have been identified. Within the first one (herein called informal), self-build projects are included, with low investments and use of recycled material. In designing these rooftop farms, particular attention must be given to issues such as the waterproof course; wind breakers that must be safely secured to the buildings and the capability of the farm to visually blend with the surrounding buildings;
- The formal type utilizes green roof technologies and requires higher investments. Consequently, specialized labour or direction of works can be used, which generally results in a more rational use of the space for production as well as the exploitation of local resources which include greywater and rainwater. In some cases, on-site power generation can be included;
- The technological type refers to all those projects that use hydroponic and aquaponic technologies, with greenhouses and enclosed spaces. These are generally implemented by commercial enterprises. Existing projects are located on commercial buildings, with a footprint sufficiently large to host high-yield medium to large farms.

References

Barnett L (2011) The supermarket growing food on its roof. The Guardian, 9th March 2011

Berardi U, Ghaffarian Hoseini AH, Ghaffarian Hoseini A (2014) State-of-the-art analysis of the environmental benefits of green roofs. Appl Energy 115:411–428

Berndtsson JC (2010) Green roof performance towards management of runoff water quantity and quality: a review. Ecol Eng 36:351–360

British Standard Institute (1988) BS 6399-3: loading for buildings – Part 3: Code of practice for imposed roof loads. BSI, London

Bundesministerium fur Bildung und Forschung. There's something growing on the roof – Rooftop greenhouse: idea, planning, implementation. Leibniz Centre for Agricultural Landscape Research. Available at: http://www.stadtentwicklung.berlin.de/bauen/oekologisches_bauen/download/broschueren/Rooftop_greenhouses_Freisinger_et_al_2015.pdf. Accessed on 30 Sept 2015

Carson TB, Marasco DE, Culligan PJ, McGillis WR (2013) Hydrological performance of extensive green roofs in New York City: observations and multi-year modelling of three full-scale systems. Environ Res Lett 8:1–13

Carter T, Keeler A (2008) Life-cycle cost–benefit analysis of extensive vegetated roof systems. J Environ Manag 87(3):350–363

Caputo S, Schwab E, Tsiambaos K (2016) Emergent approaches to urban gardening. In: Bell S, Fox-Kaemper R, Benson M, Caputo S, Kaeshavarz N, Noori S, Voigt A (eds) European urban allotments. Routledge.

Castleton HF, Stovin V, Beck SBM, Davison JB (2010) Green roofs; building energy savings and the potential for retrofit. Energ Buildings 42:1582–1591

Cerón-Palma I, Sanyé-Mengual E, Oliver-Solà J, Montero JI, Rieradevall J (2012) Barriers and opportunities regarding the implementation of Rooftop Eco. Greenhouses (RTEG) in Mediterranean cities of Europe. J Urban Technol 19(4):87–103

Chang N, Hong Y (2012) Multiscale hydrologic remote sensing: perspectives and applications. CRC Press, London

Clark C, Adrians P, Talbot B 2006 Probabilistic economic analysis of green roof benefits for policy design. In: Proceedings of the fourth North American green roof conference: greening rooftops for sustainable communities, Boston, The Cardinal Group, Toronto, 11–12 May

Clatworthy J, Hinds J, Camic PM (2013) Gardening as a mental health intervention: a review. Ment Health Rev J 18:214–225

Deplazes A (2005) Constructing architecture: materials, processes, structures. Springer, Berlin

Dvorak B, Volder A (2010) Green roof vegetation for North American ecoregions: a literature review. Landsc Urban Plan 96:197–213

Feng C, Meng Q, Zhang Y (2010) Theoretical and experimental analysis of the energy balance of extensive green roofs. Energ Buildings 42:959–965

Garden Organic (n.d.) Roof gardens in schools. Food for Life Partnership, Bristol. Available at: https://www.gardenorganic.org.uk/sites/www.gardenorganic.org.uk/files/resources/fflp/RoofGardensinSchools.pdf

Getter KL, Rowe DB, Cregg BM (2009) Solar radiation intensity influences extensive green roof plant communities. Urban For Urban Green 8:269–281

Grant G (2006) Extensive green roofs in London. Urban Habit 4:51–65

Gorgolewski M, Komisar J, Nasr J (2011) Carrot City: creating places for urban agriculture. Monacelli Press, New York

Greenroofs (2008) Green roofs for healthy cities – 2008 green roofs industry survey report. Greenroofs, available at www.greenroofs.org/resources/GRHC_Industry_Survey_Report_2008.pd. Accessed on 20 Oct 2015

Grewal SS, Grewal PS (2012) Can cities become self-reliant in food? Cities 29:1–11

Hanna SR, Brown MJ, Camelli FE, Chan ST, Coirier WJ, Hansen OR, Huber AH, Kim S, Reynolds RM (2006) Detailed simulations of atmospheric flow and dispersion in downtown Manhattan: an application of five computational fluid dynamics models. Am Meteorol Soc 87:1713–1726. doi:10.1175/BAMS-87-12-1713

Hillel D (1982) Introduction to soil physics. Academic, Cambridge

Kadas G (2006) Rare invertebrates colonizing green roofs in London. Urban Habitat 4(1):66–86

Knepper, C. A. (2000) Gardens in the sky. J Prop Manag, March/April 2000

Lamnatou C, Chemisana D (2015) A critical analysis of factors affecting photovoltaic-green roof performance. Renew Sust Energ Rev 43:264–280

Lee A (2004) Life cycle cost analysis—green roofs form an investment. In: proceedings of the second North American green roof conference: greening rooftops for sustainable communities, Portland, OR, The Cardinal Group, Toronto, 2–4 June

Litichevskaya J (2010) Reviving the world wonder: why rooftop gardens should cover urban landscapes. Rutgers Comput Technol Law J;2010 37(1):58

Mandal P, Chaturvedi MK, Bassin JK, Vaidya AN, Gupta RK (2014) Qualitative assessment of municipal solid waste compost by indexing method. Int J Org Waste Agric 3:133–139

Mandel L (2013) Eat up: the inside scoop on rooftop agriculture. New Society Publishers, Vancouver

Oberndorfer, E., Lundholm, J., Bass, B., Coffman, R. R., Doshi, H., Dunnett, N., Gaffin, S., Kohler, M., Liu, K. K. Y. and Rowe, B. (2007) Green roofs as urban ecosystems: ecological structures, functions, and services. BioScience 57(10), pb. 823–833

Papafotiou M, Pergialioti N, Tassoula L, Massas I, Kargas G (2013) Growth of native aromatic xerophytes in an extensive Mediterranean green roof as affected by substrate type and depth and irrigation frequency. Hortscience 48:1327–1333

Relf D (1995) Gardening in raised beds and containers for older gardeners and individuals with physical disabilities. Virginia Coop Ext 426:1–16

Sanyé-Mengual E, Orsini F, Oliver-Solà J, Rieradevall J, Montero JI, Gianquinto G (2015a) Techniques and crops for efficient rooftop gardens in Bologna, Italy. Agron Sustain Dev. doi:10.1007/s13593-015-0331-0

Sanyé-Mengual O-S, Montero JI, Rieradevall J (2015b) An environmental and economic life cycle assessment of rooftop greenhouse (RTG) implementation in Barcelona, Spain. Assessing new forms of urban agriculture from the greenhouse structure to the final product level. Int J Life Cycle Assess. doi:10.1007/s11367-014-0836-9

Scott TL, Masser BM, Pachana NA (2015) Exploring the health and wellbeing benefits of gardening for older adults. Ageing Soc 35:2176–2200

Thrive (2008) Carry on gardening: raised beds. Thrive, Reading

Vancouver Convention Centre (2015) Environment: functional, beautiful and 'green'. http://www.vancouverconventioncentre.com/about-us/environment. Accessed on 29 Oct 2015

VanWoert ND, Rowe DB, Andresen JA, Rugh CL, Fernandez RT, Xiao L (2005) Green roof stormwater retention: effects of roof surface, slope, and media depth. J Environ Qual 34:1036–1044

Whittinghill LJ, Rowe DB, Cregg BM (2013) Evaluation of vegetable production on extensive green roofs. Agroecol Sustain Food Syst 37(4):465–484

Williams NSG, Lundholm J, MacIvor JS (2014) FORUM: do green roofs help urban biodiversity conservation? J Appl Ecol 51(6):1643–1649

Soil Based and Simplified Hydroponics Rooftop Gardens

Alfredo Rodríguez-Delfín, Nazim Gruda, Christine Eigenbrod, Francesco Orsini, and Giorgio Gianquinto

Abstract By using rooftops for food production, new cultivation areas can be exploited. Soil-based cultivation beds and simplified hydroponics make use of low-input systems, offering accessible food production with low maintenance and low costs. Simplified hydroponic systems are particularly suitable for the installation on rooftops due to their low weight. For low-income families, soil-based and simplified hydroponic systems represent a promising approach to tackle poverty and food insecurity. In this chapter, the potential of these cultivation systems on rooftops will be reviewed. New adaptations and developments in equipment needed for different growing solutions over the last 20 years will be discussed. Emphasis on their special features for rooftop agriculture will be addressed.

Introduction

Rooftops of private houses or public buildings can be used for plant production either using soil-based, water based or substrate systems. Soil-based systems use, as substrate, translated soils and/or organic ones, enabling cultivation of a wide variety of crops, with a high level of intensity. Alternatively, growing systems without soil

A. Rodríguez-Delfín (✉)
Hydroponics and Mineral Nutrition Research Center, La Molina Agrarian National University, Lima, Peru
e-mail: delfin@lamolina.edu.pe

N. Gruda
Federal Office for Agriculture and Food and University of Bonn, Bonn, Germany

Federal Office for Agriculture and Food, Bonn, Germany
e-mail: ngruda@uni-bonn.de

C. Eigenbrod
Royal Agricultural University, Gloucestershire, UK
e-mail: christine.eigenbrod@gmail.com

F. Orsini • G. Gianquinto
Department Agricultural Sciences, Alma Mater Studiorum, University of Bologna, Bologna, Italy
e-mail: f.orsini@unibo.it; giorgio.gianquinto@unibo.it

F. Orsini et al. (eds.), *Rooftop Urban Agriculture*, Urban Agriculture,
https://doi.org/10.1007/978-3-319-57720-3_5

in situ, are called soilless cultures (Gruda 2009), which include plants grown either on a solid rooting media – called substrate or growing media – directly in a nutrient solution without any solid phase, in so-called hydroponic systems, or directly exposed to the atmosphere and periodically nebulized with the nutrient solution in the so-called aeroponic systems. For the sake of this publication, the term "hydroponic" will be used as an equivalent to "soilless culture". Using these techniques makes it possible to grow vegetables of excellent quality and to ensure a more efficient use of water and fertilizers. Yields per unit of cultivated area are generally higher as compared to traditional soil-grown plants, due to both increased plant productivity and planting density.

Soilless systems may be classified according to the management of the leachate (drain solution), as either open- or closed-loop systems. In an open system, any excess irrigation is allowed to go to waste and is not recycled. In a closed system, any drainage is captured and recycled. Most hydroponic systems are inherently closed, but many systems based on solid materials, until recently, were open. Most recent installations are closed systems, which will likely become mandatory in the future as nutrient management planning is implemented in more countries. The nutrients used in these systems are applied through irrigation water as a complete nutrient solution (Gruda and Tanny 2014).

For years, hydroponics has been used for research in the plant mineral nutrition field. Today, hydroponics is the most intensive horticultural production method; it is generally of high technology and strong investment, and it is being successfully applied commercially in developed countries. The main reason for using soilless cultures, however, is the reduction of soil-borne pathogens and the improved control over water and nutrient supplies (Gruda and Tanny 2014). Other advantages, such as better environmental protection with closed systems and improved product quality through precise dosage of nutrients, are becoming more and more important (Gruda 2009). A considerable decrease of agricultural land in developing countries makes hydroponics an interesting production alternative in urban and peri-urban areas. Within the context of the so-called urban agriculture, hydroponics can be very well applied in cities with simple and inexpensive technologies, mainly in extremely poor areas, as a way of encouraging vegetable consumption and to support the family income through self-employment in their own homes or in community centers. In these contexts, hydroponics may enable to grow a wide variety of plants in roofs and terraces of houses. With hydroponics it has been demonstrated that it is possible to cultivate a large number of plants, such as leafy vegetables (e.g. lettuce, celery, Swiss chard, basil), fruits (e.g. tomato, pepper, cucumber, eggplant), roots (e.g. carrot, radish, turnip, beet), tubers and bulbs (e.g. potato, onion, garlic), aromatic, medicinal and ornamental plants (Fig. 1). Hydroponics is also widely used to produce fresh forage to feed farm animals (Izquierdo 2005). Among the advantages of hydroponic rooftop farming are:

- Locations unsuitable for traditional agriculture are exploited.
- Yields obtained with hydroponics significantly outperform traditional soil-based production. This is due to both faster and vigorous plant growth, as well as increased number of harvests per year.

Fig. 1 A variety of flowers and aromatic crops grown in wooden and plastic containers. *Top left*, *green* roof project in the City Hall of San Miguel, Lima, Peru (Photo: A. Rodríguez-Delfín). *Bottom left*, rooftop garden in Bologna, Italy (Photo: F. Orsini). *Right*, rootop herb garden at Cookies & Cream restaurant, Berlin, Germany (Photo: F. Orsini)

- Less water and fertilizer consumption. The technique is especially suitable for areas with water scarcity.
- No fertilizer leakage into the environment.
- High-quality and nutritive crops are obtained which are good for a healthy diet.
- Hydroponic rooftop farming can also be used for social purposes to improve the incomes of disadvantaged populations to generate self-employment in their own homes.

Soil-Based Cultivation

Rooftop farming may be pursued by integrating cultivation beds and walking paths over the whole surface in the roof design. These rooftop farming systems generally use soil from agricultural fields that is mixed with organic matter (e.g. compost, humus) and/or other substrates (e.g. perlite, peat) to improve drainage and promote root growth. Cultivated beds can either be flat or ridged, and soil may be bare or covered with mulch, in order to reduce water evaporation, improve micro-climate of the root zone and limit weed outbreak. Common mulching in rooftop farming makes use of organic materials, e.g. straw or bark, but also plastic or biodegradable films may be adopted. Irrigation is generally applied through drip lines or micro-sprinklers (see chapter "Water Management and Irrigation Systems"). Excess water is eliminated by a drainage system placed below the soil and that shall be integrated into the garden design. As detailed in the chapter "Water Management and Irrigation Systems", irrigation scheduling shall be adapted to growing system and crop characteristics. Consistently, soil depth shall be defined balancing building structural limits, soil water holding capacity and crop requirements. If the roof garden is integrated into the original building plan, structural allowance can be made for deep soils, able to accommodate not only vegetables, but eventually also trees and shrubs. On the other hand, when integrating rooftop agriculture into already existing buildings, weight becomes a serious concern (see chapter "Integrating Rooftop Agriculture into Urban Infrastructure"). When saturated with water, a cubic meter of soil may weigh up to 1.6 tons (Fairholm 1999). Most buildings are built to support only the roof structure and a minimum live load to accommodate snow and occasional maintenance. Consistently, unless either the snow or the soil are removed in winter, the live load-bearing capacity of the roof may be potentially exceeded. The adoption of lightweight substrates alternative to soil is a feasible strategy to maximize the depth:weight ratio. Growing media or substrates are defined as all those solid materials, other than soil, which alone or in mixtures with other materials can guarantee better conditions than agricultural soil (Gruda et al. 2013). Light and porous growing media should be chosen preferably among those locally available. The growing media should allow maximum growth and root development, resulting in a vigorous plant. Hence, media of different origin take on the role of soil and provide anchorage for the root system, supply water and nutrients for the plant, and guarantee adequate aeration in the root area (Gruda and Schnitzler 2006). Growing media systems can be used to grow a large number of plants on rooftops, including leafy vegetables, fruits, roots, tubers, bulbs, stems, aromatic and medicinal plants, ornamental and flowers. Organic growing media and others, such as perlite and pumice are usually used in containers. However, sometimes they are used in form of bags and slabs (peat-based substrates and rock wool, respectively), mats (polyurethane foam) and troughs (e.g. rock wool) (Gruda et al. 2013). Containers of different sizes and materials can be used as long as they are waterproofed (e.g. by using thick black polyethylene film). Containers with growing media must however facilitate the drainage of the excess nutrient solution and prevent anoxia and disease outbreak in the root environment. The depth of the container depends on the type of crop. For leafy crops, the depth varies between 10 and 15 cm. For root, tuber and bulb crops, the depth can fluctuate from 20 to 25 cm. For seedlings, a minimum height of 3–5 cm is required (Rodríguez-Delfín and Chang 2014). A cultivation technique that may find applicability on rooftop is the so-called organoponics, where crops are hosted in containers filled with compost or organic matter of various origins. This technique has been commonly used in Venezuela and Cuba (Tixier and de Bon 2006). The most famous Cuban organoponic system uses a mixture of soil and organic matter (residual of the sugar production chain) in measure of 50:50 v:v and produce up to 16 kg m^{-2}) (Murphy 1999). These technologies of production are strictly related to ecologically friendly agronomic practices, and particularly to the improvement of fertility through the use of micro-organisms and the adoption of integrated and organic control systems (Van Veenhuizen et al. 2001).

Simplified Soilless Cultivation

Growing Systems

For food production on rooftops, simplified hydroponic systems, with manual or automatic operation are suitable. As not all systems are effective for all crops, it is necessary to choose the most appropriate system for the respective crop.

Hydroponic systems can be divided into (a) water and (b) substrate cultures (Savvas et al. 2013). The first group is based on water only, and includes the so-called true hydroponic systems, in which roots are partially or completely dipped in the nutrient solution. In substrate cultivation, plant roots grow and develop in a growing media or substrate, where a nutrient solution flows to water the roots. Among the most well-known and employed simplified hydroponic systems in home gardens most common ones include nutrient film technique (NFT), float hydroponics and column systems.

Nutrient Film Technique (NFT)

The NFT is a water-based system, which consists of the continuous flow of a nutrient solution through culture medium channels where the plant roots grow and develop. A thin (few millimeters) film of nutrient solution is sufficient to supply the roots with water and mineral nutrients essential to the plants and, in addition, enables to ensure good root oxygenation. The system may be modified according to available material and growing conditions, as long as the principle of the circulation of the nutrient solution is preserved. The main crops that are produced with this simplified system are various varieties of lettuce, celery, basil, strawberries and other crops. The main advantages of the NFT system are significant savings in water and fertilizer consumption in relation to the number of plants that are produced and the reduced environmental impact and costs related to the disposal (Savvas et al. 2013). Among other advantages offered by the NFT system, are the reduced manpower required, the earlier harvest (due to a shortening of the growing period of the crop), and the better product quality and hygiene. However, if automatization is considered, installation costs may be relatively high and risk of crop loss due to power cuts may emerge. Another drawback of NFT may be associated with the lack of hygiene in the handling of the system, which can result in pathogen contamination of the nutrient solution, eventually affecting the entire production. Electricity costs may be reduced by up to 75% if the nutrient solution is circulated intermittently (e.g. the pump runs for 15 min. per hour thus saving 18 h of electricity per day). In the system, adapted by the Hydroponics Research Center of La Molina University (Peru), when the pump is switched off, the height of the nutrient solution in the channels remains at 2 cm, so that the roots stay covered by the nutrient solution and get enough oxygen without suffering from stress (Valverde et al. 2009; Rodríguez-Delfín and Chang 2014). The NFT system counts on the following components:

- ***Tank***: Containers storing the nutrient solution. The volume depends on the area and the number of plants grown, but also on what can be locally found. It is recommended to use polyethylene tanks generally available for drinking water storage. Tanks that are proper for food storage are recommended, in order to avoid contamination of the water with toxic substances. For small NFT modules, tanks from 70 to 100 l can be used. For larger sized modules, tanks from 500 l or more are used. Before placing on rooftops, weight load and building collapse risk shall be considered and the tank eventually placed at ground level.
- ***Electric pump***: it brings the nutrient solution from the tank to the growing channels through a PVC or PE distribution pipe (Fig. 2). The pumps need a minimum horse power of 0.5 for this system. The nutrient solution flow for each growing channel should be approximately 2–3 l per minute. This flow allows an adequate supply of oxygen, water and nutrients. The time of operation of the electric pump can be controlled via a clock timer, so the flow can be intermittent. There are also NFT system models that do not use electric pumps, but they are of small dimensions (Gianquinto et al. 2006). In this case, the nutrient solution is applied through a small elevated tank and, using gravity, the nutrient solution travels a short distance and is received in small containers placed on the drain side. The nutrient solution is reused for a new cycle, which is done manually. A minimum of 3–4 manual cycles daily are required to recirculate and to maintain the level of oxygenation of the nutrient solution.
- ***Culture Channels***: They support the plants and allow the growth and development of their roots. It is recommended that the length of the channels does not exceed 12 m since it can decrease the oxygen levels of the nutrient solution. For leafy vegetable cultivation, such as lettuce and basil the culture channels (PVC pipes) need to have a diameter of 7.5 cm. Crops that produce a higher root volume, such as celery and strawberries need channels with a larger diameter (10 cm), as the circulation of the nutrient solution through the culture channels would be much slower otherwise. On the upper side of the culture channels, plastic pots with a diameter of 5–7.5 cm are placed in holes of the same diameter. The plastic pots are used to hold the plants in the culture channels. For lettuce, celery and other leafy crops, the distance between the holes needs to be at least 20 cm. The shape of the culture channel determines the height of the nutrient solution. The culture channels are PVC pipes with a concave section (Fig. 3). The culture channels are placed on tables or stands of timber or corrugated iron designed and built especially for this purpose.

Fig. 2 Modified NFT system that works with a pump in the children playhouse at Monterrey, Lima, Peru (Photo: F. Orsini). *Top*, views of the hydroponic gardens, *bottom*, *left* to *right*, tank for the nutrient solution, electric timer and hydraulic pump

– ***Recollection or Drainage Pipe:*** Enables collecting the nutrient solution from the culture channels and taking it back to the tank. This pipe is located on the output side and at a level slightly lower with respect to the culture channels.

These systems may allow growing up to 25 lettuce plants per square meter on a monthly basis (Orsini et al. 2009; Valverde et al. 2009).

Float Hydroponics

In this system, plant roots are grown on polystyrene trays floating in tanks filled with nutrient solution. Plant roots are partially immerged and the tray acts as a mechanical support. It has been widely adopted in rooftop agriculture projects, although constrains may come from its weight (dependent on the nutrient solution depth) and from extreme climatic conditions (mainly heat) that may be experienced on the rooftop. This system has been adapted to be used in hydroponic social projects in different developing countries, usually for growing leafy vegetables (e.g.

Fig. 3 Pak choi plants grown in a modified NFT system, adapted by the Hydroponics Research Center, La Molina Agrarian National University, Lima, Peru. When the pump is switched off, the height of the nutrient solution in the channels remains at 2 cm, so that the roots stay covered by the nutrient solution and get enough oxygen without suffering from stress (Photo: A. Rodríguez-Delfín)

lettuce, basil, celery, leaf beet, arugula). The main advantages of this system are that (a) installation costs are low, (b) plants have constant access to water and nutrients, (c) the water volume provides a large buffering capacity for pH and nutrients, and (d) crops can be moved easily. This results in fast growth and an early harvest with more production cycles during the year (Gruda et al. 2016a). To achieve a good production it is very important to aerate the nutrient solution. According to Savvas et al. (2013), the oxygen concentration should range between 5 and 6 mg per liter. This can be pursued by injecting air with a compressor or manually by using a clean plastic whisk (Fig. 4), at least once or twice a day. While white color characterizes healthy roots, the presence of dark brown colored roots is an indicator of poor oxygenation, and results in limited water and nutrient absorption, affecting the growth and development of the whole plant. This system promotes the growth and development of healthy and disease-free plants, shortening the growing season with a low consumption of water and fertilizers.

The main elements of the system are:

- ***Growing tank***: built in cheap, easily available materials (e.g. wood or bricks), waterproofed by means of a plastic (e.g. PE) film.
- ***Floating panels***: generally made of commercial Styrofoam trays (those generally used for transplant production). When not available locally, insulation Styrofoam panels or sheets can be adapted and used as shown in Fig. 5. Generally, the growing methods are: (1) direct sowing of the crop in the trays holes filled

Fig. 4 Installation and management of a floating root system in mother groups (Clubes de Madres) of Trujillo, Peru (Photo: F. Orsini). *Top left*, sowing in polystyrene trays. *Top right*, proper root development. *Bottom left* to *right*, EC-meter, manual aeration of the nutrient solution and harvest of radish

Fig. 5 Step-by-step procedure for building up a simplified floating system. *Top row*, construction of the growing table. *Middle row*, insulation and waterproofing with a black PE film. *Bottom row*, preparation of the styrofoam trays by making holes with a hot iron rod. Images from Lima and Trujillo, Peru (Photo: F. Orsini)

with a fine granular substrate (e.g. rice hulls) or (2) transplanting of already germinated seedlings folded in foam rubber after careful washing of the previous growing media residue. After harvest, the Styrofoam trays need to be washed with water and disinfected by dipping them in a 10% sodium hypochlorite water solution. Instead of Styrofoam sheets, PVC sheets can also be used, enabling to reduce algae outbreak, and providing higher mechanical resistance to breakage and therefore longer life. PVC sheets are supported on the edges of the containers, with two galvanized wires placed in the center to withstand the weight of both PVC sheet and plants. This provision allows an air space between the nutrient solution and the plant support which contributes to the root aeration, reducing oxygenation needs.

The nutrient solution remaining in the container at the end of the growing season can be recycled for watering indoor plants or the garden, thus avoiding resource depletion and environmental pollution. Afterwards, the interior of the container is cleaned with a sponge. Finally, it needs to be rinsed, and is then ready for a new production cycle. Sowing, transplanting and harvesting should be coordinated to achieve continuous production. Depending on the crop type, reported yield are of about 25–230 plants per square meter on a monthly basis (Orsini et al. 2010a).

Column System

The column or vertical cultivation system is a hydroponic system that is characterized by the vertical growth of plants in stacked pots, or in columns that contain light growing media (Rodríguez-Delfín and Chang 2014). The infrastructure must support all the weight of the pots, media, plants and the drip irrigation system. The structure must be very firm to avoid the collapse of the columns, even avoid the fall of the columns by strong winds. That is why each column has a central axis (PVC tube of 0.5 inches in diameter) that allows to pass the pots through the central hole they have. The central axis allows to hold the entire column in a beam that goes at the top of the columns. In Table 1 some advantages and disadvantages of this system are presented. Although it allows a high plant production per unit of area, its application is restricted to plants with habits that tolerate being hung and have a rather small root system. The system is widely used for strawberry production; it is also useful for producing some vegetables, flowers, aromatic and ornamental plants of small proportions. Plants that grow in a vertical system should be well lit by natural sunlight; otherwise they will have lower photosynthetic rates, affecting their growth and performances. To achieve good lighting, the distance between rows is recommended to be between 1.0 and 1.2 m and around 1.0 m apart. Using these plant densities on a roof of 50 m^2, it is possible to have approximately 50 columns with each column of 1.7 m high, maximum, each yielding around 32 plants. The columns can be PVC pipes of a 20 cm diameter, plastic sleeves of 8 microns in thickness and 25 cm in diameter, or Styrofoam pots of a 3.5-liter capacity or more, stacked one above the other, supported by a shaft or tube that goes through the central part of the

Table 1 SWOT analysis of the column system for growing plants

Strength • High production per unit area, close to 5 times with respect to traditional soil production • Efficient use of water and fertilizers • More uniform crops, resulting in fruits of higher quality • Lower incidence of root diseases • Less effort in the collection of fruits	**Weaknesses** • The initial investment for the installation can be significant • Not having a suitable growing media • The lack of knowledge of the agronomic management of the crop can cause a severe loss of plants • The lack of knowledge of the nutrient solution management can also affect the nutrition of the plant and the production • An oversight due to lack of hygiene can contaminate the nutrient solution and, in turn, all of the plants in a closed system
Opportunities • High potential for commercial production in small spaces • Pesticide-free production of quality vegetables • Reduced labor costs • Suitability to arid, water-scarce environments.	**Threats** • If not properly trained, growers may lose production and pollute the environment • Pest outbreak in case of crop mismanagement • Non availability of proper system components (e.g. nutrients or substrate)

Adapted from Caso et al. (2009) and Rodríguez-Delfín and Chang (2014)

pot. In column systems, the frequency of watering (activated manually or with a timer) will depend on the growing media, the climatic conditions, and the age of the plants. A form of watering by gravity is placing plastic containers to store the nutrient solution in the upper part of a column. Plants are then grown in pots or plastic containers, which can be connected with hoses with a small diameter (Fig. 6). The nutrient solution moistens all the pots or containers that are connected and is collected in a bottom container for further use.

Growing Media

The international trend for substrate development tends towards the use of natural resources and renewable raw materials (Gruda 2012). In countries where perlite and vermiculite are available, they are mixed with peat moss and used for rooftop farming. This is a mixture that is quite light and retentive. In other countries, such as Latin American countries, these growing media are not easily available and must be

Fig. 6 Column systems. *Top*, herb production in simplified vertical hydroponic systems in the hydroponic school garden of San Luis Gonzaga Public School, San Juan de Miraflores, Lima (Photo: A. Rodríguez-Delfín). *Below*, column growing systems at the Hydroponics Research Center, La Molina Agrarian National University, Lima, Peru (Photo: F. Orsini). The nutrient solution is either stored in the *upper* plastic tank, which drains into the *lower* pots, or occurs through drippers placed on the top pot. Watering is done once per day, and it is controlled by a key placed in the tank

imported. Alternative growing media that comply with the above terms and conditions are rice hulls, coconut fiber and pumice. Tables 2 and 3 present some of the growing media used alone or in mixture with others and their respective advantages and disadvantages.

There is no substrate that meets all the desired characteristics. In some cases, it is necessary to mix the growing media with other materials in different proportions, to improve the water retention and aeration and to obtain a lighter mixture. The mixture of substrates is performed according to the volume and not the weight. Suitable substrates need to be easy to access, low-cost, retentive, not saline and durable (does not break down or degrade easily) (Gruda et al. 2016b).

Table 2 Main advantages and constraints of inorganic materials used as growing medium or as growing media constituents (Gianquinto et al. 2006; Gruda et al. 2016b)

Material	Origin	Advantages	Constrains
Perlite	Siliceous volcanic mineral sieved and heated to 1000 °C	Long-lasting	Industrial product, may be expensive;
		Very low volume weight (90–130 kg m^{-3});	Low nutrient-and water holding capacity;
		Sterile and neutral in pH (6.5–7.5)	Energy consuming product;
		Total pore space (50–75% V/V)	
Pumice	Light silicate mineral of volcanic material	Cheap and long-lasting;	High transport costs;
		Environmentally friendly;	pH may be high;
		Low volume weight (450–670 kg m^{-3});	
		Good total pore space (55–80% V/V).	
Rock wool	Melted silicates at 1500–2000 °C	Very low volume weight (80–90 kg m^{-3});	Disposal problems;
		Totally inert;	Energy consuming during manufacture.
		Nutrition can be carefully controlled.	
		High total pore space (95–97% V/V)	
Sand	Natural origin, with particles 0.05–2.0 mm	Relatively inexpensive,	Low nutrient- and water holding capacity;
		Good drainage ability.	High volume-weight (1400–1600 kg m^{-3})
			Low total pore space (40–50% V/V)
Vermiculite	Mg+, Al + and Fe + silicate sieved and heated to 1000 °C	Very low volume-weight (80–120 kg m^{-3});	Compact when too wet;
		High nutrient holding ability;	Energy consuming product;
		Good water holding ability;	Expensive.
		Good pH buffering capacity;	
		Total pore space (70–80% V/V)	

– ***Physical Properties***. The physical properties of substrates give important information concerning numerous parameters, such as water:air ratio that are required for proper regulation of irrigation and volume weight or bulk density. Based on such parameters, it is possible to make further calculations of the substrate's mineral content (Gruda and Schnitzler 2004a). Furthermore, it is important to know the water distribution and movement at root level. The physical properties of the substrates are mainly dependent on the size of their particles. A good sub-

Table 3 Main advantages and constrains of organic materials used as growing medium or as growing media constituents (Gianquinto et al. 2006; Gruda et al. 2016b)

Material	Origin	Advantages	Constrains
Bark (well aged)	By-product or waste of wood manufacture	Good air content and water holding capacity (WHC). Sub-acid-neutral pH (5–7) and good cation exchange capacity.	Highly variable depending on plant species and age, and site of origin;
		Good total pore space, TPS (75–90% V/V);	Must be composted to reduce C:N ratio and terpenes concentration, increase WHC
		Average volume weight (320–750 kg m^{-3});	Increasing cost since used as an alternative to fuel and in landscaping
		Long lasting;	
Biochar	Solid material derived from biomass pyrolysis	Production is energy neutral;	Properties vary dependent on feedstock.
		Helps with carbon sequestration;	High production costs;
		Biologically very stable.	High pH.
Coconut coir	By-product of fiber coconut processing	Physical stability,	May contain high salt levels;
		Good TPS (94–96% V/V) and WHC;	Energy consumption during composting and transport;
		Sub-acid-neutral pH (5–6.8);	Water needed for rehydration.
		Low-volume weight (65–110 kg m^{-3}).	
Green compost	Composted plant residues	Good source of K^+ and micronutrients;	Variable in composition (risk of salinity);
		Good WHC;	High volume weight (600–950 kg m^{-3});
		Urban waste reduction.	Becomes easily waterlogged.
Peat	Natural anaerobically processed plant residues, with variable features according to type, origin and degradation degree	Physical stability;	Finite and expensive resource;
		Good TPS (85–97% V/V) and WHC;	Environmental concerns and contribution to CO_2 release;
		Low microbial activity;	
		Low volume weight (60–200 kg m^{-3});	Shrinking may lead to substrate hydro-repellence;
		Low and easily to adjusted pH;	May be strongly acidic.
		Low nutrient content.	

(continued)

Table 3 (continued)

Material	Origin	Advantages	Constrains
Rice hulls	By-product of rice processing	Very low weight volume (70 kg m^{-3});	In areas where rice is not grown, necessary to consider the freight cost for transport.
		Abundant and cheap where rice is grown;	
		Carbonization/fermentation makes it pathogen free, inhibits algae formation, avoid further fermentation and germination of viable rice seed, improves drainage and high aeration.	Not advisable to re-use the rice hulls on a long term; this growing media should be renewed after each growing season.

strate should be a mixture of different particle sizes to allow water and air availability and to achieve a better development of the crop. The selection of particle sizes can be performed by sifting the material using sieves or meshes of different openings. The water needed for the development of the plant roots and the air necessary for their respiration must be supplied to the substrate.

- ***Chemical Properties***. Salinity refers to the concentration of the soluble salts present in the substrate. Due to the small volume of growing media available to the roots of plants grown in substrates, the risk of accumulation of high levels of dissolved salts increases. The electrical conductivity (EC) of a suspension or an aqueous extract of the substrate is measured to determine the levels of salinity of a substrate; the higher concentrations of dissolved salts, the higher the EC, therefore leading to salt stress in plants. The pH plays an important role in plant substrates, determining the availability of various nutrients (Gruda et al. 2013). pH values above 7.5 cause a decrease in the availability of iron, manganese, copper, zinc and boron ions. pH values below 6.0 produce a lower solubility of phosphorus, calcium and magnesium. Although plants can survive in a wide range of pH of the substrate without suffering from physiological disorders, their growth and development is reduced markedly in conditions of extreme acidity and alkalinity. Keeping the pH of the substrate within a reduced range through the application of slightly acidic nutrient solutions is recommended. The optimal value of the pH of the substrate should be between 5.5 and 6.5.

Nutrient Solution

There are various factors that should be considered for appropriate control and management of the nutrient solution, which have direct implications on the plant growth and development.

- ***Electrical Conductivity (EC) of the nutrient solution***. The EC indicates the total salts content in the nutrient solution. The range of EC required for good crop growth is between 1.3 and 2.3 dS m^{-1} (Resh 2001). It is recommended that this evaluation is performed at least once a week in the stages of post-seedling and final transplant. If the nutrient solution exceeds the limit of the optimal range of EC, water should be added. In the event that the EC decreases, it indicates that the plants are consuming the nutrients, which must be replenished through concentrated solutions. The use of a conductivity meter allows the determination of the EC values.

- ***Oxygenation of the nutrient solution.*** Lack of oxygenation produces fermentation of the nutrient solution and may result in root rot. A healthy and well oxygenated root should be light-colored or whitish; otherwise the root becomes dark due to death of root tissue. Oxygenation can be done manually (by shaking the solution manually for a few seconds at least two times a day, especially when temperatures are highest), or mechanically (by a compressor or an air-pump injecting air throughout the day).
- ***Preparation of the nutrient solution.*** The nutrient solution is usually prepared from two high concentration solutions (A and B), which are added to the water in volumes determined for each liter of water; for larger volumes of water it is added in proportions according to the size of the container. After preparing the nutrient solution, pH and the EC should be checked, in order to avoid salinity problems.
- ***Nutrient solution duration and periodical checks***. The volume of the nutrient solution must be kept constant to ensure a good development of the crop. The plants absorb more water, and at a higher rate, than the mineral elements, which will result in an increase of the EC. So it is recommended to add water to reach a value of EC that is appropriate for the plants. This will help reduce the concentration of the nutrient solution as the plants absorb water. The life span of the nutrient solution depends mainly on the content of ions that are not used by the plants. Measuring the EC weekly will indicate the level of concentration of the nutrient solution (if it is high or low). The nutrient solution half-life which has been adjusted on a weekly basis is around 30 days (Rodríguez-Delfín and Chang 2014). If nutritional adjustments are not applied, it is recommended to replace the nutrient solution every 15 days. In order not to pollute the environment, it is recommended to use the discarded nutrient solution to water other plants, such as trees, green fences and home gardens.
- ***Watering frequency.*** Watering depends basically on the size of the substrate particles. In those substrates of fine mesh (below 0.5 mm) it is necessary to reduce the watering frequency, while, on substrates with coarse grain size (greater than 2 mm) a higher number of irrigations is recommended to maintain water availability at all times. However, the crop age and the weather conditions are also indicators for the watering frequency. As addressed in chapter. "Water Management and Irrigation Systems", days with high temperatures and excessive solar radiation create more need than cloudy days with low temperatures. Manual watering of the substrate is performed to saturate it to its capacity to retain water in such a way that the excess will drain out immediately, which allows for the determination of the water or nutrient solution amount in volume per plant and the interval between irrigations (Fig. 7). In systems that are watered manually, it is important to recover the drainage of the nutrient solution to prevent water and nutrients loss.

Fig. 7 Watering of spinach plants grown on river sand with nutrient solution in the hydroponics garden of a Public School in Villa María del Triunfo, Lima, Peru (Photo: A. Rodríguez-Delfín)

In hydroponic cultivation, drip irrigation is the world's most widely used nutrient solution delivery system (Hickman 2011). The nutrient solution is supplied to each plant through emitters or drippers. Outlets may be pinched on the pipes or come preinstalled along the pipe. The irrigation is done by applying small amounts of nutrient solution directly to the root zone. This system is widely used for fruit crops production such as tomato, pepper, melon, cucumber and watermelon. In simplified systems, drip irrigation is applied by a pressure gradient, so that the tank that stores the nutrient solution is placed to take advantage of its height above the system, generating enough force so that the nutrient solution flows from the drippers, moistening the substrate and watering the plants. In this case, the system is not large, only watering small areas with few plants, and the size of the tank shall take into consideration the weight load on the rooftop.

Conclusions

With the population shift from rural to urban areas and the increasing resource scarcity, our food production needs to be adapted to the changing demands. Efficiency of food production systems is already an important factor for today's food production. Simplified hydroponic or soil-based systems offer a reliable production of high-quality vegetables with a minimum of inputs and technology. This combination makes them especially attractive for low-income families in developing countries. Considering the population growth especially in urban areas, and the associated increasing poverty in cities, the importance of these systems is likely to rise further. If the above mentioned aspects are considered, simplified hydroponics and soil-based systems have a strong potential to improve cities food security in the future. By installing these systems on unexploited areas like rooftops, problems like land loss and a decreasing amount of land per capita in cities, can be counteracted. This ensures a maximum resource use, which is likely to play a key role in future food production. The growing demand for fresh and healthy food in cities with a rapidly growing urban population will shape world food production systems of the future. Resource efficient systems like simplified hydroponics and soil based systems are a promising method to increase cities food security sustainably and will most likely gain more popularity in the future.

Bullet Points

- Closed systems like simplified hydroponics have a variety of features which are important for future food production. The precise dosage of nutrients makes them extremely resource efficient. This is significant both economically and environmentally, as resources are saved and there is no surplus that can leak into the environment. Furthermore, these production systems do not increase the loss of arable land, which is one of the main concerns regarding the food supply in the future.
- Although the described systems are simplified, appropriate know-how and suitable technologies are vital. The systems hereby proposed are suitable for different production types and need to be seen in the context of the respective circumstances. Proper training as well as the careful selection of the right growing system is significant for a successful crop production. It is also important to consider that rooftop production requires adapted growing systems with lightweight materials. The growing media play a key role in terms of weight as well as the nutrient content, the water holding capacity but also the environmental impact. Using finite resources such as peat or rock wool is not sustainable due to both their energy cost and environmental impact. Choosing the right growing media should therefore not only be an economic choice but also contribute to environmental sustainability.
- Considering the rising urbanization and increasing poverty among urban population, food production needs to become more accessible to all urban residents. The special features of simplified hydroponics and simplified soil based systems

offer sustainable long-term solutions for low-income families. Here, rooftop production offers the extra advantage of unused space, as poor people usually have limited access to land, which limits their opportunities to grow food. However, proper training (e.g. Fig. 8) and secured access to the growing area are vital for a successful implementation. Furthermore, it is important that participants of

Fig. 8 Different simplified aquaponic and hydroponic systems built and designed by students at Rufus King High School (*top* and *bottom left*) and at Bradley Tech High School (*bottom right*) in Milwaukee, United States. Students learn about the characteristics of aquaponic and hydroponic vegetable production and maintain the systems throughout the school year. These classes give them access to healthy and nutritious food which is often unavailable in local stores. The classes further create awareness for healthy food and give the students the opportunity to grow their own food in the future (Photos: C. Eigenbrod, N. Gruda)

projects receive guidance for the system maintenance as well as appropriate input access beyond the duration of the project. Only if these recommendations are followed, the projects may become sustainable in the long-term.

References

Caso C, Chang M, Rodríguez-Delfín A (2009) Effect of the growing media on the strawberry production in column system. Acta Hort 843:373–379

Fairholm J (1999) Urban agriculture and food security initiatives in Canada: a survey of canadian non_governmental organizations, (IDRC). [Online] Available at: https://idl-bnc.idrc.ca/dspace/bitstream/10625/8568/24/117781.pdf. Accessed 5 Feb 2016

Gianquinto G, Orsini F, Michelon N, Ferreira Da Silva D, Damasio De Faria F (2006) Improving yield of vegetables by using soilless micro-garden technologies in peri-urban area of north-east Brazil. Acta Hort 747:57–65

Gruda N (2009) Do soilless culture systems have an influence on product quality of vegetables? J Appl Bot Food Qual 82:141–147

Gruda N (2012) Current and future perspectives of growing media in Europe. Acta Hort 960:37–43

Gruda N, Schnitzler WH (2004) Suitability of wood fiber substrates for production of vegetable transplants. I. Physical properties of wood fiber substrates. Sci Hortic 100(1–4):309–322

Gruda N, Schnitzler WH (2006) Wood fiber substrates as a peat alternative for vegetable production. Eur J Wood Prod 64:347–350

Gruda N, Tanny J (2014) Chapter 10: protected crops. In: Dixon GR, Aldous DE (eds) Horticulture – plants for people and places, Production Horticulture, vol 1. Springer, Dordrecht, pp 327–405

Gruda N, Qaryouti MM, Leonardi C (2013) Chapter 11: growing media. In: Baudoin W, Nono-Womdim R, Lutaladio N, Hodder A, Castilla N, Leonardi C, De Pascale S, Qaryouti M (eds) Good agricultural practices for greenhouse vegetable crops – principles for mediterranean climate areas, Plant production and protection paper 217. Food and Agriculture Organization of the United Nations (FAO), Rome, pp 271–302

Gruda N, Gianquinto G, Tüzel Y, Savvas D (2016a) Soilless culture. In: Lal R (ed) Encyclopedia of soil science, 3rd edn. Taylor & Francis, Abingdon-on-Thames, Oxfordshire

Hickman GW (2011) Greenhouse vegetable production statistics. Cuesta Roble Greenhouse Consultants, Mariposa

Gruda N, Prasad M, Maher MJ (2016b) Growing media. In: Lal R (ed) Encyclopedia of soil science, 3rd edn. Taylor & Francis, Abingdon-on-Thames, Oxfordshire

Izquierdo J (2005) Simplified hydroponics: a tool for food security in Latin America and the Caribbean. Acta Hort 742:67–74

Murphy C (1999) Cultivating Havana: urban agriculture and food security in the years of crisis. Development report No.12 [Online] Available at: http://foodfirst.org/wp-content/uploads/2013/12/DR12-Cultivating-Havana-Urban-Agriculture-and-Food-Security-in-the-Years-of-Crisis.pdf Accessed 5 Feb 2016

Orsini F, Michelon N, Scocozza F, Gianquinto G (2009) Farmers-to-consumers pipeline: an associative example of sustainable soil-less horticulture in urban and peri-urban areas. Acta Hort 809:209–220

Orsini F, Fecondini M, Mezzetti M, Michelon N, Gianquinto G (2010a) Simplified hydroponic floating systems for vegetable production in Trujillo, Peru. Acta Hort 881:157–161

Resh H (2001) Hydroponic food production, Seventh edn. Taylor & Francis, Abingdon-on-Thames, Oxfordshire

Rodríguez-Delfín A, Chang M (2014) Manual práctico de hidroponía. Centro de Investigación de Hidroponía y Nutrición Mineral. Universidad Nacional Agraria La Molina, Lima

Savvas D, Gianquinto G, Tüzel Y, Gruda N (2013) Chapter 12: soilless culture. In: Baudoin W, Nono-Womdim R, Lutaladio N, Hodder A, Castilla N, Leonardi C, De Pascale S, Qaryouti M (eds) Good agricultural practices for greenhouse vegetable crops – principles for mediterranean climate areas, Plant production and protection paper 217. Food and Agriculture Organization of the United Nations (FAO), Rome, pp 303–354

Tixier P, de Bon H (2006) Urban horticulture. cities farming for the future, urban agriculture for green and productive cities. [Online] Available at: http://www.idrc.ca/EN/Resources/Publications/openebooks/216-3/index.html Accessed 5 Feb 2016

Van Veenhuizen R, Prain G, De Zeeuw H (2001) Appropriate methods for urban agriculture research, planning implementation and evaluation. In: Urban Agriculture Magazine 5, pp 1–5

Valverde K, Chang M, Rodríguez-Delfín A (2009) Effect of the light quality on the nitrate reductase activity in lettuce plants grown in NFT. Acta Hort 843:89–96

Technology for Rooftop Greenhouses

Juan I. Montero, Esteban Baeza, Pere Muñoz, Esther Sanyé-Mengual, and Cecilia Stanghellini

Abstract Rooftop greenhouses (RTGs) can generate significant advantages provided RTGs and buildings are connected in terms of energy, water and CO_2 flows. Beyond the production of high-value crops, environmental benefits such as re-use of waste water, application of residual heat and absorption of carbon dioxide are derived from urban RTGs. Social benefits viz the creation of employment, social cohesion and so on are also important assets of RTGs. This chapter is focussed on RTGs technology. RTG share many common aspects with conventional greenhouses, but at the same time RTGs show attributes that should be discussed separately. Synergies such as using residual heat, rain water for irrigation, CO_2 exchange, etc. are part of the common metabolism greenhouse-building. This chapter will concentrate on the available technology from conventional greenhouses which is more suitable for RTGs, particularly concerning greenhouse structure, covering materials, climate control and soilless cultivation systems.

Rooftop Greenhouse Structure and Access to Light

The present chapter illustrates how greenhouses can be integrated onto building rooftops and the main elements to be considered when designing a rooftop greenhouse. Concerning crop cultivation in urban environments, the greenhouse primary function is protecting the crop against hostile conditions, such as unfavourable temperature, rain, wind, disease and pests. RTGs share the same requirements as

J.I. Montero (✉) • P. Muñoz
Environmental Horticulture Programme, IRTA, Centre de Cabrils,
08348 Cabrils, Barcelona, Spain
e-mail: JuanIgnacio.Montero@irta.cat; Pere.Munoz@irta.cat

E. Baeza • C. Stanghellini
Greenhouse Horticulture Group, Wageningen University & Research,
Wageningen, The Netherlands
e-mail: esteban.baezaromero@wur.nl; Cecilia.Stanghellini@wur.nl

E. Sanyé-Mengual
Research Centre in Urban Environment for Agriculture and Biodiversity, Department of Agricultural Sciences, Alma Mater Studiorum, University of Bologna, Bologna, Italy
e-mail: esther.sanye@unibo.it

F. Orsini et al. (eds.), *Rooftop Urban Agriculture*, Urban Agriculture,
https://doi.org/10.1007/978-3-319-57720-3_6

conventional on-soil greenhouses, but they also have some peculiarities that are discussed below. Greenhouses must have maximum transmission of natural light. This consideration applies to greenhouses in countries with low light conditions, but it is also very important in areas such as the Mediterranean basin, where winter radiation, although not scarce, may limit production in that season. Roof top greenhouses (RTGs) have generally more structural parts than conventional greenhouses to comply with building construction codes, which are stricter than agricultural codes. Besides, RTGs are sometimes shaded by building's climatisation equipment, ducts, etc., as well as cast shadow from surrounding building. Therefore optimising light transmission could be even more important in RTGs than in conventional greenhouses. Greenhouse orientation, roof slope, covering material and structural parts (which can cast shadow on the crop) are to be taken into account for maximum light transmission.

Orientation

With regard to greenhouse orientation, east to west orientation (E-W) is preferable over north to south orientation (N-S), according to different studies conducted on conventional greenhouses. For instance, in a Venlo glasshouse in The Nederland (latitude 52° north), Bot (1983) reported 45% daily transmission for E-W orientation against 35% for N-S orientation at December 21st (winter solstice). Major differences were due to the transmission of the glass cover, which was maximum at noon (nearly 60% in E-W and 40% in N-S), while shades from the constructing parts (gutters, purling, etc.) were not very different in terms of light transmission. Nevertheless, it is known that shadows produced by north to south parts move along the day while shadows from east to west parts remain (nearly) in the same position all day long; therefore wide structural parts should be avoided, and if possible such wider parts should follow the N-S orientation for better light uniformity in the greenhouse.

As season progresses the angle of incidence of solar radiation on the greenhouse changes and overall transmission increases. According to the aforementioned study, E-W and N-S greenhouses have nearly the same light transmission at March 22nd (equinox).

In Southern latitudes such as the Mediterranean basin, E-W orientation is also preferred over N-S orientation, as shown in the study by Castilla (2005). Simulations for 37°N latitude showed a significant advantage of E-W over N-S, up to 15% increase in transmission at the winter solstice when there is less availability of solar radiation. Such result refers to a greenhouse with 30° roof slope. As the roof slope decreases the effect of orientation decreases, and for greenhouses with 10° roof slope the effect of orientation on light transmission is minimal.

Roof Slope and Shape

The greenhouse roof slope has also an effect on light transmission. In Central Europe the prevailing structure is the Venlo structure, with 22° symmetrical roof slope. It is not easy to change the slope of glass-covered greenhouses unless specific architectural designs are used. Plastic covered greenhouses are more versatile in terms of roof shape. Soriano (2002) used a computer simulation programme to study the effect of roof slope on light transmission for greenhouses located in latitudes 25° N, 37°N and 45°N. At the winter solstice, for latitude 25°N the greenhouse with slope 10° had the worse light transmission (67%) while the one with slope 40° had the best (77%). Nevertheless, differences were small for roof slopes 20° or higher. For other latitudes the greenhouse with 30° roof slope gave 73% and 68% light transmission, for 37°N and 45°N respectively. A 30° roof slope is a good compromise between light transmission and construction cost; therefore, it is a recommended figure in the design of plastic covered urban greenhouses.

Shadows

Structural parts which are opaque to solar radiation are always a major source of loss in light transmission. For instance, Bot (1983) reported transmission of the ridge, gutter and bar system ranging from 70 to 76% (that is, 30–24% light losses due to the greenhouse structure) depending on greenhouse type, latitude and orientation. Modern glasshouses have reduced such losses by increasing the glass size and the typical span width thus reducing the number of glass-supporting frames.

In RTG's shadows are more abundant than in conventional greenhouses. Apart from possible shadow cast from neighbouring buildings, there are two other major reasons that limit light availability in urban greenhouses: on the one side RTGs must comply with the local buildings codes. They are by far more demanding than agricultural codes in terms of coefficients of security against wind load and snow load, load combination, displacement allowances and so on. As a consequence, RTGs frames are stronger and so more opaque. On the other side, building's equipment and requirements (e.g. acclimatisation pipes) are essential parts of the building and sometimes it is unavoidable that they are located inside RTGs. As an example, Fig. 1 shows a wide opaque duct running along the major axis in the ICTA RTG (Barcelona, Spain).

Preliminary reports have shown very significant light losses in this ICTA RTG. Differences on light transmission in ICTA RTG against that of a conventional greenhouse at the same latitude were striking. For instance, at the end of January, transmission at noon in the conventional greenhouse was 62%, and in the RTG was on the average close to 35% depending on the greenhouse spot; such poor light transmission makes questionable growing most fruit crops in the winter season. It is therefore mandatory that building designers, developers as well as other collectives

Fig. 1 View of ICTA_RTG with a hydroponic lettuce crop (Photo: Sostenipra Research Group)

with weak background on crop physiology, pay attention to the greenhouse requirements (particularly light requirements) when erecting new buildings if successful urban agriculture is to be implemented. Nevertheless, some actions can be taken to mitigate the poor light transmission issue. Reflective surfaces (white painted or aluminised screens) can be incorporated to surrounding walls and opaque areas that usually are part of RTGs sides. Moreover, RTGs are regularly narrow compared to conventional on-soil greenhouses; therefore RTGs benefit more from light gains through the side walls than wider greenhouses do.

Covering Materials for Rooftop Greenhouses: Optical Properties and Materials Selection

This chapter evaluates the properties that greenhouse materials must accomplish in rooftop greenhouses. According to their agricultural purposes, materials must show optimal optical properties for ensuring proper sunlight levels and maximal crop yields in the greenhouse. Furthermore, laws are more restrictive in urban buildings and environments than the rural ones regarding the properties that materials must ensure, mainly due to safety requirements (e.g., fire-resistance). According to these specifications, we here evaluate the appropriateness of the materials that are commonly employed in rural greenhouses to be considered in the design of rooftop greenhouses.

Table 1 Desired radiative properties of materials for greenhouse covers

Property	Optimal
Absorbance	Minimum absorbance to solar spectrum
Colour	As a general rule colour must not act as a light filter.
Photosensitivity	Material must be UV resistance, to avoid polymer degradation
Reflectivity	Minimum reflection of solar spectrum
	Maximum reflection of FIR, though this is hard to achieve in commercial transparent materials
Transmittance	Maximum transmittance of PAR (photosynthetically active radiation)
	Minimum transmittance to FIR
Solar radiation diffusion (haze)	High power of diffusion (haze), as far as it does not limit PAR transmission in the solar spectrum. Most plastics have haze > 30%

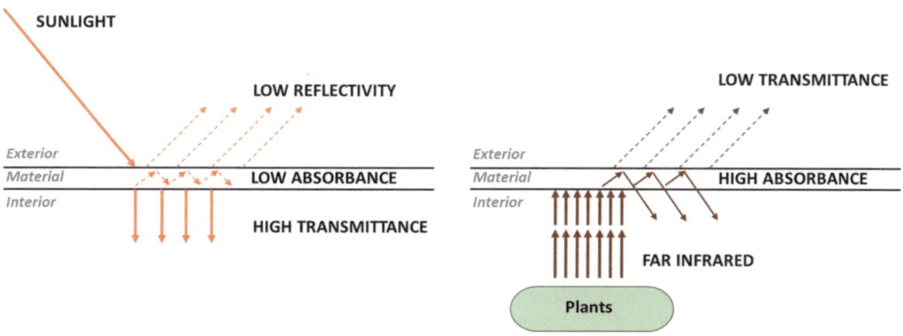

Fig. 2 Desired behaviour of materials against sunlight (*left*) and far infrared (FIR) (*right*) to maximize crop yields (Source: E. Sanyé-Mengual, J.I. Montero)

Optimal Optical Properties of Materials for Agricultural Production

Sunlight availability is a critical element in greenhouse production since photosynthesis and, thus, crop yield can be constrained. As a result, optical properties are limiting factors for the selection of materials for the covering of greenhouses. In the case of sunlight, materials must have the maximal possible transmittance in order not to limit the photosynthetic activity of the plants. In the case of thermal radiation (far infrared), which would be lost by the relatively warm greenhouse environment to outside, a greenhouse cover that is not (or as little as possible) transparent to it is a must for un-heated greenhouses. Since high absorption of FIR means high emission of FIR (and so more energy losses by radiation), maximum reflection of FIR is desired, though this is hard to achieve in commercial transparent materials. Table 1 summarizes the desirable radiative properties of materials. In particular, high transmittance to solar spectrum and low transmittance to thermal radiation (FIR) are the most limiting properties. Figure 2 illustrates the desired behaviour of materials against sunlight and FIR to maximize crop yields.

Material Properties to Ensure Safety Regulations in Urban Environments

According to Sanyé-Mengual et al. (2015), rooftop greenhouse must comply with certain laws within the urban environment, which are more constraining with the use of materials than regulations in agricultural areas. To comply with the Spanish Technical Code of Edification (CTE) (RD314/2006) and urban fire safety laws (RD 2267/2004 and Law 3/2010), the rooftop greenhouse of the ICTA-ICP building in Bellaterra (Spain) had to adapt common greenhouse structures. In particular, the steel structure of the greenhouse was oversized to ensure resistance (e.g., wind) and the covering was made of polycarbonate, as light density polyethylene (LDPE) was not permitted. In this sense, materials must comply with certain characteristics that differ from the limitations in rural environments, as summarized in Table 2. Appropriate materials for rooftop greenhouse applications will comply with fire safety laws (no inflammable, pierceable by fire) and will be more resistant (e.g., loadings, hail-resistance). Beyond policy requirements, cost, maintenance and weight can be limiting factors in the selection of the materials depending on the typology of the business model and the characteristics of the building on which the greenhouse is implemented.

Polycarbonate and PMM can be used as corrugated single layer or double-wall panels. The benefit of single layer sheets is the higher light transmission. Twin-wall panels lose up to 10% light transmission compared to single layer panels. On the positive side, double layer panels save energy for heating (between 25 and 40%, according to Stanghellini et al. 2016) and keep unheated greenhouse warmer at night.

Other materials such as glass or plastic films can also be used as a double layer cover. Double layer plastic films are generally air inflated to keep both layer separated and tensioned. Double layer covers have the same trade off mentioned for semi-rigid plastics; there is a loss in light transmission and a gain in energy saving. A fixed single layer cover and a movable double layer would be desired, since it would allow good light transmission and good energy saving. This is quite often done commercially but it requires further investments not always profitable in unheated greenhouses.

Films have the benefit of being lighter than other materials, however structural and wind loads must be taken into account in the design. Concerning film properties, FAO (2013) has recently published a manual on good agricultural practises that summarises beneficial greenhouse film properties.

- Multi-layer films are recommended over single layer films since they allow addition of the positive properties of each of the components that form the film.
- Diffusive films are preferred over clear films because they improve light uniformity and increase light interception by the crop, in so far as the overall transmissivity is not/hardly reduced.

Table 2 Evaluation of the appropriateness of materials to be employed in rooftop greenhouse, main advantages and limitations

Material	Appropriate	Advantages	Limitations
Glass	Yes	Good light transmittance; Good heat retention (i.e., particularly at night); Low transmission of UV light; Durability (long lifespan); Low maintenance costs	It needs to be hail-resistant (hardened glass, also for labour safety); Higher costs of the structure: Higher weight
Semi-rigid plastics			
Polycarbonate (PC)	Yes	Good light transmittance; Lightweight; Fire-law-compliant; Impact resistance; Hail resistance	Lifespan 10 years – It becomes brittle; Aging reduces transmissivity long before then, maintenance requirements, algae formation in the cells
Polymetacrilate (PMMA)	Side walls only. Under fire it melts and drips	Good light transmittance; Strong and lightweight; Good light transmittance; UV filter (300 nm); High corrosion resistance	Poor resistance to chemicals; Higher fragility than PC; Lifespan: up to 30 years; Fairly low hail resistance. Use high impact grades is mandatory.
Plastic films			
Polyvynilchloride (PVC)	Yes	Strong and lightweight High resistance It does not propagate the flame	Lifespan 10 years – It becomes brittle (20% transmittance lost); Environmental toxicity
PE-based films (multilayer)	Side walls only. Under fire it melts and drips	Law-compliant only for walls; Strong and lightweight; High resistance; Cheap	Maintenance (change) is required every 3–4 years; Additives for complying fire safety laws can be added (e.g., fire-retardant)
Ethylene Tretafluoroethylene Copolymer (ETFE)	Yes	Fire-law compliant; Long lifespan; Lightweight; UV filter; High corrosion resistance; High melting temperature; Flexible; High light transmittance	High cost (expensive)

Data sources: Briassoulis et al. (1997a, b), Kittas and Baille (1998) and Sanyé-Mengual et al. (2015)

Table 3 Compilation of global heat transfer coefficients for 1 Ha greenhouses, accounting for an estimate of infiltration and radiative losses

Cladding material	U-value (W $m^{-2o}C^{-1}$)
Single glass	8.8
Double glass in sidewalls	7.9
Thermopane glass	3.0
All double glass	5.2
Double acrylic	5.0
Double polycarbonate	4.8
Single PE-film	8.0
Double PE-film	6.0

From Stanghellini et al. (2016)

- Anti-drip films improve transmission and reduce dripping on the crops, but usually lose their anti-drip properties before the end of their lifespan.
- In many climates, a Near Infra-Red (NIR) film filter part of the solar spectrum and may have useful applications during the summer, but could be detrimental during the winter since they could reduce daytime temperature.

Heat Transfer Coefficient of Covering Materials

Independently that the greenhouse is heated by passive means (thermal inertia and so on) or active means (greenhouse heating with external energy) the greenhouse heat losses are directly linked to the properties of the covering material. For practical purposes, the heat losses can be quantified by an overall heat transfer coefficient U (W m^{-2} K^{-1}). Typical values for different covering materials are given in Table 3. The values given are mean values, since heat losses are also a function of wind speed and sky cloudiness, but Table 3 is also useful to compare covering materials. Obviously, the best materials are those with low U values, provided they satisfy the other requirements mentioned earlier (particularly light transmission)

RTG Climate

Currently, there are very few greenhouses in the world that are connected with a building in terms of energy, water resources and gases such as CO_2 and O_2. The aforementioned ICTA-RTG is one of the few examples of such integration and can be taken as an example to discuss the RTG climate. Compared to conventional greenhouses, recent evaluations of ICTA-RTG have shown interesting peculiarities in terms of the greenhouse climate performance (Montero et al. 2016).

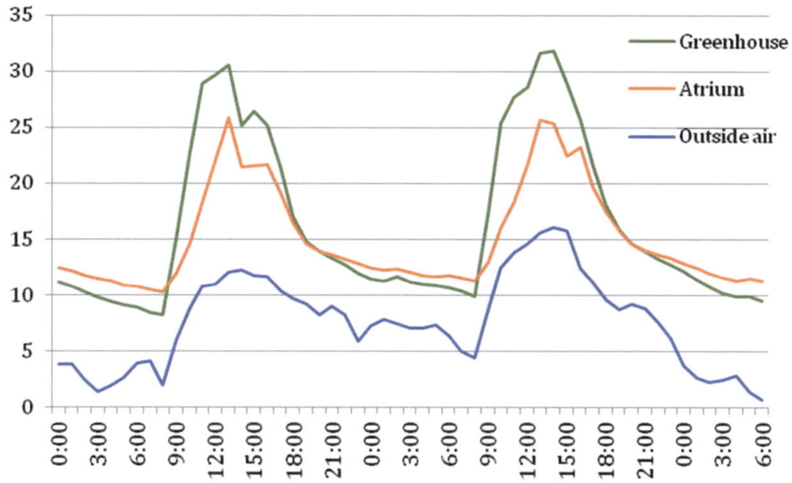

Fig. 3 Time course of ICTA greenhouse, atrium and outside air temperature, days 24–25 January 2015. Greenhouse without plants (Source: Montero et al. 2016)

Heat Transfer from the Building to the Greenhouse

Figure 3 shows the evolution of greenhouse temperature, atrium temperature and outside air temperature for days 24–25 January 2015. At this time, the greenhouse had no plants and was not heated, so it is a good representation of the natural climate of the passive RTG. It can be seen that at night the greenhouse was always warmer than the outside(average 6.3 °C, maximum increase in temperature 8.8 °C). Most of the time, the greenhouse air temperature was above 10 °C. The RTG had a thermal screen, which was not deployed during the nights shown in Fig. 3; higher increases in temperature (not shown here) were measured when the thermal screen was extended. Night-time average of 8.7 °C and maximum increase in temperature of 9.5 °C were registered with the thermal screen. This greenhouse response is in clear contrast with conventional unheated greenhouses, where in areas such as the Mediterranean thermal inversion occurs frequently (Piscia 2012) The major reason that can explain this different behaviour is that there is a significant air exchange of the RTG with the atrium, which, as displayed in Fig. 3, is slightly warmer than the greenhouse air and clearly higher than the outside air. Accordingly, the greenhouse benefits from the common areas of the building, which due to the thermal inertia of construction elements remains warmer than the outside air at night.

Moreover, the greenhouse floor is made of concrete which also could play a relevant role in storing solar energy during the day and releasing it at night. Concerning the RTG thermal inertia, the concrete floor surface temperature at night was typically up to 6 °C higher than the greenhouse temperature at the beginning of the night and around 4 °C higher at sunrise. Heat transfer calculations show that the heat release from the concrete floor to the RTG air is on the average near to 30 Wm^{-2}, while in commercial on-soil unheated greenhouses 20 Wm^{-2} have been previously reported (Montero et al. 2013). The concrete floor acts as a thermal storage, which is more efficient than most soils; it collects heat during the day and releases it at night. Such higher efficiency may be due to the fact that thermal conductivity of concrete (0.93 Wm^{-1} $°C^{-1}$ according to ASHRAE 1989) is generally higher than that of soil, which depends on its water content but can be between 0.5 and 0.8 Wm^{-1} $°C^{-1}$ (J.I. Montero, unpublished data). Nevertheless, the role of greenhouse thermal inertia is currently under investigation. Only preliminary results are available up to now.

This first evaluation has shown that, at least in Mediterranean climates, it is possible to grow winter crops on the RTG without any external source of heating due to the air exchange with the buildings as well as the thermal inertia of construction elements. Nevertheless, winter night-time temperatures in RTG were below optimal (which are 13–16 °C for tomato. Castellanos 2004) and so additional heat sources may be needed. Very recently, a study by Nadal et al. (2017) has quantified the benefits of integrating RTG and the building underneath in terms of waste energy exchange. Based on simulation results, the study reported that the ICTA-RTG recycled 341.93 $kWh/m^2/year$ of thermal energy from the main building in 2015. This figure can help us to compared integrated RTG with non-integrated or free-standing greenhouses. Assuming 100% energy conversion efficiency, compared to a non-integrated greenhouse heated with gas, the i-RTG delivered an equivalent carbon saving of 82.4 kg $CO_2/m^2/year$, and economic saving of 15.88 €//m^2/year. In Mediterranean conditions, the cost of energy makes that the vast majority of greenhouses are unheated, and so, production has to be stopped during the coldest months of the year (in Northern Mediterranean areas) or suffers from low temperature (in Southern Mediterranean areas). The i-RTG allows extending the crop cycle and increasing yield in comparison with non-integrated RTGs or on soil free standing greenhouses.

Waste heat is abundant in most buildings, particularly public buildings, retail parks, restaurants and facilities which have centralised climatisation equipment. It is a free energy source, that should be used as much as possible. Limited information on waste heat applied to RTGs show that the quality of waste heat is rather low (temperature in the range of the low twenties) which means a very generous surplus of low temperature heat is needed to cover the greenhouse heating requirements. Besides, air distribution of high-volume low-temperature heat must be carefully studied to avoid excessive air streams around the crop canopy and guarantee climate uniformity.

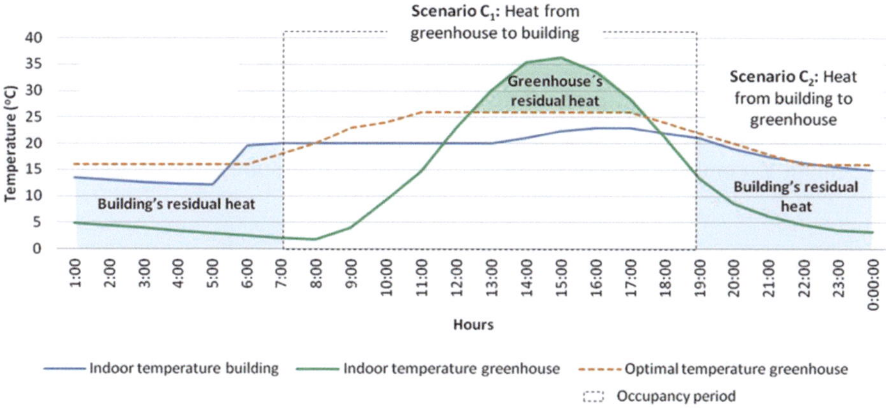

Fig. 4 Graphical representation of residual heat in RTG-Building system (Source: Cerón Palma 2012)

Heat Transfer from the Greenhouse to the Building: The Greenhouse as a Solar Collector

Figure 3 shows the time-course of temperature during the day at ICTA-RTG. Since the greenhouse was closed, the air temperature was up to 31.8 °C, nearly 16 °C more than the outside air. This means that potentially the greenhouse has a surplus of energy during the day that can be used to heat the associated building.

The greenhouse can act as a thermal solar collector provided extreme precautions are taken on crop sanitation and pest and disease management to avoid dispersion of harmful products on the building space. The greenhouse has also a potential as a building's air purification and oxygenation system, but to make sure this is possible further studies on allergens and air quality needs to be conducted. The amount of residual heat from the greenhouse to be given to the building is not well known yet. This is being evaluated by modelling the thermal behaviour of the greenhouse and its associated building. Figure 4 shows the time course of measured building's indoor temperature and greenhouse temperature in a winter day in Cabrils (Barcelona). In this case, the greenhouse was unheated. Figure 4 also shows the set point greenhouse temperature which would be desirable to maintain, provided the greenhouse had climate control equipment. During the night the greenhouse temperature is clearly below the set point temperature (16 °C) and the greenhouse could benefit from the building's heat. During the day, after 12:30 pm approximately, the greenhouse temperature is above the set point temperature (25 °C), so the greenhouse has to be ventilated or, alternatively, such surplus heat can be delivered to the building underneath.

Cerón-Palma (2012) conducted simulation studies for a RTG located in coastal Barcelona. The building was a single storey building and the greenhouse covered the whole surface of the roof. Simulation results showed that surplus heat from the greenhouse can save up to 79% of the heating building's requirements in a one to one greenhouse/building surface basis. Simulations show the potential of the greenhouse as a solar collector in a favourable area with mild winter climate. Further studies under different climate conditions are needed to fully assess the potential of greenhouse heat recovery. Besides, to make such energy saving possible, specific equipment to control greenhouse air and building temperature must be developed.

Ventilation

Most conventional greenhouses rely on natural ventilation, which is the result of pressure differences created by wind or temperature differences. It is generally accepted that wind driven ventilation prevails over thermally driven ventilation even for small wind speed such as 2 ms^{-1} (Baeza et al. 2009). It is also known that, concerning wind driven ventilation, the air exchange rate is depending on the ventilator's size and is directly proportional to the outside wind speed. Ventilation studies based on numerical simulations have enlighten the flow pattern in and around the greenhouse structure and have produced a set of recommendation to increase wind driven ventilation. Such studies show schematically the air velocity field of greenhouses with roof ventilators open to the upcoming wind (windward ventilation) and opposed to it (leeward ventilation). Windward ventilation produces higher air exchange and is generally preferred. Nevertheless, care should be taken to reduce excessive air stream impinging over the crop or dismembering the wind panel, so under moderate wind conditions it may be advisable to partially close the ventilators.

Simulations results also show the importance of combining side-wall ventilation and roof ventilation. The greenhouse slope also helps to increase the air exchange; a minimum roof slope of 25–30° is recommended for proper ventilation. The greenhouse size is also a factor to be taken into account. Good agricultural practises recommend to build greenhouses no more than 50 m wide, in order to avoid hot spots in central areas.

All these recommendations developed for commercial on-soil greenhouses are valid for roof top greenhouses, but a major difference is that wind speed changes with height, so that the wind speed is stronger on top of the buildings than at the ground level. Figure 5 illustrates how the wind profile follows the well-known logarithmic law which depends on the terrain roughness. For large cities with high buildings and skyscrapers the Swiss Wind Power Data Website (http://wind-data.ch/tools/index.php?lng=en) suggest using a roughness length of 1.6 m for calculating the wind profile with height above ground. Accordingly, if the wind speed at 10 m high is 5 ms^{-1}, at 30 m above ground would be 8 ms^{-1}, and at 50 m above ground would be 9.5 ms^{-1} approximately.

The different wind regimen of RTGs compared to conventional on-ground greenhouses have positive consequences in terms of greenhouse ventilation; as mentioned earlier, ventilation is directly proportional to wind speed and, so RTGs are generally

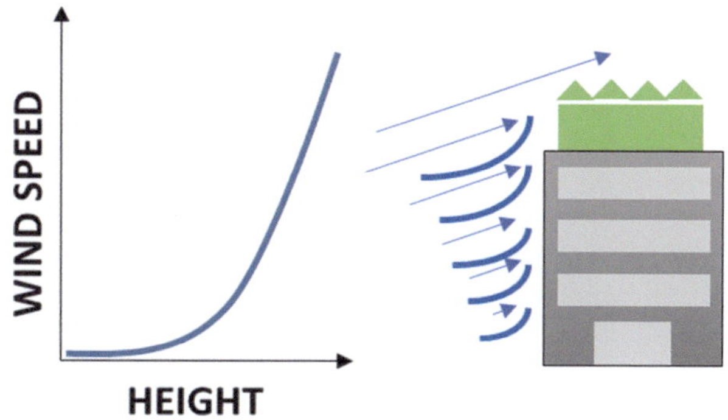

Fig. 5 Scheme of wind speed profile above ground (E. Sanyé-Mengual, J.I. Montero)

better ventilated. Actually, RTGs may need less ventilation surface than conventional greenhouses to achieve the same air exchange rate. The implication is that the design of ventilation systems for RTGs has to be as careful as for conventional greenhouses, but the vents surface can be less. Also specific climate control equipment to manage ventilation should be adapted to RTGs to prevent mechanical damages due to strong gusts as well as avoiding excessive air movement around the plants. Previous experience at ICTA-RTG has shown unexpected high crop water demand probably associated to very high wind speed in the crop canopy.

Humidity

As reported by Montero et al. (2016) relative humidity in a RTG can be very low, at least in Mediterranean climates. On the one side, the soil could be a source of water vapour in conventional greenhouses, while the concrete floor of a building is not. Besides, the specific humidity of a building's air can be low as well, as shown in the aforementioned study. Moreover, air in urban areas may also have low specific humidity compared to rural areas, where vegetation and surface water may be relevant sources of water vapour. Low humidity in RTGs can create problems in terms of pollination and fruit development. It can also have negative effects on plant growth and particularly on water consumption. Perhaps future RTGs may need additional water vapour sources and (again) specific climate control and irrigation control equipment to account for the low humidity regime.

Soilless Cultivation Systems and Management

Generally speaking, RTG crops are not grown in soil, due to the extra weight soil adds to the buiding's structure (particularly when it is humid as required for crop development) as well as difficulties associated to soil transportation and dirt.

Fig. 6 Fertigation system with a domestic programmer, concentrated nutrient solution tanks and hydraulic injectors in an urban orchard in Spain (Photo: P. Muñoz)

Therefore, soilless cultivation is generally preferred, as described in chapter "Soil Based and Simplified Hydroponics Rooftop Gardens"). Aeroponics systems (crop without soil or substrate) may be implemented in RTGs but only for some specific crops and very high technology greenhouses. Nutrient film techniques (NFT) may be used in crops such as lettuce, escarole, chard etc. Nevertheless, in some areas with hot periods, NFT is not very used due to the potential risk of crop wilting if electric power or water supply fails. Besides, excessive root temperature has been reported in crops irrigated with NFT in Southern Mediterranean conventional greenhouses.

Most RTG crops are grown in substrates (organic or mineral) similar to the ones used in greenhouse soilless production, since substrates have a buffer capacity in terms of water and nutrients. Logically, RTGs should use fertigation systems (i.e. the application the nutrients with irrigation water); optimizing the use of water and nutrients is mandatory to ensure sustainability and productivity. In RTGs production, fertigation equipment can range from the very simple to more complex ones that allow the use of concentrated solutions and injectors (Fig. 6).

Closed-loop irrigation systems are strongly recommended in RTG production; as well as enabling good environmental results, closed systems can reward the growers with saving in water and fertilizers costs (Ehret et al. 2001), with a 30% minimum saving to be expected as compared with open loop systems. Pardossi et al. (2011) provided a set of guidelines for best management of growing medium and fertigation in closed soilless cultivation, with the aim to reduce the consumption of water and fertilisers (and then production costs). This study also contemplates the

Fig. 7 Lettuce grown in perlite bags in a closed-loop irrigation systems. ICTA-RTG lab. (Barcelona, Spain) (Photo: Sostenipra Research Group)

reduction of environmental impact associated to the disposal of spent substrates and the emission of nutrients and other agrochemicals with drainage water.

Closed-loop irrigation in RTGs requires uprising the crop rows above the floor to allow lixiviate collection and pumping back to the fertilization equipment where the solution is reformulated; this is not needed in conventional greenhouses where crop stand on the soil surface and lixiviate is collected in a tank that can be dug in the soil. Cheap methods to up rise crop rows have to be developed. Figure 7 shows a lettuce crop grown in a closed-loop irrigation system (ICTA RTG, Barcelona, Spain). Perlite sacs are on top of metal benches with enough slope to collect lixiviates.

Recirculation systems can foster the transmission of root diseases, particularly under high temperature. For this reason disinfection treatments such as filtration, ozone application, UV light and so on are recommended. (Raviv et al. 1995; Ehret et al. 2001).

It must be pointed out that full recirculation is only possible if irrigation water is free of ions such as Cl and Na since they are poorly absorbed by the plants and so accumulate in the irrigation solution. Previous work in conventional greenhouses (Jeanequin et al. 1998) pointed out the difficulty in using recirculation systems if Na^+ concentration and Cl^- concentration is above 2.0 meq L^{-1} and 1.5 meq L^{-1}, respectively. Partial recirculation (lixiviate dejection when some ions accumulate above undesirable levels) is a solution for this case (Marfà 2000); one has to bear in mind that local regulations concerning lixiviate dejection and treatments have to be respected. Wetlands and biological filtering has proven to be an effective method for lixiviate treatments and partial recuperation of nutrients (Fig. 8).

Fig. 8 Wetland for lixiviate treatment in ICTA Building, Barcelona, Spain (Photo: Sostenipra Research Group)

Concerning the irrigation schedule, standard recommendations established for conventional greenhouses (FAO 2013) apply; nevertheless, as mentioned earlier RTG crops may be affected by lower humidity and higher wind speed than those in conventional greenhouses, and this has a direct consequence on crop water requirements. Shades are also more common than in conventional greenhouses, and so less energy is available for plant transpiration; all this means that the general equations for crop irrigation should be adapted and calibrated to RTG production.

An important point to be taken into account is the water source. Rain water is the obvious first source of water for urban agriculture since it is high quality water; harvesting rain water is fairly easy from the greenhouse roof, but storing it is neither easy nor cheap, particularly in greenhouses retrofitted to existing building in highly populated areas where there is no room for water reservoirs. Besides, rain water is in most regions not enough to satisfy crop requirements (Sanjuan et al. 2015). In many cases tap water from the urban network has to be used as the primary water source or as a supplement to rain water. Here below are listed some consideration concerning tap water use for urban agriculture:

- Tap water is expensive and is expected to become more and more expensive with time. Therefore optimising irrigation and reducing water consumption is a top priority in the development of any type of urban agriculture
- Tap water may be too salty for recirculation systems; diluting it with rain water is a solution

- If rain water and tap water are to be combined, the fertilisation solution has to be quickly adapted to the water source to avoid abrupt changes in electrical conductivity and pH.
- Some buildings have water softener equipment that strongly reduce Ca concentration in tap water. It is common knowledge that Ca plays a relevant role in plant nutrition, so the fertilization programme must include Ca solution to avoid physiological pathologies.
- Although tap water pipes are pressurized, experience shows that it is recommended not connecting the pipes directly to the irrigation systems, since direct connection may cause a drop in pressure when irrigation water is required. It may also happens that the pipe lines are not dimensioned to carry the water flow required for watering crops, particularly in old buildings. To avoid pressure fluctuations it is better to have a small tank to store tap water at atmospheric pressure and use a pump to regulate the pressure for the irrigation system. This also guarantees that water will be available even if the supply is cut down for a short period of time.

Conclusion

In addition to the potential benefits common to urban horticulture, the joint climatisation of buildings and roof-top greenhouses may have the advantage of a reduced input requirement. A roof-top greenhouse can take advantage of waste heat and carbon dioxide from the building and at the same time work as a solar collector and source of oxigen and water vapour for the building.

Bullet Points

There are several aspects that need to be taken in good consideration for planning an efficient rooftop greenhouse:

- ensure maximal light transmissivity of the cover (minimise structural elements)
- select the cover material with the maximal possible transmissivity
- if necessary ensure good thermal insulation through extensible screen
- carefully design the ventilation openings in view of the siting
- design the growing system in view of the crop(s) that will be cultivated and the desired sink/source effect for CO_2 and H_2O
- dimension and place any buffer (heat/water) that can be required
- consider that the intensity of management requires automatic (computer) control of climate and irrigation

References

ASHRAE Handbook. Fundamentals (1989) Atlanta. Ga. American Society of Heating, Refrigerating and Air Conditioning Engineers

Baeza E, Pérez-Parra J, Montero JI, Bailey B, López JC, Gázquez JC (2009) Computational analysis of the effect of sidewall vents on buoyancy driven natural ventilation in parral type greenhouses. Biosyst Eng 2009:1–11

Bot GPA (1983) Greenhouse climate: from physical processes to a dynamic model. PhD dissertation. 240 pp. Wageningen University

Briassoulis D, Waaijenberg D, Gratraud J, Von Elsner B (1997a) Mechanical properties of covering materials for greenhouses part 1: general overview. J Agric Eng Res 67(3):81–96

Briassoulis D, Waaijenberg D, Gratraud J, Von Elsner B (1997b) Mechanical properties of covering materials for greenhouses part 2: quality assessment. J Agric Eng Res 67(3):171–217

Castellanos JZ, Gálvez F et al (2004) Manual de producción Hortícola en Invernadero. INTAGRI. 2da Edición

Castilla N (2005) Invernaderos de plástico. Tecnología y manejo. Ediciones Mundi-Prensa, Madrid, p 462

Cerón-Paima (2012) Integración de invernaderos de producción intensiva en las cubiertas de edificación de Barcelona. Tesis doctoral. Universitat Autónoma de Barcelona

Ehret D, Alsanius B, Wohanka W, Menzies J, Utkhede R (2001) Disinfection of recirculating nutrient solutions in greenhouse horticulture. Agronomie 21:323–339

Food and Agriculture Organization of the United Nations (2013) Good agricultural parctices for greenhouse vegetable crops. FAO, Rome. ISBN 978-92-5-107649-1

Jeannequin B, Fabre R (1998) Pilotage de la fertirrigation de la tomate hor-sol à drainage reciclé. PHM-Rev Hort 396:17–20

Kittas C, Baille A (1998) Determination of the spectral properties of several greenhouse cover materials and evaluation of specific parameters related to plant response. J Agric Eng Res 71:193–202

Marfà O (2000) La recirculación en cultivos sin suelo. Elementos básicos.En Compendios de horticultura n° 14: Recirculación en Cultivos sin suelo

Montero JI, Stanghellini C, Castilla N (2009) Greenhouse technology for sustainable production in mild winter climate areas: trends and needs. Acta Hortic 807:33–44

Montero JI, Muñoz P, Sanchez-Guerrero MC, Medrano E, Piscia D, Lorenzo P (2013) Shading screens for the improvement of the night time climate of unheated greenhouses. Span J Agric Res 11(1):32–46

Montero JI, Muñoz P, Llorach P, Nadal A, Sanyé-Mengual E, Rieradevall J (2016) Development of a building-integrated roof top greenhouse in Barcelona (Spain). Acta Horticulturae (in press): greensys 2015. Evora (Portugal)

Nadal A et al (2017) Building-integrated rooftop greenhouses: an energy and environmental assessment in the mediterranean context. Appl Energy 187:338–351

Pardossi A, Carmassi G, Diara C, Incrocci L, Maggini R, Massa D (2011) Fertigation and substrate management in closed soilless culture. Deliverable 15. EUPHOROS project. http://www.wageningenur.nl/en/Research-Results/Projects-and-programmes/Euphoros-1.htm

Piscia D, Montero JI, Baeza E, Bailey B (2012) A CFD greenhouse night-time condensation model. Biosyst Eng 111(2):141–154

Raviv M, Reuveni R, Krasnovsky A, Median S (1995) Recirculation of rose drainage water under semi-arid conditions. Acta Hortic 401:427–433

Sanjuan-Delmás D, Muñoz P, Gabarrell X, Josa A, Rieradevall J (2015) Environmental assessment of adapting nutritious solution for rainwater use in the irrigation of hydroponic crops. Conference: life cycle management 2015, At Bordeaux, France

Sanyé-Mengual E, Oliver-Solà J, Montero JI, Rieradevall J (2015) An environmental and economic life cycle assessment of rooftop greenhouse (RTG) implementation in Barcelona, Spain. Assessing new forms of urban agriculture from the greenhouse structure to the final product level. Int J Life Cycle Assess 20(3):350–366

Soriano T (2002) Validación de un modelo de cálculo de transmisividad a radiación solar directa en invernadero mediante maquetas a escala y determinación del prototipo óptimo para la costa mediterránea. Tesis Doctoral. Universidad de Almería

Stanghellini C, Baptista F, Eriksson E, Gilli C, Giuffrida F, Kempkes F, Muñoz P, Stepowska A, Montero JI (2016) Sensible use of primary energy in organic greenhouse production. BioGreenhouse COST ActionFA 1105, www.biogreenhouse.org

Rooftop Aquaponics

Beatrix W. Alsanius, Sammar Khalil, and Rolf Morgenstern

Abstract Multitrophic water-based production systems, such as aquaponics, are a means to supply animal protein, fish lipids and high-quality horticultural produce rich in fibre, minerals and bioactive compounds in urban areas. In this chapter, we describe the specific demands of fish and crop production and technological solutions. However, for long-term economic viability, financial investment in such systems must be met by sustainable economic output from the systems. For methods such as rooftop aquaponics, further system development and capacity building are therefore essential preconditions for wider establishment in urban areas.

Introduction

In work to secure provision of animal protein, fish lipids and high-quality horticultural produce rich in fibre, minerals and bioactive compounds for the world's rapidly expanding urban areas, multitrophic water-based production systems are highly interesting. In such systems, fish rearing (aquaculture) are integrated with production of horticultural produce (hydroponic systems) in so-called aquaponics. The reclaimed fish tank water is reconditioned by recycling through a hydroponic unit. To optimise nitrogen provision and use efficiency, a biofilter for conversion of ammonia to nitrate is installed within the aquaculture subsystem. Irrespective of the use of low- or high-tech approaches, aquaponics systems are highly engineered. These multitrophic water-based culture systems have a high level of complexity and – apart from fish/seafood and plants – can also incorporate units for alternative green fish feed production.

B.W. Alsanius (✉) • S. Khalil
Department of Biosystems and Technology, Microbial Horticulture Unit, Swedish University of AgriculturalSciences, PO Box 103, SE-23053 Alnarp, Sweden
e-mail: beatrix.alsanius@slu.se; sammar.khalil@slu.se

R. Morgenstern
University of Applied Sciences of South Westfalia, DE-59494 Soest, Germany
e-mail: morgenstern.rolf@fh-swf.de

F. Orsini et al. (eds.), *Rooftop Urban Agriculture*, Urban Agriculture,
https://doi.org/10.1007/978-3-319-57720-3_7

Fish and Crop Demands in Aquaponics

Fish and crops grown in aquaponics systems have different demands on their environment for maximum biomass production. Fish are heterotrophic and dependent on a good supply of organic energy sources and oxygen. In contrast, plants are autotrophic and need light energy, CO_2, water, inorganic nutrients and oxygen. Water quality (oxygen content, load of organic compounds, electrical conductivity, nutrient content, pH) and water temperature are other important features for fish and plants, as is the microbiota associated with the biofilter responsible for conversion of nitrogen.

Water fulfils different functions in aquaponics systems. For plants, water is a physical growth medium but also a carrier of nutrients and oxygen, enabling plant growth. Furthermore, in classical hydroponics, water serves as a solvent and carrier for additives and pesticides. Contrary to belief, the nutrient solution in hydroponics is not sterile, but microbially colonised. For fish, water has a habitat function, but also acts as a carrier of fish feed and feed additives, as well as of fish faeces. Demands on water quality vary between the fish and plant compartments of aquaponics systems, but also between different fish species. In this context, it should be noted that plant and fish requirements involve different measures that are sometimes not comparable. When discussing water quality in aquaponics systems, the load of organic and inorganic compounds, electrical conductivity, pH, temperature and oxygen content are of critical importance. The microbial community structure in the different compartments is a function of the microbial colonisation and the physiology of the macroorganism grown, while the water quality parameters also influence the microbial load and microbial activity in the different compartments. Table 1 shows physio-chemical threshold values for water properties associated with aquaponics compartments (Alsanius 2014).

Temperature

Adequate temperature regime in the water recirculating between the fish and plant compartments is a vital prerequisite for high-quality food production.

Water temperature is critical for fish rearing, with different temperature requirements applying for cold-water and warm-water fish (Table 1). Plants respond to both suboptimal and supraoptimal temperatures by reducing growth, but plant growth is generally impaired at root zone temperatures <15 °C (Martin and Wilcox 1963; Wilcox et al. 1962). The temperature optimum for biomass production and yield varies between different plant species. However, root zone temperature also affects the crop's tolerance to pathogens and influences disease development caused by both fungi-like and fungal root pathogens (O'Brien and van Bruggen 1993; Panova et al. 2004; Sopher 2012). Temperature levels in the root environment must thus be optimised in order to maximise plant yield. To optimise the integrated

Table 1 Physico-chemical threshold values for water properties associated with aquaponics compartments

Parameter	Compartment	Threshold value
Temperature (°C)	Warm-water fish	24–30
	Cold-water fish	12–23
	Crop	>15
		20–24[a]
Oxygen content (mg L^{-1})	Warm-water fish	>2
	Cold-water fish	>5
	Crop	BOD <10
		COD <60
pH	Fish	7.5–8.5
	Crop	>5; <7
Electrical conductivity (dS m^{-1})	Fish	30
	Crop	<3[b]
Alkalinity (meq)	Fish	50–300
	Crop	<3
Carbon content (mg L^{-1})	Fish	<12
	Crop	<20–40 (TOC)[c]

Source: Alsanius (2014) and references therein
[a]Considering plant physiology and pathology factors
[b]Differences occur between different crops and developmental stages
[c]Value highly dependent on the degradability of the organic carbon source

aquaponics system, the technical design of the system needs to be adjusted to meet the temperature demands within the water cycle.

Oxygen

Oxygen content in the water loop of aquaponics systems is a function of physical factors, such as temperature, chemical factors (i.e. the load of organic compounds) and the associated macro- and microorganisms. Differences in terminology and choice of methods for measurement of oxygen content limit the possibility to obtain a general picture and thereby draw conclusions on optimisation of oxygen conditions in aquaponics culture systems. However, within the three fundamental compartments in an aquaponics system (fish, biofilter, plant), the following demands must be considered:

- In the fish compartment, oxygen content should be >6 mg O_2 L^{-1}
- In the biofilter, oxygen is a limiting factor for ammonification and nitrification
- In the plant compartment, biological and chemical oxygen demand (BOD and COD) levels of 10 mg O_2 L^{-1} and 60 mg O_2 L^{-1}, respectively, are crucial thresholds for water quality in hydroponics.

In the root zone, aerobic conditions need to be maintained for adequate water and nutrient uptake processes (Flannery and Lieth 2008) and for microbial organic carbon utilisation and degradation. However, oxygen-depleted microloci may be present, impairing plant physiological parameters and promoting certain plant pathogens.

Electrical Conductivity

The parameter electrical conductivity (EC; dS m^{-1}) is commonly used in horticulture and hydroponic contexts, but lacks precision when optimising plant nutrient supply. In the fish compartment of aquaponics systems, the EC level should not exceed 1.2 dS m^{-1}. In the crop compartment, the EC is dependent on the crop grown and its different developmental stages, as well as product quality (lettuce: 2.3 dS m^{-1}, tomato: 2.2 dS m^{-1}, cucumber: 2.2 dS m^{-1}).

Nutrients

The presence of both autotrophic and heterotrophic organisms in aquaponics systems is a particular challenge. The quality and quantity of the basic and major nutrient input to the fish compartment (fish feed) must be matched to the fish species, size and age, but since all fishes are carnivorous at early developmental stages high protein inputs are required. Herbivorous and planktivorous fishes (e.g. grass carp) have a protein requirement of 18–23%, while omnivorous fishes (e.g. *Nile tilapia*) require 24–33% and carnivorous fish (e.g. trout) require 35–50%. The excretion rate of nitrogenous waste is correlated with fish feeding behaviour and feed quality, with carnivorous, omnivorous and herbivorous fish having high, medium and low nitrogen excretion capacity, respectively.

Several factors govern influence biomass production rate and nutrient excretion in reared fish, and thus the fish feeding regime required. These include: fish species, feed integrity and quality, especially with respect to energy density, amount of fish feed provided in relation to stocking density, feeding level and fish age, as well as feed conversion ratio (FCR = kg feed/kg fish meat gain). According to Schneider et al. (2005), FCR in fish can vary between 0.71–3.00 and the level of excreted nitrogen (N) and phosphorus (P) between 14–48 g N feed and 3–5 g P per kg feed, respectively. To meet the crop's demands and optimise crop growth and development, yield and quality and the timing of crop growth, the ambient nutrient content in the water after passage through the biofilter needs to be supplemented with readily available nutrients. Plant nutrient requirements vary between plant species and phenological stages, but in general, during the vegetative stage of the crop the N to potassium (N:K) ratio has to be higher, whereas during the generative stage (fruit set) it can decrease.

Apart from the mineral composition of the solution, the pH is also important for plant nutrient uptake. The nutrient solution in hydroponic systems is adjusted to a pH level of 5.5–6, depending on the plant species grown and its developmental stage.

The nutrient load and pH supplied to plants through fish tank effluent normally does not meet the demands of the crop. Most of the data in the literature refer to N, which of course is the dominant element released by the fish and a macronutrient for plants, but high N levels alone do not ensure high biomass production (Dediu et al. 2012; Marschner 1995). Due to fluctuations in nutrient release and demand in the fish and plant compartments, nutrient adjustments need to be made continuously. Problems arise especially with respect to K, a plant macronutrient that is not needed by the fish and thus normally not administered through the fish feed (Graber and Junge 2009). The plant micronutrients zinc (Zn), copper (Cu), boron (B) and manganese (Mn) are also elements that are likely to be present in too low amounts in aquaponics systems. Accumulation of nutrients to levels that are toxic either to fish or to plants is another problem encountered in aquaponics systems (Treadwell et al. 2010).

System Design for Rooftop Aquaponics Systems

The differing requirements for the accommodation of fish and plants in a combined system, as detailed above, necessitate the establishment and connection of different instruments. To maintain high water quality in a recirculating aquaculture system, processes such as solids removal, nitrification and biofiltration, degassing, aeration and water conditioning need to be considered. Aquaculture technology has a number of tools available to achieve these tasks, some of which can perform several functions, e.g. degassing, biofiltration and aeration can be achieved using a tricking filter.

Water treatment in RAS follows an established chain of events, starting with solids removal and followed by biofiltration, degassing, aeration and conditioning. The integration with a hydroponic cultivation system can be achieved with different system designs, all of which exhibit different system behaviour and performance.

Early system designs, generally termed “one-loop systems”, pipe the effluent from the fish tanks directly into the growing beds filled with growth medium (Goddek et al. 2015), removing solids, acting as a biofilter and thus providing a habitat for the nitrifying bacteria. The cleaned and nutrient-depleted effluent is then pumped back to the aquaculture system. The one-loop system is very tightly coupled, which leaves little room for intervention and control. Long-term operational problems like sludge accumulation in ebb and flow growth beds have been reported. However, this design could still be suitable for extensively operated smaller systems without a focus on high yield. This simple design can be amended by assigning some or all of the water treatment processes to dedicated units. Pumping water in one loop still leads to sub-optimal control over the different units, which may require different mass flows and retention times.

Two- or multi-loop systems are variants referred to as “decoupled systems” (Goddek et al. 2016a). In this design the aquaculture is operated similarly to a stand-alone rooftop aquaponics system. At an integration point, nutrient-rich process water is diverted into the hydroponics system, which usually circulates the water multiple times through the growing beds via a catch tank system before returning it to the aquaculture system at a second integration point.

Recirculation of the process water in the rooftop aquaponics loop allows nutrient accumulation, raising nutrient concentrations to match plant requirements. The fish nutrient tolerance sets the threshold concentration for nutrients before discharge to the hydroponic system. The threshold levels have to be low enough to avoid negatively affecting fish welfare and yield and vary between species. *Tilapia* and different catfish species can tolerate high nutrient concentrations, making them popular choices for aquaponics systems. Preliminary results indicate that European catfish (*Silurus glanis*) can be reared at 150 mg L^{-1} nitrate-N (NO_3-N), while African catfish (*Clarias garepinius*) has been reported to perform well up to concentrations of 200 mg L^{-1} NO_3-N.

These elevated nutrient levels are still relatively low in comparison with those used in the nutrient solution of commercial hydroponic systems, which leads to suboptimal plant yield. However, by adding additional nutrients to the rooftop aquaponics system effluent water, yield increases exceeding those obtained in traditional hydroponics with comparable nutrient levels are possible (Delaide et al. 2016). Due to the limited nutrient concentration tolerance of fish, the enriched effluent from the hydroponic compartment cannot be directed back to the aquaculture compartment, which causes a system design dilemma: the desired nutrient flow and required water flow between the aquaculture and hydroponic compartments do not match the requirements of high-performance systems. Concepts for decoupled aquaponics systems to alleviate this situation have been proposed and are currently under investigation. This dilemma is commonly resolved by discharging the surplus aquaculture process water into the sewage system and thus reducing the overall sustainability of the system.

The sludge processing unit in rooftop aquaponics systems can be discharged directly into the sewage system, but this represents a significant nutrient loss from the system (Goddek et al. 2016b). In particular, it results in losses of P from the fish feed, which accumulates in the solid faeces. Thus remineralisation of these nutrients in an aerobic or anaerobic sludge treatment system is recommended. Aerobic remineralisation can be achieved with a simple technical set-up, but leads to comparatively low remineralisation rates and significant bacterial proliferation within the set-up, which counteracts the purpose of reducing sludge volume. Multi-stage anaerobic remineralisation for aquaponics systems in which the predicted sludge removal and remineralisation rates exceed 95% are currently under development. The first results are expected to be published in spring 2017.

Practical considerations regarding the integration of an aquaculture system into a rooftop farm need to consider assembly, structural calculations and logistics. Since the system will be constructed in an existing building with all its constraints, the tanks and processing units have to either be small enough to fit through elevators, staircases and doors, or assembled from parts on-site. This limits the choice of material for the tanks, as e.g. polyethylene parts can be assembled on-site using plastic welding, while with fibreglass parts this is not possible.

Placement of the aquaculture unit within the building has to be evaluated from one project to the next. Due to space restrictions, the rooftop itself is not an attractive site for the aquaculture unit. Moreover, while the plants need light, the fish do not. The floor directly beneath the roof is a good place to host the aquaculture unit and, since it usually requires much less floor space than the hydroponic unit, other areas of the floor can be utilised for nutrient and fish food storage, nutrient solution preparation, fish and plant processing and product cold storage.

When choosing the aquaculture site, the load bearing capacity of the floor has to be considered. Buildings with support pillars are most likely necessary. Ideally, a load bearing capacity of at least 100 kg m^{-2} of the floor itself and a minimum of 5–6 t in each pillar, depending on their density, must be guaranteed. Industrial manufacturing buildings and storage buildings, usually located in dockland areas, are good candidates for suitable locations.

Accommodation of the aquaculture unit on the ground level or in the basement is possible. The load bearing capacity of these floors is often more suitable, but greater electricity consumption in pumping the nutrient solution from the integration point through the riser pipe to the hydroponics compartment on the roof has to be considered.

Operation of an aquaculture system requires logistics pathways, since for every kg fish yield that has to be transported from the system, a similar (or greater) amount of fish feed has to be transported to the system. Industrial elevators with high load capacity are recommended to avoid laborious transportation through elevators for people.

Optimised System Design for Rooftop Aquaponics

Rooftop aquaponics provides countless opportunities for urban horticulture and can help to save space, secure food production and produce environmentally friendly food, as well as contributing to creation of green cities. However, the design of the aquaponics system is critically important. For the plant compartment, hydroponically cultivated plants can be grown in solid hydroponics using pots and growing medium such as peat, rockwool or pumice. Liquid hydroponics, with no growing medium to support the plant root and to anchor the plant, is another alternative. The hydroponic unit may be either horizontally or vertically organised. The latter form may optimise the use of floor space, but calls for artificial lighting to optimise light interception in all parts of the crop stand. Furthermore, optimisation of other environmental factors in vertical cropping systems differs from the approach needed in horizontally organised units. Reuse of water and nutrients is a fundamental requirement to achieve sustainable integrated production systems for fish and high-quality plant produce. As water acts as a carrier for dissolved and particulate compounds, plant pathogenic organisms causing root diseases may also be moved by the water through the system (Stanghellini and Rasmussen 1994). Different technological and filtering approaches to prevent the dispersal of root pathogens via the nutrient solution are commercially available (Ehret et al. 2001) and should be integrated to secure the viability of rooftop aquaponics. To ensure adequate nutritional conditions for the plant crop, sensors and control devices to maintain relevant pH and nutrient composition should be considered if the plant unit not only acts as a wetland, but also as a production unit for high-quality horticultural produce.

Biofiltration and bacteria in aquaponics systems are other crucial factors that play an important role in converting fish waste into plant fertiliser (Tyson et al. 2008). Using a biofilter provides a large surface area, appropriate temperature, pH and oxygen level. As discussed by Nelson (2008), raft and growing media-filled bed aquaponics systems do not require the separate biofilter needed in liquid hydroponics, since the raft and solid medium provide sufficient surface area to which the bacteria can attach. The flow rate in such systems is of great importance for the biofiltration effect, e.g. N removal has been shown to decrease with increased flow rate (Endut et al. 2010). Nitrogen removal from the system can also be adjusted by the use of an appropriate fish to plant ratio. Temperature demands for the different compartments need to be considered also from the perspective of the biofiltration system.

Requirements and Limitations

Construction stability and carrying capacity are important features for installation of rooftop aquaponics systems but, especially when separate floors for the aquaculture and hydroponic components are proposed, the economics of the process need to be considered. Space limitation is a decisive characteristics in urban areas which is taken into account in property transactions. Thus there needs to be financial incentive to build rooftops with high carrying capacity and/or to sacrifice a floor in the building for fish production. There is an obvious conflict between sustainable urban food production and economics, which limits the use of rooftop aquaponics to supply high-quality products in high quantities. High-tech hydroponics rather than "wetland" plant production approaches are the primary choice for the hydroponics compartment, compromising sustainable use of resources as discussed above. Existing horticultural knowledge and technology therefore need to be fused to create innovative sustainable alternatives that encourage investment in rooftop aquaponics. This in turn means that assets necessary to secure such production goals need to be met, e.g. multidisciplinary capacity building and training must be promoted.

Conclusions

Rooftop aquaponics is an interesting alternative for efficient use of urban spaces for sustainable food production. However, the complexity of dual aquaculture-hydroponics production systems puts high demands on the construction and on optimising the conditions for both fish and plant crop production. There is a need for further development to ensure the environmental, economic and social sustainability of rooftop aquaponics systems.

Bullet Points

- Aquaponics systems can contribute to food security in urban areas. In such systems, fish and seafood rearing (aquaculture) is integrated with production of horticultural produce (hydroponics). Reclaimed fish tank water is reconditioned by recycling through a hydroponic unit.
- Fish and plants have different demands for optimal growth that need to be reflected in the system design. The economic conflict on use of land and/or surfaces in urban areas needs to be resolved by providing strong incentive to allocate assets to food production rather than residential or commercial purposes.
- Output of high-quality fish and plant-based products, along with sustainable city development, must be promoted. There is a good basis for further innovations in which multidisciplinary teams reach out to society to build the necessary capacity.

References

Alsanius BW (2014) Sustainable systems for integrated fish and vegetable production — new perspectives on aquaponics. In: Glynn C, Planting M (eds) The SLU global food security research and capacity development programme 2012–2014; a Swedish government initiative. SLU-global report vol 2014:6. SLU, Uppsala, pp 60–65

Dediu L, Cristea V, Zhang XS (2012) Waste production and valorization in an integrated aquaponic system with bester and lettuce. Afr J Biotechnol 11(9):2349–2358

Delaide B, Goddek S, Gott J, Soyeurt H, Jijakli MH (2016) Lettuce (Lactuca sativa L. Var. Sucrine) growth performance in complemented aquaponic solution outperforms hydroponics. Water accepted

Ehret DL, Alsanius BW, Wohanka W, Menzies JM, Utkhede RS (2001) Desinfestation of recirculating nutrient systems. Agronomie 21:323–339

Endut A, Jusoh A, Ali N, Wan Nik WB, A. H (2010) A study on the optimal hydraulic loading rate and plant ratios in recirculation aquaponic system. Bioresour Technol 101:1511-1517 doi:10.1016/j.biortech.2009.09.040

Flannery RJ, Lieth JH (2008) Effects of plant root oxygen deprivation in relation to water and nitrate uptake for roses. Acta Hortic 766. doi:10.17660/ActaHortic.2008.766.4

Goddek S, Delaide B, Mankasingh U, Ragnarsdottir KV, Jijakli H, Thorarinsdottir R (2015) Challenges of sustainable and commercial aquaponics. Sustainability 7(4):4199–4224. doi:10.3390/su7044199

Goddek S, Espinal CA, Delaide B et al (2016a) Navigating towards decoupled aquaponic systems: a system dynamics design approach. Water 8(7):303. doi:10.3390/w8070303

Goddek S, Schmautz Z, Scott B et al (2016b) The effect of anaerobic and aerobic fish sludge supernatant on hydroponic lettuce. Agronomy 6(2):37. doi:10.3390/agronomy6020037

Graber A, Junge R (2009) Aquaponic systems: nutrient recycling from fish wastewater by vegetable production. Desalination 246:147–156

Marschner H (1995) Mineral nutrition of higher plants, 2nd edn. Academic, London/San Diego
Martin GC, Wilcox GE (1963) Critical soil temperature for tomato plant growth. Proc Soil Sci Soc 27:565–567
Nelson R (2008) Aquaponic equipment - the biofilter. Aquaponic J 48:22–23
O'Brien RD, van Bruggen AHC (1993) Effect of temperature on corky root disease of lettuce and growth of the pathogen Rhizomonas suberifaciens. Can J Plant Pathol 15(2):85–90. doi:10.1080/07060669309500831
Panova GG, Grote D, Kläring H-P (2004) Population dynamics of Pythium aphanidermatum and response of tomato plants as affected by root-zone temperature. J Plant Dis Protect 111(1):52-63
Schneider O, Sereti V, Eding EH, Verreth JAJ (2005) Analysis of nutrient flows in integrated intensive aquaculture systems. Aquac Eng 32:379–401
Sopher C (2012) High temperature predisposition of sweet pepper to Pythium root rot and its remediation by Pseudomonas chlororaphis. University of Guelph
Stanghellini ME, Rasmussen SL (1994) Hydroponics - a solution for zoosporic pathogens. Plant Dis 78:1128–1138
Treadwell D, Taber S, Tyson R, Simonne E (2010) Foliar-applied micronutirnets in aquaponics: a guide to use and sourcing. IFAS Extension (HS1163)
Tyson RV, Simonne EH, Treadwell DD, Davis M, White JM (2008) Effect of water pH on yield and nutritional status of greenhouse cucumber grown in recirculating hydroponics. J Plant Nutr 31(11):2018–2030. doi:10.1080/01904160802405412
Wilcox GE, Martin GC, Langston R (1962) Root zone temperature and phosphorus treatment effects on tomato seedling growth in soil and nutrient solution. J Am Soc Hortic Sci 80:522–529

Integrating Rooftop Agriculture into Urban Infrastructure

M. Gorgolewski and V. Straka

Abstract Rooftop agriculture projects need to integrate with the social, economic and environmental infrastructure of the surrounding city. Their design needs to consider numerous technical and practical factors that affect the host building, site and neighbourhood. There are various potential synergies with other urban necessities such as water management, energy use, air quality, provision of green space, employment, and community support. This chapter will focus on the relationship between rooftop agriculture and the technical infrastructure of the building and its neighbourhood.

Introduction

The design of a rooftop for food production will be influenced by some key aspects of the building, including: location, size and design of the roof, height, structural loads that it can accept, access for people and produce, safety, and water availability. In addition, rooftop agriculture can affect urban systems such as water supply and disposal, waste disposal, energy demands, urban heat island effect, and ecological diversity. Many of these issues are governed by local planning and building codes, which need to be addressed by any rooftop agriculture project. In addition to the technical characteristics, the use of the host building (e.g. residential, office, industrial) must be considered so as to exploit synergies and to avoid conflicts.

Clearly, the local bio-climate is a major factor in food production, and also affects building performance. So it is important to explore synergies between the performance of the host building and its food producing roof. Rooftop agriculture is usually more exposed to wind, sun and rain than ground level agriculture, but detail design may be able to provide some wind shelter, while green roofs can create thermal buffers to reduce cooling loads inside the building, and can protect the waterproofing layer from exposure to damaging solar radiation.

The relationship between the activities on the roof and the building's occupants should be considered at an early stage of any project so that opportunities can be

M. Gorgolewski (✉) • V. Straka
Department of Architectural Science, Ryerson University, Toronto, ON, Canada
e-mail: mgorgo@ryerson.ca; vstraka@ryerson.ca

F. Orsini et al. (eds.), *Rooftop Urban Agriculture*, Urban Agriculture,
https://doi.org/10.1007/978-3-319-57720-3_8

explored, and problems anticipated. These can be limited by providing separate access and entry to the roof, so that other building users are separated from the rooftop activity, or they can be more or less integrated. This will depend on the building type, and the requirements of other users and the owner.. There are several examples where an organisation wishing to farm on a roof have agreed a lease or other long-term arrangement with a building owner to use the roof for food production. Such agreements usually need to be for a significant period (10 years minimum) to be worthwhile for the farmers, and need to clearly set out responsibilities, access arrangements, water availability, insurance, energy use, maintenance, and liability issues.

Municipal Requirements: Building and Planning Codes

As with any construction activity, rooftop agriculture projects must address local planning and zoning requirements and building codes. These vary with each location and with the type of building and its use. Every city will have some areas that are more suited to rooftop agriculture due to the nature of the zoning. In general, zoning is concerned with location of appropriate activities and building forms within the city, so, for example, a major food producing initiative that will attract a lot of truck movements may be resisted in a residential area without the appropriate transport infrastructure. Local zoning requirements also typically restrict building heights, maximum floor areas, and setbacks from the property lines. These may particularly affect proposals that include greenhouses, pergolas, pavilions or other structures on the roof. There may also be local limitations for buildings that have been classified as historical monuments, or are part of a historical neighbourhood. Zoning requirements in some locations may not anticipate a demand for agricultural activity in the city and so clear policies may not exist for such projects, and bylaws may seem obstructive. In such cases, the municipality may need some educating and reassurance. Most municipalities have an appeal process, so it may be possible to request an exemption from local bylaws to allow the project to go ahead even if it does not meet municipal zoning requirements. However, this is likely to cause delays and increase costs, and there is no guarantee about obtaining an exemption.

Building codes generally regulate technical and safety issues and use of materials. Any project will have to demonstrate that the building can cope with the loads imposed by the rooftop activity, and meet the safety, insulation, access and fire escape requirements. These are discussed further in various sections below.

On the other hand, rooftop agriculture can provide benefits that can help to meet other municipal priorities and objectives. These include water management, reducing urban heat island effect, reducing food deserts, creating employment and creating green spaces (Fig. 1). For example, some municipalities such as the City of Toronto have introduced a Green Roof Bylaw that requires new buildings over 2000 m^2 gross floor area to include a proportion of green roof. In Toronto, this policy was instigated mainly to reduce rainwater run-off at times of peak rain fall,

Fig. 1 Ryerson University in Toronto converted an extensive green roof into an urban rooftop farm. (Photo: A. Throness)

which reduces the need for the city to upgrade its waste water drainage systems. Research summarised by Banting et al. (2005) shows that run-off from green roofs can be significantly reduced by between 50% and 98%, depending on a variety of factors. Also, they suggest that the substrate on green roofs has the ability to retain particulate matter in the storm water and to improve the quantity of runoff by reducing the mass of pollutants that flow into the drainage system.

Many municipalities are also interested to reduce the temperature in the city on hot summer days when the dark roof surfaces of conventional roofs heat up significantly from solar radiation and raise the overall temperature in the city. Investigations of the effect of green roofs on surrounding temperatures have generally used non-productive green roofs, and the results have varied significantly. Nevertheless, some benefits are now accepted and some municipal initiatives offer an opportunity for rooftop agriculture as this will reduce the temperature of the roof surface. It also allows the creation of secluded green spaces connected to the city but separated from the noise and commotion of the urban activities below. The City of Toronto has developed advisory documents including a Green Roof Designer Checklist intended to assist in such projects (City of Toronto 2009).

Another municipal objective in many cities is improving air quality. Research by Peck et al. (1999) and by Yok and Sia (2005) suggest that plants on the roof can help municipalities with improving urban air quality. Plants act as a filter of pollutants, removing as much as 95% of heavy metals such as cadmium, copper, and lead from runoff, 37% of sulphur dioxide and 21% nitrous oxide and improve air quality by trapping and absorbing volatile organic compounds, and airborne particulate matter. Particles are trapped by leaves and then washed off into the soil.

Building Structure Including Earthquake Design

Choosing a building type with an appropriate structural system is a crucial first step for any rooftop agriculture project involving an existing building. The building's foundation and columns as well as the roof structure must have sufficient capacity to ensure that they can support the weight of the soil, crops, equipment and people that may be on the roof at any time, as well as snow and wind loads. The foundations often have an additional capacity but if not, then this is the most expensive item to deal with. Columns commonly have an additional capacity, and if necessary, can be strengthen. However, a roof structure rarely has an additional capacity unless it has been designed for occupant access in mind or has ballast. In order to assess the existing building potential for rooftop agriculture, a structural engineer should audit the building and review structural drawings before assessing the structural capacity of various structural members. The structural requirements for rooftop agriculture vary depending on the exact use of the roof space. It may be that the rooftop design has to be adjusted according to the structural assessment. For example, positioning the heavier elements above columns or loadbearing walls. Some building types such as modern industrial sheds, and big box stores have large, flat roofs that may seem attractive spaces for rooftop agriculture due to the roof area they offer. However, these buildings usually have "lean" roof structures with little surplus capacity for additional loads. Thus, many such buildings would not be able to accommodate the soil and other loads that are necessary for agriculture without significant intervention. They require structural analysis and most likely strengthening to accommodate agriculture. This will have economic implications that may be difficult to justify financially. Nevertheless, due to the use of steel, this building typology can be adapted by strengthening major structural components and by introducing new joists between the existing with reasonably cost efficiently, provided the foundation has an adequate capacity.

Low rise residential buildings often have pitched roofs that make them difficult to use for food production. However, other existing building types such as apartment buildings, offices, and older industrial building from the early twentieth century (and earlier) which often use masonry or concrete construction may have overdesigned foundation and additional structural capacity to accept extra loads necessary for rooftop agriculture, or can be readily upgraded. There are several rooftop farming projects in New York such as the two projects by Brooklyn Grange that are located on older industrial/commercial buildings that have spare structural capacity. Concrete structural systems are more difficult to assess for the potential of additional load capacity because of the nature of the structural system. A concrete structure is usually poured on site and therefore it is dependent on the workmanship (placement of rebar, their cover) and actual strength of concrete mix. While steel structure is either exposed or easily identifiable, there is little information available about the actual make up of concrete members (monolithic construction) and where the reinforcement was placed. Even the actual concrete strength is not known. In case of steel structures, it is relatively easy to determine the capacity of the structure due to an ability to measure the structural elements and the historic information about the quality of steel and a relative simplicity to test a coupon sample of structural members. In some cases there is ballast on the roof and some roofs were designed with access in mind. Both cases result in an additional capacity to deal with loads from soil, etc. and so may be particularly suited for rooftop agriculture.

Considering a load on the roof, the saturated weight of the vegetated roof, is generally proportional to the thickness of the growing medium. One cubic metre of "wet earth" weighs approximately 1600 kg of additional load on the structure of the building but there are mixtures of expanded minerals, such as vermiculite and perlite with peat or other composting material which weigh less than 50% of the conventional soil. The growing medium can further be engineered to use no peat or soil. Common practice in hydroponics growing leads to reduction in density to 300 to 500 kg per cubic meter. As the previous paragraph indicates, it is very important to consider the type of green or vegetated roof proposed and to communicate the resulting limitations as a switch from a growing medium to a more regular soil has a significant impact on the imposed loads on the building structure.

When considering the loads on the structure, options for how to arrange the agriculture should be considered; i.e., distributed planting beds, or planters strategically placed (Fig. 2). These decisions are important for the assessment of the structural potential of the existing building; for example, locating planters over the columns or load-bearing walls may constrain the potential for growing but is likely to be less problematic structurally. There are some cases where industrial buildings were designed for an addition which was never built and as a result they have an additional capacity to support a vegetated roof. This scenario is demonstrated in the case of Lufa farms in Montreal and Laval where hydroponic greenhouses have been added to two buildings with little structural implications. In the case of Montreal, the two-storey commercial/industrial building was design for an additional floor which was never built. Therefore the building structure including the roof itself had an additional capacity to accommodate a greenhouse and hydroponic food growing operation.

In areas with seismic activity, placing of additional weight on the roof of mid- to high- rise buildings will result in an increase in the horizontal force acting at the roof level. This in turn may cause an inadequate lateral load resisting system which is in need of strengthening.

For new buildings, the increased cost of the structure to support a 150 mm extensive green roof is small provided it is incorporated into the project at the beginning. So it is important to engage green roof specialist's right from the start of the project. They play important role in education of the design team about the opportunities, concepts and design needs of green roofs and are essential in advising on the

Fig. 2 Planters on the Uncommon Ground roof in Chicago are used as barriers for safety on the roof. (Photo: M. Cameron)

additional loads associated with different schemes. The growing medium can be engineered to suite a range of acceptable loadings on the roof. It is important to consider "wet" weight for the growing medium as some are highly absorbent, and can increase 300% in weight when wet. In more conventional food growing, using primarily soil mix, it is common that peat or composting matter is added every year. Over the years this could lead to a significant load increase. Besides the additional weight of growing media, the following loads need to be considered:

- ***Weight of crops***; this is fortunately not a problem in locations with snow loads which are typically greater;
- ***Weight of equipment*** such as farming tools and irrigation systems; usually not significant;
- ***Weight of water storage tanks***; these can be significant loads but with a right placement (over columns or shear walls) they may not cause problems other than in a seismic area. Vegetated roofs are not particularly suitable for the collection of rain water but the greenhouse roofs create an ideal opportunity;
- ***Weight of composting units;***
- ***Load from staff working on the roof and visitors***, which must be considered as it impacts not only loading but also the number of exits required (this varies based on local codes).

Wind loads on the roof may result in scouring of soil. Typically, the worst condition is when the wind blows on the corner, creating the largest uplift. This may be a concern when the soil is dry or it contains very light particles and there is no vegetation.

Building Construction Issues

The primary objective of a roof is to provide protection from rain, snow and wind and to create a thermal and air barrier between inside and outside. Any rooftop agriculture proposal must not compromise these functions. As a result of the increased interest in green roofing in recent years, there have been significant developments in the technologies used, creating a whole new industry which specialized in new systems such as roofing membranes, root barriers, filter cloths, lightweight growing media, plant containers as well as education programs to improve installation standards. However, not all of this technology is suited to food production. Much of the research has focused on: 1) developing lighter, thinner green roof systems at a reduced cost that have minimal impact on the building structure, 2) addressing concerns about water leakage. Food production generally requires thicker growing media, of 150 mm to 500 mm and a more intensive approach.

Waterproofing is a major issue that building owners worry about. Digging on the roof must not reach the waterproof layer as this can lead to damage. Also, some roots can potentially penetrate and damage the waterproof layer and provide a route for water to get inside. However, appropriate technologies have advanced significantly in recent years to provide a range of options for waterproofing systems. Green roofs typically have several layers of protection including a "root barrier" to protect the waterproof membrane, and a filter membrane to prevent soil being washed away and blocking the building drainage. In some cases, thermal insulation is also located on top of the waterproof layer providing further protection from roots.

Such roofs can actually extend the lifetime of the waterproof layer since they protect it from exposure to ultraviolet radiation from the sun, and also from the worst excesses of temperature variations (roofs can fail when they are exposed to large changes in temperature over time). Canadian research has shown that the surface temperature of an exposed conventional roof membrane can reach over 60 °C on a hot summer day in Ottawa, Ontario. When under a green roof with 150 mm of growing medium and grass, the roof membrane reached below 30 °C (Liu and Baskaran 2003). In addition, the green roof lowered the daily temperature fluctuations on the roof membrane in the summer from a median value of 45 °C to less than 5 °C, greatly reducing the thermal stresses on the roof membrane (although this research was not on a productive green roof). This research suggests that green roofs can make the roof membrane last longer thus reducing disruption and cost to the building owner. This is reinforced by research from Europe that suggests that a green roof can double the life span of a conventional roof by protecting the membrane, leading to predicted life spans of over 50 years (Porsche and Kohler 2003).

Another factor is leak detection. Some municipalities will require a leak test prior to installing a green roof. Of greater concern is detection of leaks after the green roof has been installed. In the past, locating the leak and repairing it could be difficult, involving moving large amounts of growing medium to expose the membrane. However, this issue has spawned an industry in leak detection systems which can locate the source of a leak and avoid large scale replacement. A variety of specialist companies focus on leak detection equipment using electric field vector mapping (EFVM) to accurately pinpoint the leak on either a green or conventional roof.

Typically, building codes require a roof (as well as other elements of the building envelope) to provide a specified level of thermal insulation. This is achieved with thermal insulation integrated into the layers of the construction, and must be done in a way that avoids condensation occurring within the roof (interstitial condensation). The soil layer and plants also provide some insulation to the building below. This is often over-emphasised, as wet soil does not have good insulation characteristics, but some benefit does occur. In particular, during the summer months the absorption of solar energy by the plants and protection of the roof membrane surface from high temperatures can have a significant benefit for reducing cooling loads and maintaining comfort below.

Consideration also needs to be given to the construction process. A significant amount of soil will need to be elevated onto the roof and distributed. The height of the building will be a significant factor. Appropriate equipment is needed to deliver, raise and spread the soil. This is usually less of a problem on new building projects where machinery and space is often available, but existing buildings can lead to significant challenges, particularly if the building is in normal operation below. It is possible to blow soil up onto the roof through a vacuum pipe system, or raise it with cranes, hoists or conveyors. Smaller projects have used manual labour, particularly when community volunteers are available; but this has its limitations, one of which may be the capacity of the elevator or the number of stairs. It can also be quite messy. This process needs to be carefully considered with space made available at the base of the building for delivery of soil, and equipment.

Resource Use

One key challenge for the future viability of cities is enhancing the efficiency of using resources. Rooftop agriculture can help establish local cycles for the use of resources such as energy, water and waste within the building or community, due to their proximity. Due to the exposure of roofs to rain, and the relatively thin layer of soil used, roofs intended for agriculture should be designed with close attention to water availability and rainfall patterns. In many locations, irrigation systems may be required. If using clean municipal water for irrigation is to be avoided or minimised, rain-water collection and storage systems will usually be necessary. It may also be possible to integrate with building water systems by using grey water (water used for bathing and washing) from the building for irrigation. Research by Roehr (2009) suggests that locations with high summer rainfall such as Tokyo or Shanghai can often support rooftop food production using mainly direct rainwater (with a small amount of storage). However, in most locations significant irrigation may be required during the growing season as well as inter-seasonal rainwater storage. For example, Vancouver, although generally thought of as a wet climate, only has 24% of annual rain fall during the April to October growing season.

Collecting rainwater for irrigation is relatively low cost and low tech. Appropriate roof surfaces are required for the rain collection and water storage tanks are needed. It is generally not appropriate to collect water from a green roof since they retain 50 to 98% of water – see Banting et al. 2005). Storage tanks can add significant load to the building and so they must be carefully located and their impact should be considered when assessing the structural implications of the rooftop agriculture. Tanks can be on the roof or at (or below) ground level, but pumping water may be required in some cases. Also, in many climates the tanks and pipes must be protected from freezing, or they should be emptied during the winter period. Rooftop greenhouse projects provide a suitable roof for water collection, and such projects may require less water (depending on the growing method).

Some container systems designed for growing food on rooftops use a reservoir beneath the soil container and capillary action for water to seep into the soil from below. Alternatives, which is a leading Montreal-based NGO, developed such a planter with a water reservoir for low-maintenance vegetable gardening. This container system uses an adapted ordinary plastic recycling box, including a false bottom that hides two water reservoirs, which are filled through a vertical tube (a commercial version is shown in Fig. 3) (Alternatives 2008). The advantages of such systems are that they provide a good environment for growing healthy vegetables while reducing maintenance and water use. Rooftop agriculture may also come into competition with the requirements for rooftop mechanical systems, or renewable energy systems. Roofs are often locations for elements of heating, ventilating and air conditioning (HVAC) systems and are usually the most efficient location for solar energy technologies. In addition to being a space conflict it can cause security issues as access to mechanical equipment is usually restricted. For some building types, podium roofs can be a good option for agriculture as they are generally not very high up and so decrease the cost of adding access to growing space and allow for the mechanical systems to be physically separated higher up the building (Fig. 4). Rooftop greenhouses offer additional complications and opportunities. The spaces in a greenhouse may be heated to extend the growing season, and this can be very energy intensive since greenhouse have poor thermal insulation characteristics. Research in Canada suggests that the additional energy costs for such greenhouses are significant (Vickers and Gorgolewski 2014). However, it may be possible for such projects to explore synergies with the building HVAC systems. In particular, waste heat in the form of warm air from the host building which is expelled could be used to maintain minimum temperatures in the greenhouse, particularly at the coldest times of the year. Conversely, there may be other times of the year particularly

Fig. 3 Self watering planter with a water reservoir below the soil. (Photo: Courtesy of Bio Top)

in spring and fall when the greenhouse acts as a solar collector and will generate spare heat that can be used in the host building to reduce fossil fuel energy use. Such synergies are yet to be fully explored in built projects but may offer significant opportunities. Greenhouses sometimes use artificial lighting to extend the growing season. This will add to the electricity load of the building, and so any lighting system should use the most efficient LED lights, which can match the characteristics of natural light well.

A large rooftop farm will produce significant volumes of organic waste and will also require significant amounts of compost. This offers the opportunity for developing material cycles to reduce nutrient exports and losses, reducing the need to apply artificial fertilisers. Partnerships can be explored with local groups for exchange of organic resources within the neighbourhood. These can be from residential buildings, or may be specialist products such as coffee grinds from local coffee shops.

Fig. 4 Fairmont Hotel in Vancouver uses the podium roof for growing food. (Photo: Courtesy of Fairmont Hotels)

Rooftop Access and Safety

Issues of access and security will vary with the scale of the project. When significant amounts of food production are planned for a rooftop it is important to facilitate an efficient operation enabling an unhindered flow of people, plants, equipment, waste, and produce. Access for deliveries and removal and distribution of produce need to be planned, and as with any accessible roof, safety must be addressed. Vertical access to existing roofs is often limited as the expectation is that they are not intensively used. This can be a significant limitation to rooftop agriculture on existing buildings. Furthermore, access (or rather escape) is regulated by fire codes which usually require two separate exits for any larger project, and there may also be restrictions on flammable materials. Architects or code consultants may be required to deal with approvals.

Elevators are often important for moving people and materials particularly if the roof is higher than a couple of storeys. However, for a larger operation using an elevator designed for people may not be appropriate. Moving soil, plants and produce can be a messy process that leads to dirt, which may not be acceptable in the main elevators designed for people in many buildings. Some buildings have service or freight elevators, which can be useful as they are usually intended for more messy uses, often can take larger items, and do not exit through the public lobby. However, these may not go all the way to the roof level, and it may be necessary to extend an existing service elevator shaft to the roof. Furthermore, for larger projects, at ground level the elevator needs to have access to a loading area or other suitable location (e.g. for moving the produce). In most buildings, it will not be acceptable to use the front lobby for this activity. New buildings should consider all these issues at the design stage.

Horizontal circulation requires spaces sufficiently wide for trolleys, or other appropriate, wheeled transportation systems, both at roof level and at ground level. Also, storage will be required for tools, equipment, and produce (this may require cold storage in some cases), and space may be necessary for packing. Access to water for cleaning the produce and for operatives to use washrooms is necessary. These spaces need to be appropriately located.

Important safety factors are the railings and barriers that are required to ensure safety for anyone on the roof. Building codes usually specify requirements for the design of these (often between 1050 mm and 1200 mm minimum height). In some jurisdictions railing must be set back from the front of the building so they are not visible from the street at a distance equivalent to two times the height from the edge. Many jurisdictions also may require full accessibility for those with physical difficulties. This may depend on whether the project involves commercial production or is used for demonstration/education purposes.

Conclusions

Spaces on urban roofs are now becoming sought after due to the significant potential they offer to contribute to community energy, water and food infrastructure, and to provide amenity space and build local resilience. They are being reclaimed and transformed from unused and sterile urban surfaces into productive enclaves that help contribute to urban ecology, health, and wellbeing. These new uses for roofs can reduce the ecological footprint of the building and community, and are generally recognised by green labelling systems such as LEED and BREEAM green building rating systems.

Roofs enjoy specific benefits such as exposure to sun, wind, snow and rain, which in most conventional roofing systems are seen as a problem rather than a potential resource. However, recent developments in new technologies offer a variety of opportunities to use the roof space creatively. Roofs are increasingly seen as valuable, and some building owners are now looking to lease their precious roof space to organisations that can use them.

Bullet Points

- Municipal planning and zoning requirements will impact where urban rooftop farms may be located. Zoning requirements in some locations may not anticipate a demand for agricultural activity in the city and so clear policies may not exist for such projects, and bylaws may seem obstructive. However, rooftop agriculture can provide benefits that can help to meet other municipal priorities and objectives such as water management, reducing urban heat island effect, reducing food deserts, creating employment and creating green spaces.
- The design of a rooftop for food production will be influenced by some key aspects of the building, including: location, size and design of the roof, height, structural capacity, access for people and produce, safety, and water availability. Choosing a building type with an appropriate structural system is a crucial first step for any rooftop agriculture project involving an existing building. Also the layout of planters/beds may be constrained by structural capacities.
- The relationship between the activities on the roof and the building's occupants should be considered at an early stage of any project so that opportunities can be explored, and problems anticipated. Access for deliveries and removal and distribution of produce should be planned, and safety must be addressed.
- As with any construction activity, rooftop agriculture projects must address local building codes. Any project will have to demonstrate that the building can cope with the loads imposed by the rooftop activity, and meet the safety, insulation, access and fire escape requirements. In particular safety must be addressed with appropriate protection to prevent people, soil and tools from falling off the roof, and to provide adequate escape in case of emergency.

- Consideration should be given to the construction process since a significant amount of soil will need to be elevated onto the roof and distributed. The height of the building will be a significant factor. Appropriate equipment is needed to deliver, raise and spread the soil.

References

Alternatives (2008) Guide to setting up your own edible rooftop garden. Published by Alternatives and the Rooftop Gardens Project. 80 pp. Available at: http://www.rooftopgardens.alternatives.ca/sites/rooftopgardens.alternatives.ca/files/ready_to_grow.pdf.pdf. Accessed 28 Feb 2016

Banting D, Doshi H, Li J, Missios P (2005) Report on the environmental benefits and costs of green roof technology for the City of Toronto, Prepared For City of Toronto and Ontario Centres of Excellence – Earth and Environmental Technologies (OCE-ETech) By Ryerson University

City of Toronto (2009) Green roof designer checklist Ontario building code O.Reg 350/06 & ByLaw 583–2009. Available online at: http://www1.toronto.ca/city_of_toronto/city_planning/zoning__environment/files/pdf/Green_Roof_Designer_Checklist.pdf. Accessed 28 Feb 2016

Liu KKY, Baskaran BA (2003) Thermal performance of green roofs through field evaluation, Green Roof Infrastructure Conference, Chicago, IL (NRCC-46412)

Peck SW, Callaghan C, Kuhn ME, Bass B (1999) Greenbacks from green roofs, CMHC, Canada

Porsche U, Kohler M (2003) Life cycle costs of green roofs – a comparison of Germany, USA and Brazil, RIO 3 – World climate & energy event. Brazil, Rio de Janeiro

Roehr D (2009) Rooftop agriculture: greenroofs as productive envelopes. Available at www.greenskinslab.sala.ubc.ca. Accessed 28 Feb 2016

Vickers A, Gorgolewski M (2014) The energy performance of urban rooftop greenhouses, NBS2014 Conference, Sweden

Yok TP, Sia A (2005) A pilot green roof research project in Singapore. Green Roofs for Sustainable Communities, Washington, DC

Part III
Rooftop Agriculture Management

Giorgio Gianquinto

Water Management and Irrigation Systems

Ioannis L. Tsirogiannis, Francesco Orsini, and Paulo Luz

Abstract An important urban societal challenge is to properly match citizens'needs and activities with reasonable use of the available resources. In this context, Rooftop Agriculture (RA) requires well documented guidelines to link its multifunctionality and sustainability to the water cycle, mainly when water shortage, scarcity, flooding and other difficult to handle or limiting conditions occur. Proper water management in RA involves a number of decisions and this chapter outlines the available knowledge, the options and the criteria, related to irrigation and drainage systems design and management at the micro-scale (e.g. plot, end-user) level. Irrigation practices and their impacts are approached considering site-specific conditions (soil-plant-atmosphere system) with respect to six different climatic regions. The main agro-environmental and socio-economic factors are evaluated in order to point out best strategies, comprising suitable and available technologies, to enable successful RA planning and actions regarding water management.

Introduction

The potential benefits of urban agriculture and nature interlinked activities, providing products and ecosystem services, have become increasingly dependent on sustainable relationships with water, soil and energy, aggregated in dynamic open systems. Water is a topic being thoroughly assessed in any agricultural system, regarding its value in three main aspects: (1) agro-environmental; (2) economic and (3) social. In the context of those water-related thematics, relevant challenges and

I.L. Tsirogiannis (✉)
TEIEP – Technological Educational Institute of Epirus, Arta, Greece
e-mail: itsirog@teiep.gr

F. Orsini
Department of Agricultural Sciences, University of Bologna,
Bologna, Italy
e-mail: f.orsini@unibo.it

P. Luz
INIAV – Instituto Nacional de Investigação Agrária e Veterinária, Oeiras, Portugal
e-mail: paulo.luz@iniav.pt

F. Orsini et al. (eds.), *Rooftop Urban Agriculture*, Urban Agriculture,
https://doi.org/10.1007/978-3-319-57720-3_9

strategies in RA activities, with multifunctional characteristics, may be pointed out, such as:

(1) Water cycle approach, involving its variability (resources quantity and quality level) in time and space;
(2) Implementation of reliable (preferably certified) structures and practices (e.g. design, equipment, management, water quality and reuse);
(3) Monitoring, evaluation, risk assessment and auditing procedures;
(4) Cost-benefit analysis.
(5) Water governance as a system/exercise, involving decision-makers and stakeholders, to establish a multidisciplinary framework of specific objectives and guidelines for water use.

These issues will be described and analysed through this chapter, focusing key questions commonly raised from RA applications.

Considering a soil-plant-atmosphere system associated to a defined RA scale, the hydrological cycle study is a basic step to characterize main interlinked variables, involving ecological aspects (e.g. quantity and quality of natural resources). This analysis provides data and information to define the level of rain-fed, water storage, irrigation and drainage requirements and further decisions to install and manage specific equipments.

Adressing water management issues, regarding objectives of sustainability (e.g. promoting natural resources conservation practices) and multifunctionality (e.g. providing multiple outputs as vegetable products and socio-ecological services) in RA structures, require a multidisciplinary approach. In this context, the efficient use of resources, water efficient species, and social science expertise are required, and in addition the adoption of irrigation technology and management practices must ensure high performances, supported by a culture of evaluation and auditing (Conellan 2004).

A holistic cost-benefit analysis in RA structures shall include water-related equipments, regarding design and management options, considering several engineering, operational and maintenance expenses, as well as environmental and socioeconomic values (USDA 2008).

Water governance strategies and plans implementation, in the RA sector, require specific innovative solutions, recommendations and policies to improve stakeholders knowledge and behaviour about socioecological practices and efficiencies. An important goal is also to increase responsibility for societal compromises with defined criteria and regulations related to an adequate urban water management. In addition, participatory management, training and educational opportunities are crucial development principles to achieve best results.

Water Management in the Context of Rooftop Agriculture

Water management should utilize interdisciplinary approaches, allowing a better understanding of cross-cutting water resource issues (Dziegielewski 2006), thus contributing to transfer most comprehensive information for adequate technical-economic options at the micro-level agriculture sector (from community allotments to private house plots). Consistently, the main challenges in RA implementation include interactions between infrastructure, activities and advanced technologies, involving systemic assessments and analysis, to link ecological and socio-economic conditions and factors, in order to ensure the spread of successful and sustainable water use solutions. Within the scope of promoting a more rational water management, key objectives are emerging in RA, comprising: (1) protection of urban ecosystems - focusing abiotic (soil-water) and biotic (habitats) elements; (2) climate change resilience - focusing water shortage and scarcity, storms, floods and other difficult or limiting conditions and (3) sustainable management of human activities - focusing water-energy saving based economy, pollution regulation, food production and public health (both threatened by excessive pressure on freshwater demand).

In addittion to the above mentioned points, institutional regulations and policies (EC Directives, or OECD and FAO documents etc.), have been published during the last years, integrating water management, climate change and agro-environmental issues. Focusing on Europe, several resolutions and actions to adapt to water shortages and improve the efficient use of water are included in the EU's 2020 Strategy and, in particular, to the 2011 Resource Efficiency Roadmap of which the EU's Water Blueprint is the water milestone (EC 2007, 2011). In regions subject to water scarcity periods, innovative and sustainable techniques are being adopted to save freshwater, such as the use of recycled/treated wastewater and saline waters, with low negative effects on health and environment, or deficit irrigation. Furthermore, to support a water supply planning, regarding the water balance from rain-fed and irrigation practices, water scarcity seasons and return periods (based in historical trends and probability distributions of climate data) shall be considered.

Concerning water issues, some objectives referred in recent approaches (Harrison 2013) will contribute to reduce vulnerabilities, namely: (1) water savings due to technological and behavioral change; (2) reduce diffuse source pollution; (3) improvements in irrigation efficiency; (4) adaptations (people and technology-based) in areas subject to flood and/or drought risk to improve resilience and (5) promotion of information platforms. Considering RA conditions, several approaches must be developed to improve specific potentialities and benefits and to reduce the negative impacts involved, from planning to operational stages. The integration of objective actions, such as monitoring, evaluation and risk control, will be helpful in providing comprehensive data to methodological tools (e.g. indicators systems, decision support tools, benchmarking) used to define standards, tresholds and performances of RA water management practices.

The selection process of irrigation systems in RA has to take into account the assessment of environmental and economic impacts, as a mean of fostering efficient technologies and achieving a high productivity (physical or monetary) of resources. Consistently, a technical-economic approach is the basis for proper solutions, linking issues related to: (1) the alternatives to the system design and expected costs; (2) options of irrigation scheduling to improve water application efficiency and uniformity and (3) fluctuations of energy cost. Thus, with particular focus on the water-energy nexus, skills must support advanced solutions, but taking into account expected trade offs between objectives (e.g. minimizing investment costs and maximizing energy saving; maximizing yield and maximizing water use productivity). For example, the selection of: (1) pressurized irrigation techniques (e.g. drip systems); (2) devices for automated operational support and control (e.g rain or soil moisture sensors), might result in higher initial costs, but the higher efficiency of resources use (water and energy) has the potential to improve the effectiveness of irrigation events in most conditions.

Irrigation Systems

To ensure proper water management strategies, special guidelines have to be considered when analyzing an irrigation system in the context of RA. Any decision, with regard to a properly designed and functioning installation, shall be based upon site-specific requirements and limitations. Such conditions are approached by numerous factors which concern natural resources characterization (climate, soil/substrate, plant, water and other inputs availability). In addition, the increasing risks of drought periods determine the need for technological innovations and well founded criteria. These considerations shall result in improved use efficiency of natural resources, and in solutions to tackle competing objectives concerning environmental, technical, and socio-economic issues in urban areas. Several procedures and steps must be implemented to identify the best options related to planning and management of irrigation systems. Making use of proper tools (e.g. decision support systems) to classify and rank the feasible irrigation systems according to their suitability to the input factors, the selection process will consist in evaluation stages, while meeting needs, constraints and beneficial procedures (Agritech 1990). Furthermore, the accomplishment of most suitable practices and high performances shall be observed through adequate monitoring and evaluation procedures. The following outline indicates the roadmap to provide irrigation guidelines for each site situation.

Planning and Design

The performance of an irrigation system depends upon variables specified by: (1) the system design (e.g. components' characteristics and layout) and (2) operation practices (e.g. frequency and duration of irrigation events) as their performance is affected by many constraints like high variability (spatial and temporal) of soil and microclimate, variable water/hydraulic supply and operating conditions, vegetation quality and architectural patterns (Connellan 2004).

Certain principles of planning to establish a feasible and successful project are crucial, as quality criteria for natural resources, installation of irrigation components and financial evaluations (USDA 2008). Site-specific studies regarding a soil-plant-atmosphere system are a key objective to ensure a reliable irrigation design (Luz 2013). For optimum performance parameters (regarding water and energy savings) a suitable pressurized micro-irrigation system must be ensured by a design process comprising a set of technical specifications: (a) water supply and pumping needs; (b) system capacity (flow rate) and (c) irrigation system layout and characteristics (components, pipelines, laterals and outlets etc.). Figure 1 displays a sample micro-irrigation installation layout and its main components and Table 1 presents some typical characteristics of outlets that are used in micro-irrigation systems.

The physical layout must be adjusted to the rooftop plot conditions taking into account the factors affecting the irrigation system selection. Furthermore, a proper

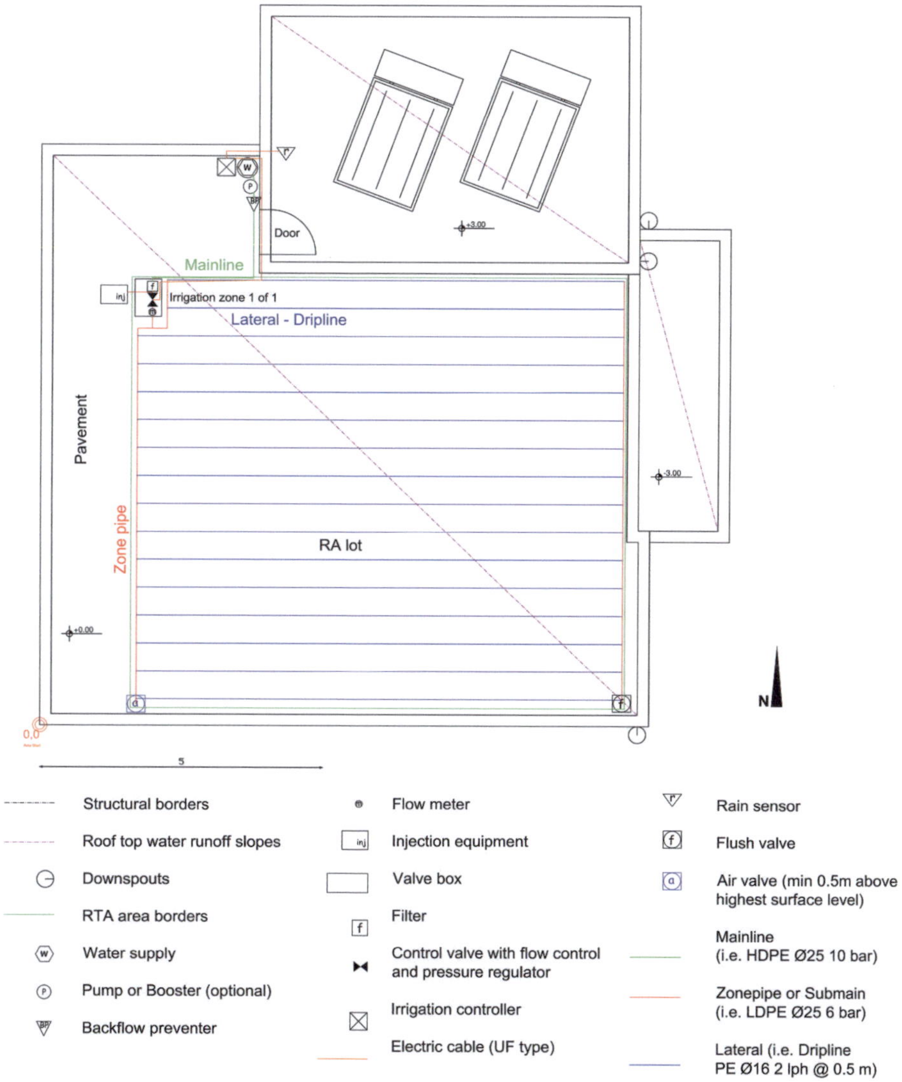

Fig. 1 Sample of a micro-irrigation system layout for rooftop agriculture

Table 1 Typical characteristics for outlets of micro-irrigation systems

Characteristic	Micro-sprinkler	Bubbler	Driplines Tape	Driplines Emitter (Dripper)	Driplines Subsurface emitter
Flow rate ($L\ h^{-1}$)	20–100	20–40	0.5–1	2–8	2–4
Spacing (m)	1.5–4	0.5–2	0.2–0.5	0.3–1	0.3–1
Application rate ($mm\ h^{-1}$)	5–20	5–20	5–20	5–20	5–20
Operating pressure (bar)	1–2	1–3	0.5–1	0.5–2	0.5–2

layout may define trade-offs between agro-environmental and economic options. When planning a plot irrigation system, the configuration to be adopted shall present numeric data, regarding the proposed layout and components (e.g. size, type, spacing, rates and pressures of outlets and pipes), in order to match the common market standards. It should also be noted that, for instance, the type of outlets used affects the size of the other components and generally, micro-sprinklers require larger filters, mainlines, and sub mains as compared to drip systems (Morris and Schwankl 2011). The system capacity determination provides a basic hydraulic knowledge, whether the irrigation system guarantees to apply the required water to meet the peak watering demand of the plants. The fact that extra pressure may be needed because of the height above ground of the RA plot should be taken into account. The pumping system features, obtained from the hydraulic and energy analysis (regarding water distribution, pressure losses, operating pressure and flow rate), must be consistent with irrigation operational considerations.

To compensate for excessive evaporation due to solar radiation and wind, installation of appropriate equipment and components (nets to provide partial shade and windscreens using plants or structures) or application of special techniques (i.e. mulching) can be used. Regarding standards for Green Roofs, the most popular one in Europe is "Forschungsgesellschaft Landschaftsentwick-lung Landschaftsbau" commonly referred as FLL (FLL 2008), which however, provide only few specifications regarding irrigation and drainage (Van Mechelen et al. 2015).

Scheduling and Operation

Concerning the variability of water needs, successful irrigation decisions shall point out solutions to prevent water shortage, ponding/runoff occurrence or drainage losses, regarding particular conditions of RA. Scheduling methods are developed to answer "when", "how often" and "how much" irrigate. As described by Jensen (1983) there are several criteria for irrigation scheduling, related to the soil-plant available water, soil water potential, evapotranspiration deficit etc. Generic climatic schedules can be developed on weekly to monthly basis using widely approved procedures and relevant free softeware like CropWat (FAO 2016). These schedules can provide a good basis for the actual watering of plant regardless if this will be applied manualy or using an automatic controller (timer). The need for irrigation can be assessed by a number of techniques which could be based on soil moisture (e.g. hands feel or indication from a soil moisture sensor) or plant condition (e.g. visual inspection, leaf temperature) indications, computed water balance (e.g. water balance models that estimate water inputs and losses). In the case of micro-irrigation for RA, small water application amounts and frequent irrigation event are suggested because in most cases the soil layer is relatively thin. The initial irrigation schedule can be established with a regular amount, and intervals (cycles) and timing based on estimated water needs of the soil-plant system, along different crop stages, and weather conditions. To ensure a proper irrigation control and performance, interactive skills and instruments shall provide update information to adjust an irrigation event to changing and actual conditions. Then, at the indicative water deficit level, or, when a certain range of thresholds is surpassing (e.g. sudden climate extreme conditions), a start or shut off action (either manually or automatically) is activated.

Moreover, irrigation systems operate along a period of time based on the scheduling information, in accordance with the water application amount (WA) and rate (WR). Important soil (or substrate) physical parameters, measured or estimated, define basic conditions for scheduling and operation procedures. Thus, the WA shall not -except if salinity issues exist- exceed the available water capacity (AWC) of the soil and the WR must be lower than the infiltration rate. Considering these, the soil must be characterized by the profile depth, main texture classes (concerning coarse, medium or fine soil types) and soil hydraulic parameters, which comprise the saturation, field capacity, wilting point and infiltration capacity (Phocaides 2007). In order to compute the WA it is common to use the plant (or soil readily) available water (PAW) in the root-zone, as a fraction of the AWC. With most plants when the PAW approaches 50%, water stress begins, but great issues occur once PAW declines below 30% (Stewart and Lawford 2011). Also, to plan a permanent adequate water level in the root zone, a management allowed deficit of the available soil water (MAD) may be selected. Generally, recommended MAD values vary between 50 and 60% of AWC, along growth stages (with exception of sensitive periods for soil water stress such as flowering and initial yield formation). MAD depth values (mm) incresase from coarse to fine-textured soils, which indicate that a larger water supply (as well as the interval between irrigation events) in loam and clay soils than in sandy soils tends to be an adequate practice.

Maintenance

A well-designed micro-irrigation installation needs also frequent and reliable maintenance of equipment, in order to enable a well-operating system to apply the water uniformly and with a high efficiency. Commonly, emitters and filters could be clogged by particles resulting in pressure and flow variability, thus, limiting the system ability to deliver the water with uniformity. Maintenance items shall include checking (e.g. discharge rates, operating pressures), cleaning (e.g. filters, laterals, emitters) and flushing procedures (USDA 2008).

Monitoring, Evaluation and Auditing

Effective monitoring in irrigation systems aims to collect data to check and evaluate potentialities and vulnerabilities of actual practices, thus, promoting a correct diagnosis to support and improve the manager's decisions. Many parameters/variables can be selected to a specific monitoring objective, for instance, related to relationships between soil moisture level, plant growth stages, flow rates and irrigation time. Adequate devices (e.g. probes, data loggers) are a key-component of required tasks. An evaluation process, regarding site-specific conditions, involves various steps and measurements to obtain the required parameters and a classification framework with standard values. Following this purpose, the manager shall be able to verify the effectiveness of the irrigation system or the magnitude of water application problems. For instance, main performance parameters, as actual application efficiency or distribution uniformity, of a micro-irrigation systems effectively operated, shall range between 80 and 90% (Jensen 1983; Smajstra et al. 2002). To improve the actual irrigation performance depend not only on promoting the implementation of a well-designed system and recommended scheduling methods, but also on monitoring and evaluation procedures to determine how effectively the system is operating. Irrigation Association (2012) has published a very practical and widely used, guide regarding audits of irrigations systems.

Drainage Systems

The protection of the building -that hosts the RA plot- against water load and moisture is a major issue. In every case roofs are not expected to pond water, which shall drain properly. The plumbing section of the applied building code will provide design information about the minimum slope, number of drains and other requirements (NRCA 2009). Any leakage will eventually cause significant damage, and any obstacle in the way of water to find its way to the downspouts could generate even structural failures as water is a very heavy element (water in excess is considered part of the live load of the structure). The adequacy of roof slope and routing of the water to the downspouts shall be carefully checked before the design of any RA system. Any event that leads to ponding or very low flow rate will cause problems to be solved from a technical point of view. Also, the waterproof membrane must be checked before all the components of the assembly are installed (there are even electronic leak detection systems to help in this) (Luckett 2009; Snodgrass and McIntyre 2010).

Rainfall reaching the rooftop may be retained, detained or drained. Retention refers to rainfall that is held within the roof system and does not leave the roof as runoff. Consistently, retained rainfall may subsequently leave the roof as evapotranspiration. Detention refers to the temporal delay that occurs between rainfall that is not retained hitting the roof and emerging as runoff. In general RA provides retention and detention, both affected by hydrologic conditions due to the characterisitics of soil/substrate profile (e.g. horizons texture and depth) and vegetation (e.g. cover type/composition and percent) (Neitsch et al. 2005; Stovin et al. 2015). In conventional green roof systems, waterproofing and root repellent membranes are placed on the roof surface and above them there is the drainage layer. The drainage layer should be very porous to permit water to pass easily through it. It must be permanent and continuous over the entire roof surface and strong enough to support the weight of the plant materials and hardscape above it. This layer must be kept free of any materials that could prevent the free flow of water to the downspouts. Roof drains shall avoid growth media from entering the building's plumbing system. They need to be regularly checked and eventually cleaned.

Aggregate drainage layers and synthetic cups (also refered as pegs) matrix drainage panels are the most typically adopted solutions. These cups may vary in size, height and spacing, or in the size of their drainage holes. Above them, a fabric holds the growing medium, while allowing for water and root penetration (Luckett 2009; Snodgrass and McIntyre 2010; Van Mechelen et al. 2015). In addition to the substrate, water can also be retained in the drainage layer or, if present, in the mat constituting the water retention layer (WRL) (Van Mechelen et al. 2015). When used as kitchen gardens or for more intensive agricultural activities, substrate depth is at least 20 cm of growing medium and includes blended organic matter. FLL (2008) includes a standard test to determine the coefficient of discharge from a green roof.

In cases where RA is performed in containers it is recommended that the waterproofing of the roof is checked and precaution to be made against the possibility that the drain route is blocked by structures or clogged by debris (substrate, leafs etc.). Containers should also have at their base an adequately layered drainage mass and their effluents should be properly channelled (Weiler and Scholz-Barth 2009).

Other Water-Irrigation Issues and Strategies

Water Quality

The quality of water used for irrigation has implications for agricultural yields, products quality and human health. Regarding abiotic parameters, Table 2 presents the acceptable limits for water pH and EC (Ayers and Westcot 1994). Using a salt tolerant plant does not solve every problem when it comes to the application of high salinity water for irrigation, as salts in the water can build up and damage both

Table 2 Water quality for agriculture (Ayers and Westcot 1994)

Parameter	Unit	Degree of restriction on use		
pH	–	normal range: 6.5–8.4		
Salinity / EC	dS m^{-1}	< 0.7, none	0.7–3.0, slight to moderate	> 3.0, severe

plants and soil. In cases like these, special salt can be leached –by applying more water that needed to cover plant needs – away from the rootzone.

When discussing about water quality, that of the effluents from the RA project should be mentioned too. The leaching of contaminants (from fertilizers, plant protection substances etc.) should be controlled too. In the same framework, the possibility to pollute drinkable water through a backflow event should be avoided and thus adequate backflow preventers should be placed in the irrigation system.

Alternative Sources of Water

Rain Water Harvesting (RWH) systems should be incorporated in a holistic approach as the collected water can have various uses in a building and irrigation of an RA lot could be among them. Rainwater harvesting systems range from simple to complex and are considered as low impact development practices for an urban environment and a way to lower the urban environmental footprint. Such a system is composed of the following basic components: the supply (rainfall), the rainfall catchment (precipitation surface and conveyance pipes), the irrigation/distribution system that discharges water to the plants, and the demand system (substrate water holding capacity and landscape water requirement). Storage is an additional element, which may be optionally integrated, if it is not included, rainwater is distributed immediately to the planted areas. An RA project is by itself a kind of RWH system. Once maximum storage capacity is reached, runoff water can be channeled into a grey water system and returned to the roof as irrigation (Chang et al. 2011). If the rainwater harvested at the rooftop level exceeds the rooftop plant requirements, it can be also used for irrigation of landscapes at lower floors or ground level, given its latent pressure which is very useful in case drip-lines are used (every 10 m of height difference corresponds to about 1 bar or 100 kPa). Rainwater, when compared to other alternative water sources (e.g. grey and recycled water) has the advantage to contain less contaminants. For some urban environment this condition, is an issue for discussion. The BS (2009) for rainwater harvesting refers that, in a RWH system, all pipework should be in contrasting color (not blue but green or black with green stripes) from main pipeworks and properly labelled.

The pressure on water resources has encouraged more active consideration of using alternative water sources. Typical regenerated alternative sources of fresh water are recycled, grey water and saline water. In a very recent European Commission's JRC Science and Policy Report (Sanz and Gawlik 2014), the need to find sustainable solutions to water challenges in urban, industrial and agriculture sector was highlighted. In the same publication, a model for wastewater reuse potential in European countries up to 2025 was presented. Recycled water may be primary, secondary, or advanced (tertiary) treated municipal or industrial wastewater. The characterization "recycled" refers in general to any water that has undergone one cycle of human use and then received sufficient treatment at a sewage treatment system in order to become suitable for various reuse purposes, including irrigation. Grey water refers to soft-treated or even untreated water that has gone through one cycle of use, usually in households or office buildings. Grey water by definition does not include the discharge from toilets or other uses that may contain human waste or food residues (which make up the sewage or blackwater). Grey water usually passes through appropriate filters before it can be used. As it contains many fewer pathogens than blackwater, it is more easily treated and recycled onsite for a number of purposes among which is landscape irrigation. These types of water pose the risk of toxicity to plants because of dissolved salts. They may also contain a wide array of hazards including microbial, chemical and physical agents that could pose a risk to human health and environmental matrices. In order to implement irrigation with alternative water sources, these risks must be mitigated. In this framework the use of such water sources is subjected to legal limitations (Sanz and Gawlik 2014).

Sensors and Decision Support Systems

The expected increase in application efficiency of an automated irrigation system provides the opportunity to conserve water resources while maintaining the plants in good condition. Nevertheless, automated residential irrigation systems tend to result in higher water use than non-automated systems (McCready and Dukes 2011; Cárdenas-Lailhacar and Dukes 2012). Olmsted and Dukes (2014) reported a too frequent and/or too long operating time, among the most frequent problems in the irrigation of residential gardens. This can be attributed to several factors, including a tendency to improperly program the irrigation controllers (timers) or to apply a flat programme during changing weather conditions. To improve water productivity, it is important to irrigate based on plant water needs and take account of rain; and for that modern controllers include special features such as multiple programs and start times, water budget and rain delay functions. Generic climatic schedules (like these developed using CropWat (FAO 2016)) can provide basic watering frequency and duration values which can be fine tuned by using information on actual crop water needs. Free services around the world provide this kind of weather based information and the California Irrigation Management Information System (CIMIS 2016) is one of the best examples.

Another alternative is to attach a sensor to the controller in order to provide information for triggering or halting irrigation events or even to adjust the running program to the actual weather conditions. Rain sensors can suspend irrigation for a certain period after rain, they are quite cheap, save water but typically they feel rain, after a certain amount of rainfall. Soil tension and moisture sensors allow irrigation only when soil moisture is below a set threshold, however, many soil-mixtures used in RA are too porous to allow electrical sensors to function properly, so before choosing a sensor it shall be confirmed the adaptability to the used growing medium (Luckett 2009). Other type of sensors include pyranometers (solar radiation sensors), which allow controllers to monitor solar radiation as a measure of evapotranspiration and continuously adjust the frequency of irrigation events, while evapotranspiration based controllers (ET controllers) use a series of meteorological data to calculate the actual ET at the site and use it to adjust irrigation. The latter can either have a mini meteorological station attached or receive data from a relevant provider. From 2002, Irrigation Association (IA) runs a very interesting initiative called Smart Water Application Technologies (SWAT) as a collaboration of water suppliers and the irrigation industry in USA (IA 2014). In 2006, the US Environmental Protection Agency (EPA) created a US national program called the WaterSense aimed at promoting water use efficiency. From 2012, WaterSense integrates weather-based irrigation controllers (EPA 2014).

Cost-Benefit

For the case of RA the discussion regarding the need for installing an irrigation system can be very short, as when the goal is to produce vegetables, fruits etc., plentiful availability of water is a must. An interesting discussion topic is how much money should be invested to the irrigation system. For small projects even a garden hose could be sufficient. For larger projects, manual irrigation is labor intensive and the installation of a permanent irrigation system may be the only practical means of water distribution (Luckett 2009). Among the various commercial systems, micro-irrigation is the one that fits better in the case of RA. In many cases, there will be the need to obtain larger water capacity on the roof, which means extra infrastructural costs.

Simple timers do not cost a lot of money but if the project needs more than one irrigation stations to get irrigated then a more complex controller is needed. In this case the cost of a rain sensor is not considered prohibitive and it could save a lot of water. At this point, it must be noted that in many RA projects drinkable water is used, for which the cost is sometimes significant (Sanyé-Mengual et al. 2015). The use of alternative water resources is always a good solution.

Moving to more sophisticated solutions, fertigation components can also added to a RA system when the yield goals are higher. Close loop hydroponic (with or without substrate), aeroponic and aquaponic systems are considered to be more adequate for the case of commercial projects.

Regarding drainage, only one way exists, it has to function perfectly in any case, otherwise the safety and the value of the building could be put in danger.

Irrigation Management – Case-Studies in Different Climatic Regions

The impact of climatic factors on hydrological cycle and irrigation practices are approached, concerning six (6) different regions:

(a) **Mediterranean, South Europe** (Rosário 2004; EU 2005)
(b) **Germany, North** (Riediger et al. 2014; IOW 2013)
(c) **Ireland** (Mills 2000; Dwyer 2012)
(d) **Canadian, Central Prairies** (Stewart and Lawford 2011)
(e) **China, North Plain** (Pereira et al. 1998; Shen et al. 2002)
(f) **Ethiopia, North Savana** (Cramer 2014)

The knowledge of the soil-plant-atmosphere system is crucial to control water requirements in rooftop plots. A water balance procedure must be driven by climate components, crop characteristics and stages, and physical - chemical parameters of a soil/substrate profile within the plant root zone. In *Mediterranean* (e.g. South Europe) *Semi-arid* (e.g. Northeast of China) and *Tropical* (Ethiopia, Savana) climates, drought periods resulting in water scarcity, are being associated to short recurrence intervals. On the other hand, in a *Continental* climate (e.g. Germany and Canada regions with a warm season, but also very cold winters) year-to-year changes may be observed, thus creating various wet-dry effects. In typical *Temperate* climate conditions with sea influence (e.g. Ireland) dry periods are rarely expected. Therefore, even during the summer season, water shortage is not commonly observed. In Ireland, Dublin's meteorological data show from June to September, the precipitation (P) ranging from 200 mm to more than 500 mm and the potential evapotranspiration (ETp) has a similar variation. Less than 1% of the cultivated area of Ireland is irrigated (Dwyer 2012).

In the scope of the urban water cycle management, the implementation of irrigation and drainage practices in RA is required when the dynamic equilibrium of precipitation-evapotranspiration is not reached over a given time period. This relationship may be assessed, considering typical drought conditions of some climatic groups, as well as anomalies currently referred to in various continental climatic groups, concerning seasonal variations. Statistical methods may be applied to create probability distributions of rainfall, temperature, moisture or wind patterns (e.g. duration, intensity or amount). Detailed and reliable databases obtained from long-term average characteristics are often used to comparative or risk assessment. Considering site-specific conditions, statistically significant trends of extreme events occurrence shall be consistent with return periods, namely, where high seasonal and annual variability is expected.

The analysis of Table 3 Shows significant differences among climatic regions data, mainly the mean annual temperature (T), which may range from 4 to 23 °C. However, the assessed climatic variables are not correlated. For instance, in Central Prairies of Canada (CPC), with the lowest temperature, precipitation value is similar to North Plain of China (NPC) or Mediterranean South Europe (MSE). On

Table 3 Climatic regions characterization

Location	Climatic regions	T – mean annual	P – annual	ET_p – annual	Aridity index: P/ET_p (annual)	PWDP (seasonal - s)	P (s)	ETp (s)	Crop water req. (s)[a] ($P–ET_p$)	Water storage needs (Irrigated area: 50m^2)	Water harvest: available rooftop area (P = 100 L m^{-2})
		°C	mm	mm	$mm.mm^{-1}$	Days (Period)	mm	mm	mm	m^3	m^2
Mediterranean. South Europe	Temperate Mediterranean	18	600	1200	0.50	120 (Jun-Sept)	<100	600	**500**	25	250
Germany. North	Sub-continental	9	600	500	1.20	120 (May-Aug)	150–300 (dry – wet)	250	**100**	5	50
Ireland	Temperate oceanic	9	1200	500	2.40	120 (Jun-Sept)	300	300	**0**	0	0
Canada. Central Prairies	Continental	4	500	500	1.00	120 (May-Aug)	180–300 (dry – wet)	300	**120**	6	60
China. North Plain	Semi-arid	12	500 (80% from Jun to Sept)	1000	0.50	120 (Feb-May)	< 100	350	**250**	12.5	125
Ethiopia. North Savannah	Tropical	23	1000 (80% from Jun to Sept)	1350	0.75	120 (Feb-May)	<150	550	**400**	20	200

<u>Notes</u>: Temperature (T), Precipitation (P), Potential Water Deficit Period (PWDP), Potential Evapotranspiration (ETp)

[a]In this simulation, average crop water requirements (WR) are related to seasonal ETp, but ETc is lower than ETp at crop initial stages. On the other hand, water volumes needed for storage/irrigation would be higher considering the water application efficiency. Thus, on the basis of both considerations, proposed approximations are acceptable

the other hand, temperature in NPC is three times higher and in MSE two times, than in CPC. Furthermore, the ETp from NPC is twice the CPC value, but is similar to MSE value. Excluding Ireland, five climatic regions are subject to periods of potential water deficit (PWDP). This deficit occurrence was estimated considering a seasonal length of 4 months. Potential critical dry periods shall happen in different months as observed in PWDP column. P and ETp data point out to regular expected soil water deficit in three regions (Mediterranean, semi-arid and tropical), and, in continental climates, to very likely deficit under dry summer conditions.

With respect to simulations in the scope of a hypothetical exercise, Table 3 is useful to demonstrate differences in water management, regarding the proposed climatic regions. Following an approach with P and ETp data, it is also worth noting that using the aridity index (AI) (UNESCO 1979) the regions with regularly conditions of seasonal water deficit can be classified as semi-arid to subhumid (AI values lower than one). In continental regions, with seasonal water deficit predicted in dry years, the AI shall be close to one. Thus, regarding pratical effects for the accomplishment of a proper water management, there is a strong correlation of those AI values with water requirements in rooftop plots at PWD periods. In this framework and according sample calculations (given in Table 3), irrigation needs in such periods may commonly range from around 100 mm (mm equal to L m^{-2}) in continental regions to between two and five times in drier regions. Considering water supply scenarios, provided for plot areas of 50 m^2, the required storage volume for agriculture purposes, ranges from 5–6 m^3 (continental climates) to a maximum of 25 m^3 at the worst conditions, met in the *Mediterranean* climate. It appears that an adequate water budget must be developed and ensured by some functional elements as rainwater harvesting systems, storage devices and distribution components. Before the water scarcity period, hypothetical precipitation events reach 100 mm. Therefore, in the latter case (*Mediterranean*), to overcome the magnitude of water resources vulnerability, a reservoir shall be installed to be filled with rainwater from a collection surface covering a rooftop size of 250 m^2. The overall area needed in such conditions would be 350 m^2 (plot area = 50 m^2; water harvest area = 250 m^2; reservoir = 50 m^2: if height = 0.5 m and volume = 25 m^3).

The irrigation scheduling is based upon irrigation system flow rate, crop water requirements and soil-water reservoir (root-zone). A soil/substrate profile with 0.2 m depth may provide an available water capacity of 16 to 40 mm (coarse or fine textures) and allowing a common minimum depletion level of 50% for most soil-plant systems, means 8 to 20 mm of water is required to replenish the soil to field capacity. In *Mediterranean* climatic areas, during the summer peak period and given a coarse-textured soil with a 0.2 m root-zone, irrigation amounts around 10 mm shall be planned with an interval of 1 or 2 days.

An important issue concerns buildings structural design and safety for RA. For example, recommendations for loads in buildings target a minimum limit of 200 kg m^{-2} (Appleton 2013). This value is of similar level to a soil/substrate profile of 0.20 m, saturated with water (Palha 2011). However, a reservoir, as previously determined, requires the adoption of a building structural design providing a resistance of 500 kg m^{-2} (25 t for 50 m^2).

Conclusions

Based upon results obtained from those climate scenarios, it must be stressed that adopting RA requires specific guidelines, due to variability of natural conditions, anomalies influence, and human activities and policies. For example, when evaporation transpiration demand exceeds precipitation, water scarcity periods are more likely. Thus, water requirements and irrigation schedules may vary a lot, in respect to a specific soil/substrate-plant system (e.g. profile depth and water capacity, rootzone, evapotranspiration), in particular regarding the control of water stress and salinization tolerance levels. Comprising plots, storage devices and water harvest systems, a set of principles and detailed calculations of sizes are required. Suitable and effective irrigation practices are associated to key challenges and concerns (e.g. water quality and supply, building design). In any case, by using adequate tools (e.g. indicators, classification, ranking and decision support systems) main agro-environmental and socio-economic variables may be assessed in order to improve technologies and strategies to enable an integrated and successful RA planning.

Bullet Points

- Water management in RA systems, require a multidisciplinary approach.
- Roof water proofing and efficient drainage are key elements for RA systems design.
- Suitable and cost-effective irrigation systems practices are associated to key challenges and concerns (e.g. availability of flow rate (system capacity) and pressure, system type, water saving techniques, use of alternative water resources, efficient irrigation management etc.).
- Effective monitoring of irrigation systems aims to evaluate potentialities and vulnerabilities of actual practices, thus, promoting a correct diagnosis to support and improve the manager's decisions.

References

Agritech (Ed.) (1990) 5th international conference on irrigation: Proceedings, Tel-Aviv, Israel, 26–27 March. Agritech Israel, 326 pp

Appleton J (2013) Estruturas de Betao (Vols. I e II). Orion, Lisboa, 1346 pp

Ayers RS, Westcot DW (1994) Water quality for agriculture. FAO (UN Food Agriculture Organisation) - FAO Irrigation and Drainage Paper #29. Retrieved on Sept 2015 from: http://www.fao.org/docrep/003/t0234e/T0234E00.htm#TOC

British Standard BS8515 (2009) Rainwater Harvesting Systems code of practice (+A1:2013)

Cárdenas-Lailhacar B, Dukes MD (2012) Soil moisture sensor landscape irrigation controllers: a review of multi-study results and future implications. Trans ASABE 55(2):581–590

Chang NB, Rivera BJ, Wanielista MP (2011) Optimal design for water conservation and energy savings using green roofs in a green building under mixed uncertainties. J Clean Prod 19(11): 1180–1188

CIMIS – California Irrigation Management Information System (2016) System overview. Retrieved on Sept 2015 from: http://www.cimis.water.ca.gov/

Connellan GJ (2004) Evaluating the performance of urban irrigation. Proceedings of NZWWA Conference, Christchurch, October

Cramer W (2014) Regional environmental change. Springer. Vol. 14 (1)

Dwyer N (2012) The status of Ireland's climate. Environmental Protection Agency, Wexford

Dziegielewski B (2006) Interdisciplinary in IWRA. International water Resources Association. Water Int 31(2):143–144

EC – European Commisision (2007) Addressing the challenge of water scarcity and droughts in the European Union. Communication from the Commission to the Council and the European Parliament, 18.07.07, COM (2007) 414, Brussels

EC – European Commisision (2011) A resource-efficient Europe – Flagship initiative under the Europe 2020 Strategy. Communication from the Commission to the Council and the European Parliament, COM (2011) 21, Brussels

EPA – Environmental Protection Agency USA (2014) WaterSense program. Retrieved on Oct 2015 from http://www.epa.gov/watersense/information.html

EU – European Union (2005). Soil atlas of Europe. European Soil Bureau Network, European Comission,. Luxembourg. Retrieved on Dec 2015 from http://esdac.jrc.ec.europa.eu/content/soil-atlas-europe

FAO – Food Agriculture Organisation (2016). CropWat. Retrieved on 1 Jan 2016 from: http://www.fao.org/nr/water/infores_databases_cropwat.html

FLL – Forschungsgesellschaft Landschaftsentwicklung Landschaftsbau (2008) Guide-lines for the planning, construction and maintenance of green roofing. Forschungsgesellschaft Landschaftsentwicklung Landschaftsbau e. V, Bonn

Harrison P (2013) Climate change impacts, adaptation and vulnerability in Europe: an integrated approach. CLIMSAVE Consortium. University of Oxford, UK

IA – Irrigation Association USA (2012) Landscape irrigation auditor manual. Fairfax, USA

IA – Irrigation Association, USA (2014). Smart water application technologies. Available at: http://www.irrigation.org/swat/

IOW – Institute for Ecological Economy Research (2013) Stakeholder participation in adaptation to climate change – Lessons and experience from Germany. German Federal Environment Agency. Vol 12. Dessau-Rosslau. Retrieved on Sept 2015 from https://www.umweltbundesamt.de/sites/default/files/medien/461/publikationen/climate_change_12_2013_stakeholder_participation_in_adaptation_to_climate_change_bf_0.pdf

Jensen ME (ed) (1983) Design and operation of farm irrigation systems. ASABE (ASAE) Monograph, Michigan

Luckett K (2009) Green roof construction and maintenace. The McGraw-Hill Companies, Inc, New York, USA

Luz PB (2013) Guia para a avaliação e seleção de sistemas de rega. Programa de Investigação e de Formação Pós-Graduada. Provas de Habilitação para o exercício de funções de coordenação científica. 15/1/2013. INIAV

McCready MS, Dukes MD (2011) Landscape irrigation scheduling efficiency and adequacy by various control technologies. Ag Wat Man 98(4):697–704

Mills G (2000) Modelling the water budget of Ireland – evapotranspiration and soil moisture. Ir Geogr 33(2):99–116

Morris M, Schwankl L (2011) The California microirrigation pocket guide. ATTRA - National Center for Appropriate Technology (NCAT). Butte, MT, USA

Neitsch SL, Arnold JG, Kiniry JR, Willliams JR (2005) SWAT – soil and water assessment tool. Theoretical documentation. ARS, Texas

NRCA – National Roofing Contractors Association (2009) Vegetative roof systems manual, 2nd edn. National Roofing Contractors Association, Rosemont

Olmsted TR, Dukes MD (2014) Frequency of residential irrigation maintenance problems. UF/IFAS Publication #AE472. Retrieved on Dec 2014 from: http://edis.ifas.ufl.edu/ae472

Palha P (2011) Coberturas ajardinadas: uma solução para as cidades? Revista Da Associação Portuguesa De Horticultura 106:26–31

Pereira L, Liang R, Musy A, Hann M (eds) (1998) Water and soil management for sustainable agriculture in north China plain. DER – Instituto Superior de Agronomia, Lisboa

Phocaides A (2007) Handbook on pressurized irrigation techniques. 2nd ed. FAO Water Development and Management Unit and International Programme for technology and research in irrigation and drainage (IPTRID), Rome. Retrieved on Dec 2015 from http://www.fao.org/docrep/010/a1336e/a1336e00.htm

Riediger J, Breckling B, Nuske R, Schroeder W (2014) Will climate change increase irrigation requirements in agriculture of Central Europe? A simulation study for Northern Germany. Environ Sci Eur 26:18

Rosário L (2004) Indicadores de desertificação para Portugal Continental. DGRF. Min-Agricultura, Lisboa

Sanyé-Mengual E, Orsini F, Oliver-Solà J, Rieradevall J, Montero JI, Gianquinto G (2015) Techniques and crops for efficient rooftop gardens in Bologna, Italy. Agron Sustain Dev 35:1477–1488

Sanz B, Gawlik M (2014) Water Reuse in Europe. A synoptic overview. Relevant guidelines, needs for and barriers to innovation. Joint Research Center (JRC) Science and Policy Reports. European Commission. Retrieved on Sept 2015 http://publications.jrc.ec.europa.eu/repository/bitstream/JRC92582/lb-na-26947-en-n.pdf

Shen Y, Kondoh A, Tang C, Zhang Y, Chen J, Li W, Sakura Y, Liu C, Tanaka T, Shimada J (2002) Measurement and analysis of evapotranspiration and surface conductance of a wheat canopy. Hydrol Proced 16:2173–2187

Smajstra AG, Boman BJ, Haman DZ, Pitts DJ, Zasueta FS (2002) Field evaluation of microirrigation water application uniformity. IFAS Extention Bulletin 265. University of Florida. Retrieved on Aug 2015 from: http://24benefi37g1f8w8w3ohxam1.wpengine.netdna-cdn.com/files/2011/10/AE09400.pdf

Snodgrass EC, McIntyre L (2010) The green roof manual. Timber Press, Portland

Stewart R, Lawford R (eds) (2011) The 1999–2005 Canadian prairies drought: science, impacts, and lessons. DRI - Drought Research Initiative, Manitoba

Stovin V, Poë S, De-Ville S, Berretta C (2015) The influence of substrate and vegetation configuration on green roof hydrological performance. Ecol Eng 85:159–172. Available at:http://www.sciencedirect.com/science/article/pii/S0925857415302159

UNESCO (1979) Map of the world distribution of arid regions. MAB Technical Notes 7, UNESCO, Paris. Retrieved on Aug 2015 from: http://unesdoc.unesco.org/images/0003/000326/032661eo.pdf

USDA – U.S. Department of Agriculture (2008) National engineering handbook: Part 652, Irrigation Guide. NRCS. Washington, DC. Retrieved on July 2015 from: http://directives.sc.egov.usda.gov/viewerFS.aspx?hid=21431

Van Mechelen C, Dutoit T, Hermya M (2015) Adapting green roof irrigation practices for a sustainable future: a review. Sustain Cities Soc 19:74–90

Weiler S, Scholz-Barth K (2009) Green roof systems – a guide to planing, design and construction of landscapes over structure. Wiley, New Jersey

Managing Mineral Nutrition in Soilless Culture

Alberto Pardossi, Luca Incrocci, Maria C. Salas, and Giorgio Gianquinto

Abstract In most cases, rooftop agriculture uses soilless cultivation (or hydroponics) of plants, as the yield and the quality of the soilless-grown crops are often higher than those grown in the agricultural soil. In soilless culture, the elements that are essential or beneficial for plant growth and development are supplied through: (i) the addition of organic and/or synthetic fertilisers to the substrate before and after crop plantation; (ii) the supply of a nutrient solution, which is prepared dissolving one or more soluble fertilisers in the raw water and thus is delivered with the irrigation system (fertigation). In this chapter, the basic aspects of the mineral nutrition of hydroponically-grown plants and the methods that could be used for a sustainable management of fertigation in rooftop soilless culture and to improve the organoleptic and nutritional quality of rooftop food crops are described.

Introduction

In most cases, rooftop agriculture uses soilless cultivation (or hydroponics) of plants, as the yield and the quality of the soilless-grown crops are often higher than those grown in the agricultural soils (Olle et al. 2012). The better control of weeds and root-borne diseases, the higher content of available water and air, as well as the lower specific weight are the main reasons for the large use of substrates (or nutrient solution) instead of the agricultural soil in the rooftop gardens. Soilless culture consists of growing plants outside their natural medium, which is soil, with the

A. Pardossi (✉) • L. Incrocci
Department of Agricultural, Food and Agro-Environmental Sciences, University of Pisa, Pisa, Italy
e-mail: alberto.pardossi@unipi.it; luca.incrocci@unipi.it

M.C. Salas
Department of Agronomy, University of Almeria, Almeria, Spain
e-mail: csalas@ual.es

G. Gianquinto
Department of Agricultural Sciences, Research Center in Urban Environment for Agriculture and Biodiversity, University of Bologna, Bologna, Italy
e-mail: giorgio.gianquinto@unibo.it

F. Orsini et al. (eds.), *Rooftop Urban Agriculture*, Urban Agriculture,
https://doi.org/10.1007/978-3-319-57720-3_10

consequent limited volume of water and nutrients. This type of culture requires the application of nutrients in irrigation solution and/or by direct application in the growing medium to allow healthy growth rate and acceptable yield.

In commercial rooftop greenhouses, hydroponic technology is an advanced technology integrated with climate control that requires skilful and expert growers. The basic principles and techniques applied in high-tech rooftop greenhouse hydroponics are those described in many textbooks, review articles and summarized in this and other chapters. Alternatively, rooftop soilless cultivation could be run by non-professional growers for purposes (e.g. horticultural therapy, school projects etc.) other than food production and marketing. In this context, low technology is needed for less expert growers and organic soilless culture with simplified management of plant mineral nutrition could be applied.

In this chapter, the basic aspects of the mineral nutrition of hydroponically grown plants and the methods that could be used for a sustainable management of fertigation in rooftop soilless culture and to improve the organoleptic and nutritional quality of rooftop food crops are described. More comprehensive texts on the management of mineral nutrition in soilless culture are those published by Savvas and Passam (2002), Raviv and Lieth (2008), Sonneveld and Voogt (2009), and Resh (2012).

Plant Mineral Nutrition in Soilless Culture

Mineral Nutrients

The essential elements for plant growth and development can be grouped in:

- non-mineral elements (carbon, C; hydrogen, H; oxygen, O), which are taken from air and water;
- mineral elements (nitrogen, N; phosphorus, P; potassium, K; calcium, Ca; magnesium, Mg; boron, B; iron, Fe; zinc, Zn; manganese, Mn; copper, Cu; nickel, Ni), which are taken from the growing medium (soil, soilless substrate or nutrient solution, NS) and exceptionally from foliar fertilisers.

Essential elements are constituents of organic molecules (e.g. amino acids contain N; chlorophyll contains N and Mg; DNA and RNA contain both N and P), play a direct or indirect role in enzymatic reactions and/or act as charge carriers and osmolytes. Other elements are beneficial as they promote plant growth in many species: chloride (Cl), cobalt (Co), iodine (I), selenium (Se), silicon (Si) and sodium (Na).

All the essential elements are absorbed by the plants as cation (NH_4^+; K^+; Ca^{2+}; Mg^{2+}; Fe^{2+} or Fe^{3+}; Mn^{2+}; Zn^{2+}; Cu^{2+}; Ni^{2+}) or anion (NO_3^-; $H_2PO_4^-$ or HPO_4^{2-} depending on pH; SO_4^{2-}; MoO_4^{2-}) with the exception of B, which is taken up by the roots mainly as undissociated boric acid. Nitrogen may be absorbed as either an

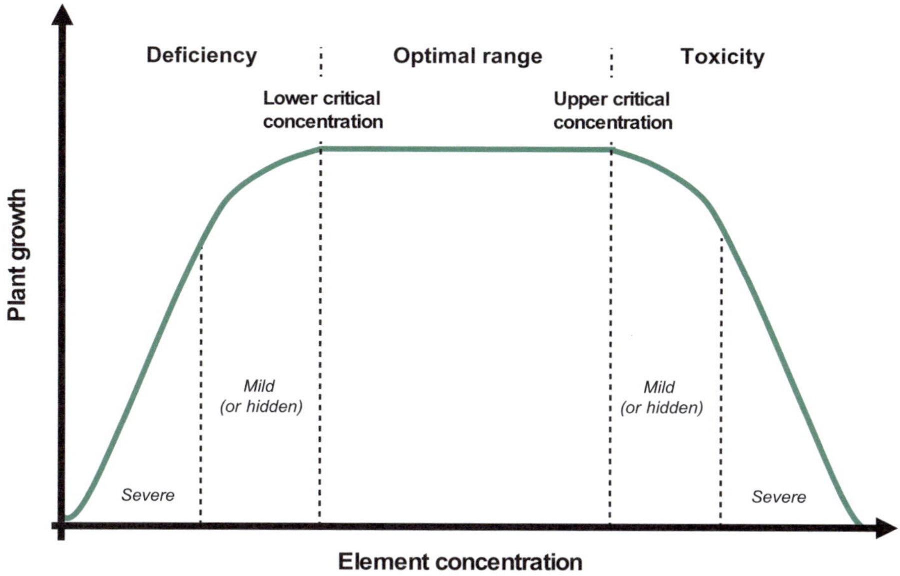

Fig. 1 Response of plant growth to the concentration of essential elements in the root zone or in plant tissues

anion (NO_3^-) or a cation (NH_4^+, ammonium). Almost all greenhouse crops prefer NO_3^- and NH_4^+ may be toxic at concentration higher than 0.5–1.0 mmol L^{-1} in NS depending on crop species and growing conditions.

Figure 1 illustrates the typical response of plant growth to the nutrient concentration in the growing medium or in plant tissues. There are different nutrient levels: deficiency, when nutrient concentration is low and plant growth is limited; optimal range, when plant growth is maximum and does not change with increasing nutrient concentration; toxicity, when nutrient concentration is excessive and plant growth is inhibited.

Essential mineral elements are divided in two groups depending upon their lower critical concentration (the concentration at the transition from deficiency to sufficiency; Fig. 1), which is 10 to 5000 times greater for macronutrients than for trace elements (Table 1).

Luxury consumption takes place when crop plants absorb nutrients without a corresponding increase in growth and yield. This condition may be responsible for the occurrence of physiological disorders such as lush growth, delayed flowering, poor fruit set, greater susceptibility to pests (e.g. aphids) and diseases, or poor product quality due, for instance, to the accumulation of free nitrates in leafy vegetables.

Nutrients are taken up from the growing medium by transporting proteins, which are located on the membrane of root cells, and transported to aerial organs in the xylem sap. Excess nutrients accumulate in the cell vacuole and the level of this pool could be determined to assess plant nutrient status. For instance, the determination

Table 1 Concentration of essential mineral elements sufficient for adequate plant growth (Epstein and Bloom 2005)

Element	Symbol	mg kg^{-1} DW
Nitrogen	N	15.000
Potassium	K	10.000
Calcium	Ca	5.000
Magnesium	Mg	2.000
Phosphorus	P	2.000
Sulfur	S	1.000
Chlorine	Cl	100
Iron	Fe	100
Boron	B	20
Manganese	Mn	50
Zinc	Zn	20
Copper	Cu	6
Molybdenum	Mo	0.1
Nickel	Ni	0.1

vof nitrate content in leaf tissues or in the petiole sap can be used to determine whether and how much N must be supplied to the crop (Peña-Fleitas et al. 2015).

Some elements can be remobilized in the phloem sap, such as N, P, K, Mg, Mo and Ni; in contrast, Ca, S, B, Fe, Mn, Zn and Cu are nutrients with a low or very low mobility in the phloem. The difference in the phloem mobility is used to recognize mineral deficiency. A deficiency of phloem-mobile element results in symptoms (for instance, leaf yellowing or scorch, interveinal chlorosis etc.) of older leaves, from which the element moves to the developing organs in the phloem. Conversely, the deficiency of phloem-immobile nutrient primarily affects younger leaves, which have a low transpiration rate and thus receive fewer nutrients via the xylem.

Mineral Supply

The goal of a fertilisation programme is matching crop demand and supply of mineral elements. This is indeed a difficult task since the rate of crop mineral uptake and the mutual ratios by which the nutritive elements are absorbed by roots are influenced by climatic conditions and considerably change during growing period, especially in long-cycle crops such as *cucurbit* and solanaceous species. Mineral nutrition depends on plant growth and is strongly influenced by the rate of photosynthesis; therefore, nutrient uptake increases with increasing irradiance. Other factors affects the uptake of nutrients, in particular their concentrations and those of other elements (due synergistic or antagonist effect), pH, salinity and moisture content of the growing medium.

The pH of the growing medium strongly affects root nutrient uptake. Optimal pH range for most of greenhouse crops is between 5.5 and 6.5. At pH higher than 6.5–7.0, some elements are less available (P, Ca, B, Fe, Mn, and Zn) and nutrient deficiency may occur. At pH lower than 4.5–5.0, the uptake of most trace elements (Fe, Zn, Mn and Cu) is stimulated and toxicity may occur with high element concentration in the growing medium. Sensitivity to sub- or supra-optimal pH largely depends on crop species.

The major reason for seasonal variation of plant mineral uptake is ontogenesis, which involves the formation of tissues and organs with different mineral composition. Major changes take place as a consequence of the transition from vegetative to reproductive growth, as fruits and vegetative organs have different nutrient content. In young tomato plants, for example, leaves are rapidly developed and N and K are absorbed at a weight ratio around one. The uptake of K increases in fruiting plants and the N:K uptake ratio drops to 0.20–0.25, as K balances the negative charges of organic acids that accumulate in the fruit cells. Therefore, K supply must be greater for fruiting plants while a reduction in N supply reduces plant vigour and improves fruit set and ripening. The same phenomenon occur for the Ca:K ratio as Ca is contained in fruit tissues at much lower concentration (<4 g kg^{-1} DM, dry matter) than in green organs (20–50 g kg^{-1} DM; Mills and Jones 1996).

Soil Culture Versus Hydroponics

Plants grown hydroponically generally have a faster growth and a higher yield than soil-bound plants as water uptake is facilitated by greater hydraulic conductivity and water content of soilless substrates and mineral nutrition is stimulated by frequent delivery (generally in excess) of a complete NS with optimal pH.

There are significant differences between soil and soilless culture as regards root morphology, the movement and the concentration of inorganic ions in the root zone, and the system buffering capacity. For instance, hydroponically-grown plants often have longer primary roots and much fewer lateral roots and root hairs compared to soil-bound plants.

Differences that are more important concern the movement of nutritive ions to root surface. In soil, which is much more heterogeneous than any soilless substrate, ion movement occurs by mass flow, driven by plant transpiration, and by diffusion, driven by root ion uptake. Moreover, in soil culture, root interception substantially contributes to mineral uptake and then root growth is essential for water and nutrient uptake. Conversely, root growth is not so important for mineral uptake in hydroponic culture and in some crops (e.g. tomato or melon) grown in nutrient film technique (see Chapter 2.b Soil based and simplified hydroponics rooftop gardens) root growth can be excessive, with the occurrence of hypoxic conditions in the root zone.

What is more important in soilless culture is the rate by which the NS on the root surface is renewed, because its volume, oxygen content and ion content rapidly change due to root uptake. For this reason and due to the low volume of substrate

Table 2 Range of ion concentrations in hydroponic nutrient solutions

Macronutrients		Micronutrients	
Nutritive element	Concentration (mmol L^{-1})	Ion	Concentration (μmol L^{-1})
N-NO_3	5–15	Fe	15–40
N-NH_4	1–3	B	15–30
P	1–2	Zn	5–10
K	5–10	Mn	5–10
Ca	3–8	Cu	1–5
Mg	1–2	Mo	0.1–0.5
S	1–5	Ni	0.1

generally available to the plants grown in soilless culture, the frequency of irrigation plays a major role in the mineral nutrition of hydroponic plants.

Another important characteristic of soilless culture is its limited system buffering capacity, in particular against changes in pH, temperature and moisture content in the root zone. Consequently, hydroponic systems are easier to control than soil but they depend on continuous control of irrigation and fertilisation with the risk of negative effects of technical mistakes or failures on crop yield and quality.

Nutrient Solution in Conventional Soilless Culture

Nutrient solution is the aqueous solution of fertilisers containing all the essential mineral elements used for plant water and nutrient replenishment. The main NS parameters are the total molar concentration and its electro-conductivity (EC), the mutual ratio between macronutrients and its pH.

Ion Concentration

Depending on many factors, such as crop characteristics (e.g. tolerance to salinity) and stage, climate and hydroponic system, total molar concentration in NS ranges between 20 and 40 mM or between 1 and 2 g L^{-1}. The ion concentration of NS is normally reported as milli- and micro-moles per litre, respectively for macro- and micro-nutrients (Table 2).

The total ion concentration of NS is often expressed as electrical conductivity (EC, dS m^{-1}), which ranges between 1.0 and 4.0 dS m^{-1}, since there is a strong correlation between the two quantities and EC can be easily and accurately measured using portable instruments.

A simple linear relationship may be used to convert equivalent concentration of cations (C^+, meq L^{-1}) in EC, assuming that the concentration of cations is equal to the one of anions (Sonneveld et al. 1999):

$$EC = 0.095C^{+} + 0.19$$

Only the macronutrients (Ca^{2+}, K^{+}, Mg^{2+}, NH_4^{+}) and possibly Na^{+} (contained in the raw water) are considered because trace elements are dissolved at low concentrations and thus have negligible effects on EC.

Nutrient Ratios

The most important ratios in the NS are the nutrient cation ratio (K:Ca:Mg) and the ratio between N-NH_4 and total N.

A K:Ca:Mg molar ratio of 1:0.5:0.25 minimizes the antagonist effects among these elements in most crops (Sonneveld and Voogt 2009) while the N-NH_4:N ratio should range between 0.07 and 0.25 (Sonneveld 1995; Savvas et al. 2009). Using both N-NH_4 and N-NO_3 is a method to control the pH in the recirculating NS and in the substrate (see the following paragraph).

pH

One of the most important parameters of NS is pH, since it influences the solubility of nutritive ions and their availability to the crop; optimal pH is between 5.5 and 6.5.

The pH of the irrigation water and NS mainly depends on its alkalinity, which is a measure of the capacity of water for neutralizing an acid solution (usually expressed as meq L^{-1}) and determines how much acid is required to adjust the pH.

The major components (>90%) of the alkalinity of irrigation water and hydroponic NS are bicarbonates (HCO_3^{-}) and, to lesser extent, carbonates (CO_3^{2-}). The acid requirement depends on the target pH and the concentration of dissolved HCO_3^{-}. For instance, adjusting pH to 6.0 requires an acid concentration corresponding to approximately 70% of HCO_3^{-} concentration.

Normally, some bicarbonates are kept in the NS in order to maintain a buffer system against excessive pH decrements, due to inaccurate acid injection and/or unbalanced root cation/anion uptake. When the uptake rate of cations exceeds the one of anions (e.g. N-NO_3^{-}), there is a net root H^{+} extrusion with consequent acidification of the outer medium; the opposite phenomenon occurs when the uptake rate of anions exceeds the one of cations. The NH^{+}_4:N ratio in the NS could be tuned in order to reduce or increase rhizosphere acidification. If nutrient solutions are prepared with distilled (or deionized) water (for instance, in experimental cultivations) or rainwater, a small amount (1 mmol L^{-1}) of sodium or potassium bicarbonate should be added to stabilize the pH.

Irrigation water and NS are acidified using strong inorganic acids, which are also fertilisers: nitric acid, sulphuric acid and phosphoric acid. Nitric acid is the mostly

Table 3 Nutrient content and pH reaction of some mineral fertilisers that are largely used in soilless culture

Fertiliser	Chemical formula	N-P-K (% w:w)	pH reation
Sources of macronutrients			
Ammonium Nitrate	NH_4NO_3	33-0-0	Acid
Ammonium Sulfate	$(NH_4)_2\ SO_4$	21-0-0-24 S	Acid
Calcium Chloride	$CaCl_2$	74-77Ca	Basica
Calcium Nitrate Tetrahydrate	$Ca(NO_3)_2\ 4H_2O$	12-0-0-23 CaO	Alkaline
Diammonium Phosphate	$(NH_4)_2HPO_4$	18-46-0	Alkaline
Magnesium sulfate	$MgSO_4\ 7H_2O$	0-0-0-16MgO-13S	Acid
Magnesium Nitrate Hexahydrate	$Mg(NO_3)_2\ 6H_2O$	11-0-0-9.5 MgO	Acid
Monopotassium Phosphate	KH_2PO_4	0-23-28	Acid
Monoammonium Phosphate	$NH_4\ H_2PO_4$	12-61-0	Acid
Nitric Acid (37%-1.23 kg L^{-1})	HNO_3	8,3-0-0	Acid
Nitric Acid (59%-1.37 kg L^{-1})		13-0-0	Acid
Phosphoric Acid (37%)	H_3PO_4	0-11,5-0	Acid
Phosphoric Acid (75%)	H_3PO_4	0-23-0	Acid
Potassium Chloride	KCl	0-0-60	Neutral
Potassium Nitrate	KNO_3	13-0-46	Alkaline
Potassium Sulfate	K_2SO_4	0-0-41-18S	Neutral
Urea	$(CO(NH_2)_2$	46-0-0	Acid
Sources of macronutrients			
Boric Acid	H_3BO3	17 B	
Sodium Molybdate	$Na_2MoO_4\ 2H_2O$	40 Mo	
Ammonium Heptamolybdate	$(NH_4)_6Mo_7O_{24}$	58 Mo	
Manganesium Sulfate Monohydrate	$MnSO_4\ H_2O$	32 Mn	
Zinc Sulfate 7-hydrate	$ZnSO_4\ 7H_2O$	23 Zn	
Copper Sulfate 5-hydrate	$CuSO_4\ 5H_2O$	25 Cu	

used, since the optimal concentration of both S and P is lower than N concentration.

Nutrient Solution Preparation

In commercial operations, NS is generally prepared by diluting stock solutions of fertilisers and acids (with a high purity grade and solubility; Table 3) with raw water at a ratio of 1:50 to 1:250. Many fertigation devices are readily available on the market with different characteristics in terms of method of fertilisers injection in the irrigation system, automation and integration with the greenhouse climate control computer (an example is shown in Fig. 2).

Fig. 2 An example of a medium-tech fertigation system. Two stock nutrient solutions and diluted acid are mixed to raw water with Venturi injectors under the control provided by EC and pH probes. The controller also schedules irrigation based on weather-based estimation of crop evapotranspiration

Normally, the minimum fertigation configuration has two stock solutions in order to separate Ca salts from $H_2PO_4^-$ and SO_4^{2-} to avoid precipitations of salts, such as calcium phosphate ($Ca_3(PO_4)_2$) and calcium sulphate ($CaSO_4$).

The determination of salt composition of the stock NS requires the knowledge of the composition of irrigation water and the target NS (recipe). Some spreadsheets and computer programs are available, also in the Internet, to calculate the exact amount of salts required to prepare stock NSs. An example is an Excel spreadsheet at the University of Pisa and available at http://www.wageningenur.nl/en/Research-Results/Projects-and-programmes/Euphoros-1/Calculation-tools/Nutrient-Solution-Calculator.htm.

Fertigation Strategies

In substrate, NS is generally supplied in excess (20–40%) respect to the evapotranspiration (ET) to prevent the difficulties associated to the unequal transpiration of individual plants and to avoid the salt accumulation and the imbalance in the nutrient solution.

Soilless growing systems can be classified in closed or open systems, if this drainage water is or not captured and reused after the adjustment of pH and nutrient concentration. Both systems can be used for drip-irrigated substrate culture while floating system, NFT and aeroponics are basically closed systems. In open soilless systems, there is a massive waste of water and nutrients, which is responsible for higher running costs and contamination of ground and surface water in agricultural areas. Therefore, the application of closed soilless systems is essential for sustainable rooftop greenhouse horticulture. Nevertheless, the adoption of closed systems has some drawbacks, such as an easier diffusion of root pathogens and a more difficult nutrient replenishment of the recirculating NS, in particular when water of poor quality (saline) is used (Massa et al. 2010).

To reduce the risk of the occurrence of root diseases, the recirculating NS is disinfected using different methods (e.g. heat treatment, UV light, chemical oxidation, slow sand filtration etc.; Postma et al. 2008) and a number of prophylactic measures (e.g. use of pathogen-free propagation materials; regular test for early detection of the presence of pathogens in the substrate and the recirculating solution; substrate or NS DNA scanning analyses; prompt removal of diseased plants etc.; Pardossi et al. 2011).

If saline water is used, there is a more or less rapid accumulation of ballast ions, which are dissolved in the water at concentration much higher than crop uptake concentration (the ratio between the ions and the water taken up by the plants). Under these conditions, the nutrient solution is recirculated till its EC and/or the concentration of some ions (e.g., Na^+, Cl^- or trace elements such as B) reach maximum acceptable thresholds value for the crop under consideration; afterwards, the nutrient solution is replaced, at least partially (flushing). During this period, EC is not a good marker of total nutrient concentration due to the build-up of ballast ions such as Na^+ and Cl^- (Fig. 3) and therefore recirculating NS must be analysed (Massa et al. 2011). In commercial greenhouse, three different procedures can be adopted, as follows:

A. NS prepared by mixing raw water and drainage NS at a ratio generally equal to the leaching fraction (the ratio between the NS applied to the crop and drainage NS) and adding nutrient stocks to reach a target EC. With this procedure, the EC of recirculating NS is kept constant but there is a progressive nutrient depletion, if some or more ballast ions are dissolved in raw water (Fig. 3). The nutrient solution is discharged when the concentration of a given ion reaches a maximum concentration or when the concentration of some polluting agents (e.g. N-NO_3; P-H_2PO_4) is lower than limits imposed by current legislation on wastewater management.

B. Water consumption due to crop evapotranspiration is compensated by refilling the mixing tank with full-strength NS (with nutrient concentrations equal or close to the corresponding uptake concentrations) and the recirculating NS is flushed out whenever its EC and/or concentration of a given ion surpasses preset thresholds. This procedure results in a relatively constant concentration of nutritive ions but leads to a progressive increase of EC due to accumulation of ballast ions. Besides, there is great oscillation in EC and this is not suitable for many crops; for instance, it may be responsible for the occurrence of blossom-end rot (Fig. 4) or cracking in tomato fruits.

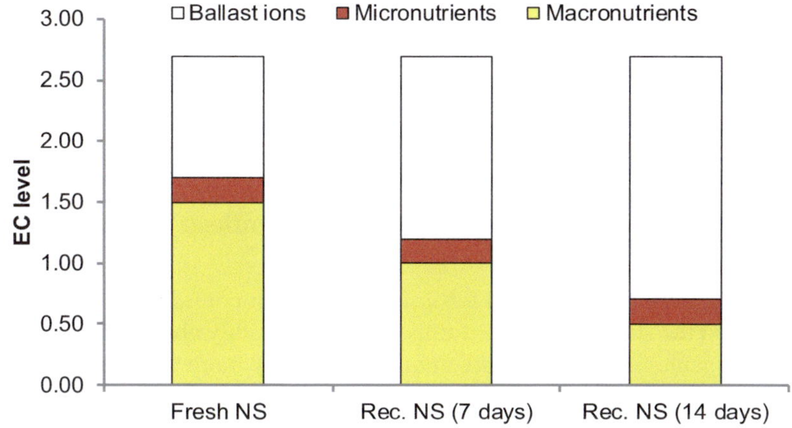

Fig. 3 Contribution of different types of ions to the EC of nutrient solution (NS) in closed substrate culture of greenhouse tomato. The values refer to newly-prepared (fresh) NS or NS that was recirculated for 1 or 2 weeks (Rec. NS). (Pardossi et al., unpublished results)

Fig. 4 Blossom-end rot (BER) of tomato. This physiological disorder is induced by a localized calcium (Ca) deficiency and also affects peppers, eggplants and melons. Fruits need Ca for healthy development and growing conditions leading to reduced root uptake and translocation to fast-growing organs of this element may results in BER. Water or salinity stress resulting from inadequate irrigation scheduling, high salinity or large oscillation in the osmotic pressure of the nutrient solution are the major factors responsible for the occurrence of BER in tomatoes grown in soilless culture

C. This procedure is similar to the first one. However, when the maximum EC or concentration of toxic ions is reached, crop water uptake is compensated with pH-controlled raw water for few days till the concentration of polluting ions drops below the maximum specified levels for wastewater discharge.

Managing Mineral Nutrition in Organic Soilless Culture

In most soilless cultures, mineral fertilisers are used in consideration of their high solubility and the stability of ionic forms, which are readily absorbed by the plants. However, organic fertilisers are becoming increasingly more and more popular, in particular in non-professional soilless cultivation in urban and peri-urban areas.

The limited volume of substrate in soilless cultures originates several drawbacks when organic sources are used to supply nutrients. Organic fertilisers have low content of nutrients readily absorbed by the plant, which means the microorganisms present in the growing media play an important role in determining the actual availability of nutrients to the crop by breaking down the fertiliser components so that they are readily absorbed by plants. The type of fertiliser source utilized for nutrient application, the physicochemical characteristics of the substrate, and the presence of microorganisms must be considered to ensure proper fertilisation.

Solid organic materials can be utilized as amendments prior to transplanting, facilitating initial availability of nutrients. If they are liquid or soluble solid form can be dissolved in irrigation water.

The most common types of organic materials of animal origin are processed manure (dehydrated, composted or liquid), or waste products such as blood, fish and bone meal. Among the available types of organic materials of plant origin are composted or fermented waste products, peat, sawdust, wood ash, vinasse, and seaweed. Seaweed extracts are widely utilized in irrigation or foliar fertilization (Fornes et al. 2002; Selvaraj et al. 2004; Haider et al. 2012; Battacharyya et al. 2015).

Rooftop agriculture could use also the compost originating from the organic fraction of municipal organic solid waste (OFMSW) as a minor component of substrate as well as a fertiliser (Hargreaves et al. 2008). The main problem on the use OFMSW compost is the possible high content in heavy metals, salinity and the possibility of the presence of some human (i.e. *Salmonella* sp. and *E. coli*) and plant (*Fusarium* spp.) pathogens as well as some organic contaminants and/or phytotoxic substances. These drawbacks could be overcome by the separation of sewage sludge from the solid biomass before to apply a proper composting process in order to obtain a compost with a low content of heavy metals and phytotoxic organic compounds (Hargreaves et al. 2008; Cesaro et al. 2015).

To evaluate the quality of compost used as substrate, it would be necessary to know the evolution of the physicochemical properties to guarantee their viability over time. Comparative trials of tomato production in containers with 4 mixtures of organic substrates and organic fertilisers versus conventional fertilization in rockwool obtained greater yields in the organic treatments without significant differences (Surrage et al. 2010). Compost is often used for organic soilless culture. This material must be mature enough to be used as a substrate component, which C:N ratio should be 25 (Dresbøll 2004).

To evaluate the quality of compost used as fertiliser it is important to consider the forms in which N is provided (organic, nitric, ammonium). For example, the content in NO_3-NH_4 is 13.3–4.8 in mg L^{-1} (w:v) for garden plant waste (Hawkins 2010), 1.300–1.000 mg kg^{-1} (w:w) for compost was obtained from seaweed washed ashore (Illera-Vives et al. 2015). A total N content higher than 0.6% (DW) with organic-N > 90% and a NO_3-NH_4 ratio in favour of the oxidized form with NH_4 < 0.04%, are considered desirable for mature compost (Senesi 1989).

In general, positive results are obtained with pre-planting applications of organic fertilisers in short-cycle culture. In trials with calendula culture in containers, two types of pelletized chicken manure (plus bedding) were compared with a synthetic controlled-release fertiliser (Bi et al. 2010). The chicken manure values were 4-0.87-1.66 and 3-1.31-2.49 (N-P-K), applied in 4 doses between 0 and 1.555 mg L^{-1} N; and the synthetic fertiliser (Osmocote, slow release fertiliser, 14-6-11) was applied in 4 doses between 0 and 831 mg L^{-1} N. The low and medium doses of pelletized chicken litter produced the largest amount of dry matter, similar to the high dose of controlled-release fertiliser. However, the high dose of pelletized chicken showed symptoms due to an excess of fertiliser. The results of Hernández et al. (2016) confirm that manure and sewage sludge composts can be used as an alternative to inorganic fertilization in lettuce crop, leading to similar or even higher yields and reducing nutrient-leaching risks. Furthermore, the leaves of lettuce grown in organically treated soils contained lower nitrate concentrations than it with inorganic fertilisers (Pavlou et al. 2007).

For long-cycle crops, the organic material incorporated as a nutrient source is normally not enough to feed the plant and needs to be supplemented with the addition of solid or liquid fertilisers over the cycle (Atiyeh et al. 2000; Mejía and Salas 2016).

A specific case of this was a study of pepper culture in organic substrate that utilized shrimp meal fertiliser, with applications of 0, 400, 800, and 1.600 mL m^{-3}. In addition, the process included the use of a commercial liquid organic fertiliser added three times per week. Five weeks after sowing, the plants fertilized with 400 and 800 mL m^{-3} were larger; however, 5 weeks later, the effects of the initial application of shrimp meal were no longer visible (Zhai et al. 2009). In contrast, there are other works which indicate that nutrients in manure compost can indeed fulfill the nutrient requirements of long-cycle crops (Márquez and Cano 2004; Raviv 2005). According to Márquez-Hernández et al. (2013), different organic fertilisers, compared to mineral forms, used in tomato culture in containers with a sand-composted manure substrate (50–50% v:v) obtained greater yield in compost treatments, requiring the application of organic top dressing fertilisers. However, there does seem to be a consensus among most studies regarding the improvement of organoleptic properties of fruits by means of increasing the content of soluble solids with organic fertilization in organic substrates (Márquez-Hernández et al. 2013; Preciado Rangel et al. 2011; Márquez and Cano 2004). Higher accumulation of soluble solids in fruits may be the result of high concentration of salts in the solution applied

(Satti et al. 1994; Wu and Kubota 2008). Extensive studies on the effect of a commercial seaweed extract, *Ascophyllum nodosum*, treatment on spinach showed that not only the storage quality was improved, but also flavonoid synthesis and nutritional quality of the spinach leaf was enhanced (Fan et al. 2013).

Liquid organic fertilisers come from different materials, such as manure (Capulín-Grande et al. 2005), compost, vermicompost (Jarecki and Voroney 2005; García et al. 2008), compost tea (Hargreaves et al. 2009; Ochoa Martìnez et al. 2009) or vermicompost tea (NOSB 2004), which can be applied in fertilizing irrigation. Compost tea is a good method for applying soluble nutrients directly to foliage or roots during the growing season. Recent works have evaluated the viability of using organic nutrient solutions versus minerals. Preciado Rangel et al. (2011) conducted a study for tomato culture in containers using sand as substrate and compared different nutrient solutions: inorganic, compost tea, vermicompost tea and vermicompost leachate. The results revealed that inorganic fertilization obtained the highest contents of foliar N-NO_3 in the petiole extract and greater yield. However, there were fewer soluble solids in the fruit than in organic fertilization. Regarding the organic treatments, vermicompost tea obtained the greatest yield. The effects of vermicompost tea and effluent were compared to Steiner (1961) nutrient solution on three plant species in a NFT hydroponic system, according to Gonzalez et al. (2013). Vermicompost tea favoured the growth in a similar way to the mineral solution. Nutrient solutions prepared with organic sources must be diluted to reduce EC to 2.0 dS m^{-1} to avoid salinity problems (García et al. 2008; Carballo et al. 2009; Olivia-Llaven et al. 2010; Ruiz and Salas 2016) and to adjust the pH with citric acid and vinegar.

We must also consider that the utilization of organic fertilisers can cause blockages in injection systems, and while solutions that have been stored prior to usage can ferment or develop a biofilm in their own containers. Furthermore, an additional drawback is the inhibition of plant growth due to the presence of phytotoxic organic components (Garland et al. 1993, 1997). Nevertheless, recent demand for organic solutions in fertilizing irrigation has inspired new innovations. Many products now undergo a preliminary processing based on the use of microorganisms that produce ammonification and nitrification of the organic-N prior to being added to the nutrient solution. However, the result of the efficiency in generating N-NO_3 from organic N is below 30% (Strayer et al. 1997). Recent works studying hydroponic in lettuce culture with an organic nutrient solution have shown promising results, the solution was vermicompost tea diluted to EC (2.0 dS m^{-1}) and injection of air getting by action of microorganisms increase in NO_3-N concentration up to 5 times compared to the beginning of the process (Ruiz and Salas 2016).

The combination of materials with different mineralization rates facilitates nutrient availability, and with supplementary applications, it is possible to achieve acceptable production (Treadwell et al. 2007; Mattson 2014). The availability of nutritional elements in organic fertilisers depends on the mineralization rate generated by microorganisms and the nature of the organic fertiliser source (Gaskell and Smith 2007). Williams and Nelson (1992) conducted tests on different materials, and after 2 weeks, half of them had 50% of N mineralized at 10 °C, and 60% at

25 °C. In general, during the initial weeks both the proportion of mineralized organic nutrients and the losses from rinses are greater (Illera-Vives et al. 2015).

In order to improve nutrient absorption the use of microorganisms is becoming more and more common. The most widely used are *Arbuscular mycorrhizal* fungi and *Sinorhizobium* spp., which have demonstrated rather inconsistent results (Carpio et al. 2005; Russo 2006). There are commercial microorganism-based products that facilitate the absorption of N, P and K, such as *Azotobacter vinelandii* (Ahmad et al. 2006; Ponmurugan et al. 2012), *Bacillus megaterium* and *Frateuria aurantia* (Cakmakçi et al. 2006; Wu et al. 2005), respectively. Additionally, in anaerobiosis conditions, *Azospirillum* spp. can be used to favor absorption of N (Ahemad and Kibret 2014). Mejía and Salas (2016) compared some commercial microorganisms Tusal (*Trichoderma* spp.), Bactel-Bioera (bacterias mix) and Bioradis-Tablet (Mycorrhiza) for melon culture in containers using organic substrate mixture (different percentage of vermicompost (V)-coconut fiber (CF)). The results showed that *Trichoderma* spp. and substrate 60%V-40% CF obtained highest yield.

Overall, it seems necessary to adjust the physicochemical properties of substrates and their evolution over time, as well as irrigation management to ensure the activity of microorganisms utilized. The adaptation of numerous factors such as aeration, pH, EC, organic matter and element concentrations to the needs of microorganisms must be studied to integrate them as active components in plant nutrition.

Improving Crop Quality Through Mineral Nutrition

Improving Organoleptic Quality by Salinity

Growing vegetable fruit crops under conditions of moderate water or salinity stress reduces the fraction of water received by fruits trough the xylem, but increases the contribution of phloem sap and its dry matter concentration (Dorais et al. 2001). This results in a significant reduction of the fresh weight gain of the fruit with no important effects on the accumulation of dry matter, thus providing the basis for higher quality.

The reduction of xylematic water influx into fast growing fruits limits the supply of Ca, thus increasing the occurrence of some physiological disorders such blossom-end rot (Fig. 4). Proper control of greenhouse climate and fertigation regime is necessary to improve fruit quality by means of osmotically induced stress without any deleterious effect on crop yield and quality.

Prevention of Nitrate Accumulation in Leafy Vegetables

Nitrate may lead to the formation of nitrite, nitric oxide and N-nitrous compounds, which may have health problems such as methemoglobinemia and carcinogenesis (EFSA 2008).

Some vegetables (e.g. lettuce, spinach, rocket salad, celery and basil) can accumulate large amount of NO_3^- in leaf tissues (up to 15–20 g kg^{-1} FW; EFSA 2008). In the European Union some limits have been laid down to the NO_3^- content of some vegetables for fresh consumption or processing (EU 2011).

Excessive nitrate accumulation in plant tissues is due to an imbalance between root uptake and nitrate assimilation (incorporation into organic compounds such as amino acids) in the leaves. Low radiation and any growing conditions reducing the rate of photosynthesis, without a concomitant reduction in root uptake of NO_3^-, inevitably results in NO_3^- accumulation in leaf tissues, where this anion is transported with the xylem sap along the transpiration stream (Santamaria 2006).

In soilless culture, a marked reduction of leaf NO_3^- content can be obtained by feeding plants for a few days before harvest with a NO_3^--free nutrient solution or by replacing part of NO_3^- with Cl^-.

Biofortification

Biofortification is the increase of the bioavailable concentration of a given element (e.g. Fe, Zn, Ca, Mg, Cu, I or Se) in food crops by means of fertilisation or plant breeding with the aim to solve the risk of micronutrient in deficiency in the humans (White and Broadly 2005).

Micronutrient biofortification is achieved by mean of foliar or soil application of specific salts (e.g. potassium selenite and potassium iodine). Crop biofortification is much simple in hydroponic culture as the element of interest can be dissolved in the nutrient solution. Micronutrient supplements may have also a positive effect on plant growth and produce quality by an indirect effect. For instance, the application of Se enhances shelf-life of fresh-cut leafy vegetables because this element interferes with ethylene synthesis (Malorgio et al. 2009).

Hydroponics has been also used to enrich vegetables with nutritional factors such as Ω-3-fatty acids (Palaniswamy et al. 2000).

Conclusions

Soilless cultivation is one of the most important components of rooftop agriculture, which can be carried out by either professional or non-professional growers. In the first case, soilless culture are generally run under greenhouse by skilful growers using high-technology devices (e.g. computer-controlled fertigation device), high-grade fertilisers and growing media with standardized characteristics. When conducted by non-professional growers, low technology is applied to reduce production costs and some difficulties may be encountered as regards, for instance, the management of mineral supply. The application of the technical principles of organic soilless culture can solve most of these problems. In both professional and non-professional cultures, simple measures can be taken to improve the nutritional quality of vegetables, such as the addition of low amount of specific salts containing beneficial elements such as iodine and selenium.

Bullet Points

- Rooftop agriculture is based on soilless culture.
- Professional soilless culture is generally conducted under greenhouse using high-technology and is not substantially different from conventional greenhouse production.
- Non-professional soilless cultivation can be run according to the principle of organic soilless culture.
- Simple measures can be taken to improve the nutritional quality of vegetables grown in soilless culture regardless of the technology level.

References

Ahemad M, Kibret M (2014) Mechanisms and applications of plant growth promoting rhizobacteria: current perspective. J King Saud Univ Sci 26(1):1–20. doi:10.1016/j.jksus.2013.05.001

Ahmad F, Ahmad I, Aqil F, Ahmed Wani A, Sousche YS (2006) Plant growth promoting potential of free-living diazotrophs and other rhizobacteria isolated from Northern Indian soil. Biotechnol J 1(10):1112–1123. doi:10.1002/biot.200600132

Atiyeh RM, Subler S, Edwards CA, Bachman G, Metzger JD, Shuster W (2000) Effects of vermicomposts and composts on plant growth in horticulture container media and soil. Pedobiologia 44(5):579–590. doi:10.1078/S0031-4056(04)70073-6

Battacharyya D, Babgohari MZ, Prithiviraj PRB (2015) Seaweed extracts as biostimulants in horticulture. Sci Hortic 196:39–48. doi:10.1016/j.scienta.2015.09.012

Bi G, Evans WB, Spiers JM, Witcher AL (2010) Effects of organic and inorganic fertilizers on marigold growth and flowering. Hortscience 45(9):1373–1377

Cakmakçi R, Dönmez F, Aydın A, Şahin F (2006) Growth promotion of plants by plant growth-promoting rhizobacteria under greenhouse and two different field soil conditions. Soil Biol Biochem 38(6):1482–1487. doi:10.1016/j.soilbio.2005.09.019

Capulín-Grande J, Núñez-Escobar R, Sánchez-García P, Martínez-Garza A, Soto-Hernández M (2005) Producción de jitomate con estiércol líquido de bovino, acidulado con ácidos orgánicos e inorgánicos. Organo Cientifico de la Sociedad Mexicana de la Ciencia del Suelo, AC, Terra

Carballo T, Gil MV, Calvo LF, Morán A (2009) The influence of aeration system, temperatura and compost origin on the phytotoxicity of compost tea. Compost Sci Util 17(2):127–139. doi:10.1080/1065657X.2009.10702411

Carpio LA, Davies FT Jr, Arnold MA (2005) Arbuscular mycrorrhizal fungi organic and inorganic controlled-release fertilizers: effect on growth and leachate of container-grown bush morning glory (Ipomoea carnea ssp. fistulosa) under high production temperatures. J Am Soc Hortic Sci 130(1):131–139

Cesaro A, Belgiorno V, Guida M (2015) Compost from organic solid waste: quality assessment and European regulations for its sustainable use. Resour Conserv Recycl 94(1):72–79. doi:10.1016/j.resconrec.2014.11.003

Dorais M, Papadopoulos A, Gosselin A (2001) Greenhouse tomato fruit quality. Hortic Rev 26:239–350

Dresbøll DB (2004) Optimisation of growing media for organic greenhouse production. In: PhD Dissertation, Royal Veterinary and Agricultural University, Denmark

EFSA (2008) European Food Safety Authority – Nitrate in vegetables: scientific opinion of the panel on contaminants in the food chain. EFSA J 689:1–79

Epstein E, Bloom AJ (2005) Mineral nutrition of plants: principles and perspectives, 2nd edn. Sinauer Associates, Inc., Publishers, Sunderland

EU (2011) Commission Regulation (EU) No 1258/2011 of 2 December 2011 amending Regulation (EC) No 1881/2006 as regards maximum levels for nitrates in foodstuffs. http://eur-lex.europa.eu/legal-content/EN/TXT/HTML/?uri=CELEX:32011R1258&from=IT. Accessed 20 Sept 2016

Fan D, Hodges DM, Critchley AT, Prithiviraj B (2013) A commercial extract of brown macroalga (*Ascophyllum nodosum*) affects yield and the nutritional quality of spinach in vitro. Commun Soil Sci Plant Anal 44(12):1873–1884. doi:10.1080/00103624.2013.790404

Fornes F, Sanchez-Perales M, Guardiola JL (2002) Effect of a seaweed extract on the productivity of' de Nules' Clementine mandarin and Navelina orange. Bot Mar 45(5):486–489

García GR, Dendooven L, Gutiérrez MFA (2008) Vermicomposting lechate (worm tea) as liquid fertilizer for maize (*Zea mays* L.) forage production. Asian J Plant Sci 7(4):360–367

Garland JL, Mackoqiak JL, Sager JC (1993) Hydroponic crop production using recycled nutrients from inedible crop residues, technical paper 932173. SAE, Warrendale

Garland JL, Mackowiak CL, Strayer RF, Finger BW (1997) Integration of waste processing and biomass production systems as part of the KSC Breadboard project. Adv Space Res 20(10):1821–1826. doi:10.1016/S0273-1177(97)00847-8

Gaskell M, Smith R (2007) Nitrogen sources for organic vegetable crops. HortTechnology 17(4):431–441

González K, Rodríguez MN, Trejo L, García JL, Sánchez J (2013) Efluente y te de vermicompost en la producción de hortalizas de hoja en sistema NFT. Interciencia 38(12):863–869

Haider MW, Ayyub CM, Pervez MA, Asad HU, Manan A, Raza SA, Ashraf I (2012) Impact of foliar application of seaweed extract on growth, yield and quality of potato (*Solanum tuberosum* L.) Soil Environ 31(2):157–162

Hargreaves JC, Adl MS, Warman PR (2008) A review of the use of composted municipal solid waste in agriculture. Agric Ecosyst Environ 123(1–3):1–14. doi:10.1016/j.agee.2007.07.004

Hargreaves JC, Adl MS, Warman PR (2009) Are compost teas an effective nutrient amendment in the cultivation of strawberries? Soil and plant tissue effects. J Sci Food Agric 89(3):390–397. doi:10.1002/jsfa.3456

Hawkins MR (2010) Consumer interest and compost substrate management of organic and sustainable plant. University of Maine Electronic Thesis and Dissertations. Paper 729. http://digitalcommons.library.umaine.edu/etd/729

Hernández T, Chocano C, Moreno JL, García C (2016) Use of compost as an alternative to conventional inorganic fertilizers in intensive lettuce (*Lactuca sativa* L.) crops – effects on soil and plant. Soil Tillage Res 160:14–22. doi:10.1016/j.still.2016.02.005

Illera-Vives M, López-Mosquera ME, Salas-Sanjuan MA, López-Fabal A (2015) Leaching techniques for saline wastes composts used as growing media in organic agriculture: assessment and modelling. Environ Sci Pollut Res 22(9):6854–6863. doi:10.1007/s11356-014-3897-1

Jarecki MK, Voroney RP (2005) Evaluation of compost leachates for plant growth on hydroponic culture. J Plant Nutr 28(4):651–667. doi:10.1081/PLN-200052639

Malorgio F, Diaz KE, Ferrante A, Mensuali A, Pezzarossa B (2009) Effects of selenium addition on minimally processed leafy vegetables grown in floating system. J Sci Food Agric 89(13):2243–2251. doi:10.1002/jsfa.3714

Márquez HC, Cano R (2004) Producción de tomate orgánico bajo invernadero. In: C Leal Ch, Garza JAG (eds) 2do. Simposium Internacional de Producción de Cultivos en Invernadero. May Monterrey, NL, Fundación UANL y Facultad de Agronomía, UANL, pp 1–11

Márquez-Hernández C, Cano-Ríos P, Figueroa-Viramontes U, Avila-Diaz JA, Rodríguez-Dimas N, García-Hernández JL (2013) Yield and quality of tomato with organic sources of fertilization under greenhouse conditions. Phyton (Buenos Aires) 82:55–61

Massa D, Incrocci L, Maggini R, Carmassi G, Campiotti CA, Pardossi A (2010) Strategies to decrease water drainage and nitrate emission from soilless cultures of greenhouse tomato. Agric Water Manag 97(7):971–980. doi:10.1016/j.agwat.2010.01.029

Massa D, Incrocci L, Maggini R, Bibbiani C, Carmassi G, Malorgio F, Pardossi A (2011) Simulation of crop water and mineral relations in greenhouse soilless culture. Environ Model Softw 26(6):711–726. doi:10.1016/j.envsoft.2011.01.004

Mattson N (2014) Comparing substrate fertilizer amendments for spring bedding plants. Greenhouse Growing Magazine 32(13):48–54

Mejía P, Salas MC (in press, Nov 2016) Uso de vermicompost de residuos hortícolas como sustrato en cultivo sin suelo de melón. V Jornadas de la Red Española de Compostaje. Sevilla

Mills HA, Jones JB (1996) Plant analysis handbook II. A practical sampling, preparation, analysis, and interpretation guide. MicroMacro Publishing, Inc, Athen

NOSB (2004) Compost tea task force final report. National Organic Standards Board. www.ams.usda.gov/nosb/meetings/Compost TeaTaskForceFinalReport pdf. Accessed 25 Oct 2012

Ochoa Martínez E, Figueroa Viramontes U, Cano Ríos P, Preciado Rangel P, Moreno Reséndez A, Rodríguez Dimas N (2009) Té de composta como fertilizante orgánico en la producción de tomate (*Lycopersicon esculentum* Mill.) en invernadero. Revista Chapingo. Serie Hortic 15(3):245–250

Oliva-Llaven MA, Rodríguez HL, Mendoza NP, Ruiz SB, Álvarez S, Dendooven L (2010) Optimization of worm-bed leachate for culturing of tomato (*Lycopersicum esculentum* Mill) inoculated with *Glomus fasciculatum* and *Pseudomonas fluorescens*. Electron J Biotechnol 13(2):1–8

Olle M, Ngouajio M, Siomos A (2012) Vegetable quality and productivity as influenced by growing medium: a review. Žemdirbystė=Agriculture 99(4):399–408

Palaniswamy U, McAvoy RJ, Bible B (2000) Omega-3-fatty acid concentration in *Portulaca oleracea* L. is altered by nitrogen source in hydroponic solution. J Am Soc Hortic Sci 125(2):190–194

Pardossi A, Carmassi G, Diara C, Incrocci L, Maggini R, Massa D (2011) Fertigation and substrate management in closed soilless culture. EUPHOROS project, Deliverable n. 15. http://www.wageningenur.nl/upload_mm/8/c/0/aa4b4486-a9db-429f-8b03-f19d4cec3ee6_Fertigation%20and%20Substrate%20Management%20in%20Closed%20Soilless%20Culture.pdf. Accessed on 25 July 2017

Pavlou G, Ehaliotis CD, Kavvadias VA (2007) Effect of organic and inorganic fertilizers applied during successive crop seasons on growth and nitrate accumulation in lettuce. Sci Hortic 111(4):319–325. doi:10.1016/j.scienta.2006.11.003

Peña-Fleitas MT, Gallardo M, Thompson RB, Farneselli M, Padilla FM (2015) Assessing crop N status of fertigated vegetable crops using plant and soil monitoring techniques. Ann Appl Biol 167(3):387–405. doi:10.1111/aab.12235

Ponmurugan K, Sankaranarayanan A, Al-Dharbi NA (2012) Biological activities of plant growth promoting *Azotobacter* sp. isolated from vegetable crops rhizosphere soils. J Pure Appl Microbiol 6(4):1–10

Postma J, Van Os E, Bonants PJM (2008) Pathogen detection and management strategies in soilless plant growing systems. In: Soilless culture: theory and practice. Elsevier Publications, Amsterdam, pp 425–457

Preciado Rangel P, Forti Hernández M, García-Hernández JL, Rueda Puente EO, Esparza Rivera JR, Lara Herrera A, Segura Castruita MA, Orozco Vidal JA (2011) Evaluación de soluciones nutritivas orgánicas en la producción de tomate en invernadero. Interciencia 36(9):689–693
Raviv M (2005) Production of high-quality composts for horticultural purposes: a mini-review. HortTechnology 15(1):52–57
Raviv M, Lieth JH (2008) Soilless culture: theory and practice. Elsevier, London, p 608
Resh HM (2012) Hydroponic food production: a definitive guidebook for the advanced home gardener and the commercial hydroponic grower. CRC Press, Boca Raton
Ruiz JL, Salas MC (in press, November 2016) Valorización del uso de bionutrientes en hidroponía. V Jornadas de la Red Española de Compostaje. Sevilla
Russo VM (2006) Biological amendment fertilizer rate, and irrigation frequency for organic bell pepper transplant production. Hortscience 41(6):1402–1407
Santamaria P (2006) Nitrate in vegetables: toxicity, content, intake and EC regulation. J Sci Food Agric 86(1):10–17. doi:10.1002/jsfa.2351
Satti SME, López M, Al-Said FA (1994) Salinity induced changes in vegetative and reproductive growth in tomato. Commun Soil Sci Plant Anal 25(5–6):501–510
Savvas D, Passam H (2002) Hydroponic production of vegetables and ornamentals. Embryio Publications, Athens
Savvas D, Ntatsi G, Passam HC (2009) Plant nutrition and physiological disorder in greenhouse grown tomato, pepper and eggplant. Eur J Plant Sci Biotechnol 2(1):45–61
Selvaraj R, Selvi M, Shakila P (2004) Effect of seaweed liquid fertilizer on *Abelmoschus esculentus* and *Lycopersicon esculentum*. Seaweed Res Utility 26:121–123
Senesi N (1989) Composted materials as organic fertilizers. Sci Total Environ 81:521–542. doi:10.1016/0048-9697(89)90161-7
Sonneveld C (1995) Fertigation in the greenhouse industry. In: Proceeding of the Dahlia Greidinger International Symposium on Fertigation. Haifa, Israel, pp 121–140
Sonneveld C, Voogt W (2009) Plant nutrition of greenhouse crops. Springer, Dordrecht
Sonneveld C, Voogt W, Spanns L (1999) An universal algorithm for calculation of nutrient solutions. Acta Hortic 481:331–359. doi:10.17660/ActaHortic.1999.481.38
Steiner AA (1961) A universal method for preparing nutrient solutions of certain desired composition. Plant Soil 15(2):134–154. doi:10.1007/BF01347224
Strayer RF, Finger BW, Alazraki MP (1997) Evaluation of an anaerobic digestion system for processing CELSS crop residues for resource recovery. Adv Space Res 20(10):2009–2015
Surrage VA, Lafrenière C, Dixon M, Zheng Y (2010) Benefits of vermicompost asa constituent of growing substrates used in the production of organicgreenhouse tomatoes. Hortscience 45(10):1510–1515
Treadwell DD, Hochmuth GJ, Hochmuth RC, Simonne EH, Davis LL, Laughlin WL, Li Y, Olczyk T, Sprenkel RK, Osborne LS (2007) Nutrient management in organic greenhouse herb production: where are we now? HortTechnology 17(4):461–466
White PJ, Broadly MR (2005) Biofortifying crops with essential mineral elements. Trends Plant Sci 10(12):586–593. doi:10.1016/j.tplants.2005.10.001
Williams KA, Nelson PV (1992) Low, controlled nutrient availability provided by organic waste materials for chrysanthemum. J Am Soc Hortic Sci 117(3):422–429
Wu M, Kubota C (2008) Effects of high electrical conductivity of nutrient solution and its application timing on lycopene, chlorophyll and sugar concentrations of hydroponic tomatoes during ripening. Sci Hortic 116(2):122–129. doi:10.1016/j.scienta.2007.11.014
Wu SC, Cao ZH, Li ZG, Cheung KC, Wong MH (2005) Effects of biofertilizer containing N-fixer, P and K solubilizers and AM fungi on maize growth: a greenhouse trial. Geoderma 125(1):155–166. doi:10.1016/j.geoderma.2004.07.003
Zhai Z, Ehret DL, Forge T, Helmer T, Lin W, Dorais M, Papadopoulos AP (2009) Organic fertilizers for greenhouse tomatoes: productivity and substratemicrobiology. Hortscience 44(3):800–809

Sustainable Pest Management

Giovanni Bazzocchi and Stefano Maini

Abstract This section addresses the application of an ecosystemic approach in pest control issues in rooftop agriculture. Biotope (e.g., physical and climatic characteristics), possible biocenosis (e.g., insect pests, plant diseases, beneficials) and related ecological relationships are described, altogether with their practical consequences.

Taken for granted that under the rooftop conditions the use of synthetic and broad spectrum pesticides is unwise, the main ecological IPM practices potentially pertinent to rooftop agriculture are described. Pest exclusion and prevention practices, biological control with beneficial arthropods, use of natural and botanical insecticides, habitat manipulation and use of functional biodiversity for pest control are discussed.

Introduction

The world will need 70–100% more food by 2050 (World Bank 2008). Smallholder farmers and rooftop agriculture can play, at least in some countries and cities, an important role in supplying urban markets and meeting the food demand of growing urban centres (FAO 2007, 2012; Orsini et al. 2014). The challenge is a new eco-productive agriculture that integrates environmental, economic and social aspects (Celli et al. 2001; National Research Council 2010).

The rooftop is a totally new environment to grow plants for food production, and it greatly differs, from a physical, climatic, social and economic standpoint, from open fields and other agricultural contexts. Pest management strategies must take into account all these aspects. An old problem in a new scenario, but also an opportunity and a challenge for researchers, practitioners and growers to devise new techniques or adapt existing ones to a different environment.

Researchers and policy makers consider unwise the use of conventional chemical pesticides in urban agriculture and urban green spaces management. Anyway,

G. Bazzocchi (✉) • S. Maini
Department of Agricultural Sciences, Research Center in Urban Environment for Agriculture and Biodiversity, University of Bologna, via G. Fanin 42, 40127 Bologna, Italy
e-mail: giovanni.bazzocchi@unibo.it; stefano.maini@unibo.it

F. Orsini et al. (eds.), *Rooftop Urban Agriculture*, Urban Agriculture,
https://doi.org/10.1007/978-3-319-57720-3_11

mode of use and specific applications of synthetic pesticides are well explained and documented, including the specific regulations by a large number of well-accessible sources. For these reasons in this chapter are only considered integrated pest and plant diseases management strategies without conventional chemical pesticides.

Agroecology (Altieri 1987), the study of ecological processes applied to agricultural production systems, has led to a rich set of scientific and practical consequences. It is not always possible to apply agroecological techniques to the RA contexts, as they are fundamentally based on the concept of ecological complexity as control method, while a rooftop is an extremely simplified system also different from cultivated open agroecosystems. On the other hand, many of the successes and applications of pest biological control properly happens in very small environments (i.e. vegetable and ornamental home or urban gardens) or particularly simplified systems as greenhouses and hydroponics. A major goal of this chapter is to describe the rooftop "ecological scenario" in which the integrated pest management (IPM) principles can be implemented avoiding the use of synthetic pesticides, and pest prevention and control methods can be applied.

Finally, the chapter specifically describes the physical, mechanical, cultural, biological means of pest prevention and management, that can be applied in RA contexts. We will not take into account the methods of pest management in rooftop greenhouses – except when they can be applied even in the open air rooftop systems – as these do not substantially differ from greenhouses placed in other contexts in which the techniques have already been widely described (Baudoin et al. 2013).

Pesticide-Free Rooftop Agriculture

There are well documented reasons, that also meet the simple common sense, to support and practice a "pesticide-free" (referring to synthetic and broad spectrum pesticides) urban agriculture. The urban environment represents an important and underestimated arena of pesticide use and the hazard of their use in these environments has been reported and highlighted by recent and less recent studies (Racke and Leslie 1993; Burns et al. 2013; Rauh et al. 2015). Urban and rooftop agriculture are conducted in close proximity to people and residences not involved in the agricultural practices. Children in particular, for reasons linked to behaviour, physiological development and body size, are more susceptible to health risks from exposure to pesticides (Makris and Rowe 1998; Alarcon et al. 2005; Nasterlack 2006). Several authors have also highlighted as people involved in farming practices in urban environment often do not use protective equipments and safety measures. In particular, unsafe use of pesticides is the rule in less developed countries (Wesseling et al. 1997; Ackerson and Awuah 2010), where, moreover, many older, non patented, more toxic, environmentally persistent and inexpensive chemicals are used extensively, creating serious acute health problems and local and global environmental contamination (Ecobichon 2001). Policies promoting agriculture without conventional pesticides in developing countries should be supported. In addition to public health arguments, there are questions of food safety and quality and relative consumer perception and demand (Grunert 2005). Furthermore, urban and RA farmers are often young and not professional agricultural people inclined to a low impact agriculture, which makes proposals of ecological pest management more feasible and easily accepted.

From the political side, after a well documented study of the California Department of Public Health (CDPH 2014) and other previous independent studies by public institutions (Valcke et al. 2004), on May 22, 2014, on the occasion of the Biodiversity World Day, the French Minister for Ecology and Sustainable Development, Mme Ségolène Royal, launched the initiative "towns and villages without pesticides" in order to eliminate pesticides in the management of green spaces and gardens in cities. Several European countries (France, Belgium, Germany, the Netherlands, Denmark), following the recommendations of the EU directive 2009/128/EC (European Parliament and European Council 2009) are planning a ban of some pesticides in all the public areas (see http://www.pan-europe.info/campaigns/towns last access 10/10/2016).

Rooftop Ecology as a Strategy for Pest Management

Pest control, in a low-impact and integrated perspective, cannot be separated by a systemic and ecological framework. The agroecological paradigm shift (Altieri 1987) led to a rich set of scientific, practical and social consequences (Wezel et al. 2009) also contextualizing Integrated Pest Management and biological control in an ecology-based perspective (Gliessman 2006). In this paragraph it is described the rooftop as a mini-agroecosystem, its main physical and biotic features and the main differences compared to an open field agroecosystem. The consequences from a pest control perspective are highlighted.

Biotope

Size

Rooftop gardens and farms are very little systems (between 1000 to 40,000 m^2 usually) in comparison with open field farming and agro-ecosystems. This has both positive and negative implications regarding the pest management. Generally, pest control should be easier in little systems: manual, mechanical and physical techniques (Olkowski et al. 1991; Baudoin et al. 2013) and handmade bio-pesticides such as macerated and decoctions (Stoll 1996) are easily applicable and often effective. In addition, in this specific conditions (low-density, isolated populations), some methods for the detection or monitoring of insects in conventional agricultural conditions, such as light, coloured sticky and pheromone traps, can be used to decrease the populations of some pests at a not harmful threshold or to eradicate them (mass trapping) (El-Sayed et al. 2006; Cocco et al. 2012; Suckling et al. 2015; Braham and Nefzaouil 2016). Nevertheless, with the exclusion of greenhouses and hydroponic cultivations, there are very limited well documented data and scientific literature on pest management in very small farm systems, so far it has been proceeded more for trial and error that with well-documented and standardized control strategies. On the other hand, well-documented methods used in open-field productive agriculture cannot be applied for economic and logistic (mechanical means) reasons associated with the scale. Size does matter… but not always in the same direction.

Environmental Conditions

The microclimate and environmental conditions on the rooftops are very peculiar and different from all the other agricultural environments, included the urban on-ground gardens. In general, an uncovered rooftop is characterized by a wider difference between day-time and night-time temperatures and also by higher fluctuations in temperature during the year with respect to lower locations of the same geographical area. This factor, together with a particular incidence of winds and sun rays with consequent relatively low relative humidity, makes rooftops an environment in some way similar to arid or semi arid-zones (relatively to geographical location). From a pest management perspective, this implies a lower incidence of diseases caused by fungal, and in some cases virus and bacterial pathogens; in this sense, air circulation assumes great importance (Jarvis 1992). Similarly, non arthropod pests such as Gasteropoda and Nematodes should not be noxious. However, some biological control practices as the use of entomopathogenic nematodes (Stuart et al. 2015) and preparations based on entomopathogenic fungi (Wraight et al. 2016) are difficult to implement in arid environments. Furthermore high relative humidity favours some very important natural enemies in relation to their prey (*Phytoseiulus persimilis* vs *Tetranychus urticae*) in greenhouse conditions with a high reduction in egg vitality under 40% relative humidity (Stenseth 1979). However, recent works had shown that the effect of RH depends very much on the mean temperature and their nocturnal and diurnal fluctuation (Audenaert et al. 2014), then the release strategy of this important predator mite (including dose and application frequency) may perhaps be adapted to the RA conditions.

On rooftop, finally, the general climatic conditions are tougher and greater is the probability of extreme weather events, such as spells of high temperature and sun irradiation, torrential rains and considerable wind gusts, which cannot be mitigated by trees, natural vegetation and buildings themselves. In general, this type of abiotic stress can make the plants more susceptible to disease and insect pests (Rosenzweig et al. 2001). Optimizing water use and fertilization is of fundamental importance in these conditions, not only in order to avoid wilting problems but also to minimize the incidence of diseases (Rotem and Palti 1969).

Biocenosis

As in all agricultural system botanical composition strongly depend on human choices. Generally, rooftop crops are characterized by poor structural diversity and a predominance or exclusivity of herbaceous species such as vegetables, herbs, floral cultivations. Biocenosis in general is extremely simplified. Species biodiversity is further lower for at least three reasons: (1) a lower number of ecological niches (no soil, no spontaneous vegetation), (2) geographical isolation of the system compared to natural ecosystems (island effect, loss of ecotones) (3) the number of species is also a function of the system size (microecosystem). In general terms, this extreme simplification can make management of pests easier, but also makes population fluctuations more extreme and increases the risk of outbreaks. Even the soil microfauna heavily depends on the choice of cultivation systems and substrates (hydroponic, soilless, etc.) and can be easily manipulated in favour of the soilborne diseases ecological management, e.g. through the use of beneficial organisms mixes.

Pests

It is just a matter of time until a new habitat will be invaded, not only by species deliberately introduced by the man, but also by unwanted species whose populations can easily assume the state of pest (Niche theory). Rooftop crops can attracts both urban and agricultural pests. As opposed to the agricultural context, in which weeds are the primary pests, in the urban environment insects represent the most important pests for which control measures are instituted (Racke and Leslie 1993). As already mentioned pathogens, nematodes, snails and slugs should be less present on RA systems because of the climatic conditions. Among the insects, except for good flyers (Lepidoptera, Hymenoptera and some Diptera), the way they can arrive on a rooftop is to be transported by wind (e.g. a few mites and aphids species) or introduced with tools, soil, seedlings and seeds, and mainly plants directly infested in the commercial nurseries.

The species of insect pests mostly likely present on the rooftop gardens and farms depend on geographical area, climate and type of crops. In Table 1 is reported a list of pests' groups potentially present on rooftops and their natural enemies.

While not pests in the traditional sense, birds (Pigeons, Sparrows, Starlings, Blackbirds, Seabirds, etc.) can eat the vegetables, little fruits and seeds, and damage the irrigation system holing it in search of water (De Grazio 1978; Furness and Monaghan 1987). Roof rats, mice and squirrel are a constant presence in most cities. A rooftop garden can be irresistible to them.

Table 1 Main groups of arthropod pests and their natural enemies that can be used in rooftop agriculture

Pests	Natural enemies	Microorganisms
Aphids	Lacewings, ladybeetles, parasitic wasps (*Aphidius*, *Lysiphlebus* spp.), syrphid fly larvae	Entomopathogenic fungi
Spider mites	Predatory mites, ladybeetles (*Stethorus* spp.), lacewings, minute pirate bugs, Cecidomyiid fly larvae	
Thrips	Minute pirate bugs, predatory mites, predatory thrips, lacewings	
Scales	Parasitic wasps (e.g. *Encarsia*, *Aphytis*, *Coccophagus* spp.), predatory mites, lacewings, ladybeetles	Entomopathogenic fungi
Moths and butterflies (caterpillars)	Parasitic wasps, egg parasitic wasps (*Thricogramma* spp.), lacewings, spiders, predatory bugs, birds	*Bacillus thuringiensis kurtstaki*, entomopathogenic fungi and viruses
Mealybugs	Ladybeetles (*Cryptolaemus montrouzieri*, *Scymnus* spp), lacewings, parasitic wasps (e.g. *Leptomastix dactylopii*)	
Whiteflies	Parasitic wasps (*Encarsia*, *Eretmocerus*, *Cales* spp.), lacewings, ladybeetles, minute pirate bugs, spiders	Entomopathogenic fungi
Weevils, root or soil-dwelling	Parasitic wasps, spiders	Entomopathogenic nematodes (*Steinernema*, *Heterorhabditis* spp.) *Bacillus thuringiensis tenebrionis*
Psyllids	Pirate bugs, lacewings, ladybeetles, parasitic wasps	
Leafhoppers	Parasitic wasp (*Anagrus* spp.), Ladybeetles, minute pirate bugs, lacewing predatory mites	

Beneficials

Generic insect predators (e.g. Carabid beetles and others) can achieve and be present on the green roof (Kadas 2006), on equal terms, however, they are much more numerous in a nearby ground-level garden than in a rooftop garden (MacIvor and Lundholm 2011; Steck et al. 2015). Among generic predators spiders are also important, their abundance and number has been shown to be positively correlated to pest control ecological service (Bennett and Lovell 2014). As noted earlier for phytophagous species, also the number of predators, parasitoids and pollinators that can independently reach the rooftops is generally lower than that present in the same areas on the ground level. However, these species are in general better fliers. Cultivated rooftops, in fact, have a great potential as green corridors "hot spots" for beneficial insects and wild bees pollinators in particular (Orsini et al. 2014). Among the insect natural enemies of agricultural interest, there is presumably a wide presence of hoverflies and lacewings, widely common in urban environments and good fliers. It is not, instead, guaranteed a natural presence of ladybirds (Coccinellidae) because of their not excellent flight propensity.

Ecological Relationship

A key feature of the rooftops is the ecological isolation (island effect), which can be more or less accentuated in relation to the distance from other agricultural sites, parks or natural ecosystems. First of all this causes a lower migration of species from and towards the outside. The biodiversity simplification is increased by the fact that the ecotones – the transition areas between two biomes – are totally absent. An ecotone is, normally, a highly biodiverse zone, because it is populated by species of both biomes, and by many species that find their specific habitat right there. It is an oversimplified ecosystem.

The growth of the population of an organism in absence of limiting factors (situation precisely similar to that of an oversimplified ecosystem) follows an exponential theoretical trend (incidentally, this is why in a new or simplified environment the pest outbreaks are so common, also after a "knock-down" treatment). Nevertheless, these are also the ideal conditions for the application of the predator – prey theory, based on the Lotka – Volterra non parametric equations, used to analyse the population dynamics of predator – prey interactions in biological systems in which only two species interact (Volterra 1926). The dynamic of both populations is characterized by oscillations around a specific equilibrium line. The oscillations are of decreasing amplitude and with a slight phase delay of the predator with respect to the prey. This should lead to a stable equilibrium point and resilient of populations (depending on the initial population, the size of the population of predators introduced and some other variables) below a damage threshold of the pest (Fig. 1).

Despite a number of assumptions about the environment and evolution of the predator and prey populations which do not always occur (i.e.: predators have limitless appetite; during the process, the environment does not change in favour of one species, etc.) the so called paradoxes of enrichment the theory and its evolutions (Berryman 1992) were applied and verified in some simplified systems such as the greenhouse. As examples: in the prey – predator interaction between the spidermite *Tetranychus urticae* and its predator *Phytoseiulus persimilis* (Sabelis et al. 1991; Kozlova et al. 2005) and in several host – parasitoid systems (Mills and Getz 1996) as *Trialeurodes vaporariorum*/*Encarsia formosa* (Rumei 1991) and *Aphis gossipy*/*Lysiphlebus testaceipes* (Bazzocchi and Burgio 2000). Right the oversimplification of the system, if from one side is a risk factor, however, allows a greater possibility of manipulation of the system by humans, favouring a species (the natural enemy) compared to other (pests). In fact, these theories have been leading to significant applications in greenhouses, the environment in which there are the most

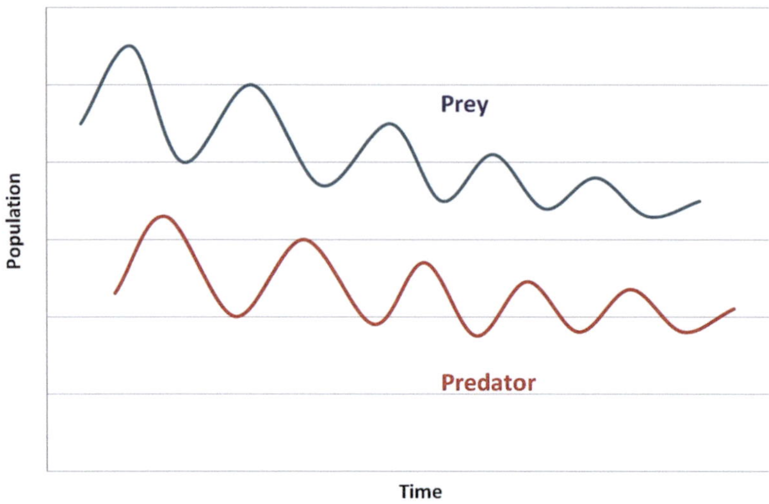

Fig. 1 Simulation of the population dynamics of predator-prey interactions in biological systems in which only two species interact, based on the Lotka–Volterra non parametric equations

biological control applications with arthropods. Most of these techniques can be easily transferred to the rooftops, such as inundative and seasonal inoculative releases of biocontrol agents (Lenteren van and Woets 1988) and, in the case of cultivation systems without production breaks, inoculative and augmentative biological control strategies. Useful from this point of view also the metapopulations ecological theory which studies extinctions and, conversely, the possibility of eradication of organisms in discrete habitat patches (fragmentation of the habitats) (Hanski 1998). As reported by Ives and Settle (1997), one of the application is the synchronous (in absence of predators) or asynchronous (in presence of predators) planting.

Final Considerations

The rooftop ecosystem is similar to the greenhouse one from the point of view of the biological communities and ecological relationships, but it is different with regard to the climatic conditions, which makes it more like arid or semi-arid agroecosystems, and geographical isolation that makes it similar to an island and a habitat fragmentation situation. Many ecological pest control applications are possible taking them from those developed in these situations and in particular from greenhouses. Rooftop agriculture is also a great opportunity for pest control ecologist and entomologist to develop and set up new agro ecological strategies in real environments.

Integrated Ecological Pest Management

Integrated Pest Management is an ecosystem-based strategy that focuses on long-term prevention of pests or their damage and emphasizes techniques such as mechanical and physical control, biological control, habitat manipulation, modification of cultural practices and use of resistant varieties. In this paragraph the main ecological pest control and management practices potentially pertinent to RA will be described, following the IPM principles, with exclusion of synthetic pesticides and less strictly link to economic thresholds.

Pest Exclusion, Sanitation, and Quarantining

Only a small percentage of the potential pests are able to reach the rooftop autonomously. Most of the pests (insects or pathogens) of the rooftop gardens and farms are transferred by operators, tools, contaminated plants. This also because "rooftop growers" are often not professional operators and little accustomed to good agricultural practices (GAP). In addition, these locations are often frequented by many people not involved in the garden management and sometimes by pets and other animals.

The physical, topological and ecological features described above, however, make that one of the strengths of "growing on the roofs" is precisely the ability to keep out potential pests with some simple preventive activities that can considerably reduce costs and product losses.

Farm Setup and Designing

In accordance with what recommended for greenhouses (Badgery-Parker 2015), it is fairly simple, compared to a open cultivated field and even a greenhouse, designing the rooftop vegetable garden/farm in order to provide for three sections: an outside zone, a clean buffer zone and a cultivation area. The first is the area thought to intercept and minimise pest and disease threats. Here people stop, for a short period of time, before entering the farm, and brush clothes, leave accessory bags, jackets, objects potentially carriers of nocive organisms, and, in some particular cases, wear specific clothes for growing. Crop debris and waste should be posed in this area (also hydroponic substrates can be a source of pests and diseases) and then disposed. Bins closed are one of the simplest, cheapest and very effective ways of reducing pest and disease problems. Vehicles, tools and all equipment need to be kept free from soil and residues before enter in the buffer and cultivation zones, and occasionally be taken out and cleaned thoroughly. It is possible to plan a "quarantine room" where to place plants and seeds from outside (also those purchased) for a period of time of at least several days, and put in place specific prevention practices (check of the phytopathological state, heat treatments for seeds and seedlings, etc.). The use of a double-door entry between outside and clean zone, or, alternatively, two plastic ‘curtains’ which overlap, is recommended. The second section is an area, all around the cultivation zone (the area where there are pots, containers, hydroponic systems, etc. and where plants are reared), that needs to be kept clean and that acts as a buffer zone (Badgery-Parker 2015). These simple arrangements of the rooftop garden, almost completely reduces the opportunity for pests to reach the crop.

A careful rooftop garden/farm design is also necessary for the plants wellness, that makes them more vigorous and resistant to pests and pathogens (see also the Ecocrop FAO database http://ecocrop.fao.org/ecocrop/srv/en/home). For many crops, the exposure to an excessive number of direct sunlight hours is not the ideal condition to grow, so that shadow areas are created through screens. Movable or easily transportable cultivation containers may be useful. Also, vegetated or artificial shelters should be set up in order to mitigate violent weather events such as rainstorm, hail, gusts that can produce abiotic convergent stresses to the plants and more vulnerability to insect and pathogen attacks (Mittler 2006). Furthermore, from a purely “pest exclusion” standpoint, putting windbreaks along the rooftop borders (e.g. hedgerow of appropriate evergreen plants) will drastically reduce the levels of pests and pathogens carried by the wind into the production area. Climbing plants, such as Algerian or English ivy, star jasmine, and honeysuckle, provide shelter and food for rats and should be avoided (Timm and Marsh 1997).

Preventing Pest Problems

Sanitation

Where the cultural cycles are not continuous (areas with cold winters or other periods of vegetative stasis) between each production cycle and, in any case, once in a while, the rooftop should be completely cleaned and disinfected. This practice, called sanitation – removal of old crop and items that will not be reused including substrate, bags, twine; cleaning and disinfecting of all equipment, tools, plant containers, bins, clips, plant hangers; sweeping down walls, floors and all internal structures – it is also a good control measure in case of persistent diseases or infestations (Hogendorp and Cloyd 2006; Cloyd 2016).

Seeds and Seedlings

Seedlings are a very common source of pests and diseases. In order to prevent pest problems in the cropping system, inspections of seedlings before they are moved inside are mandatory. For this purpose a well isolated quarantine chamber is very useful. Accurate visual inspections should be carried out to check general plant health and identify signs of diseases and pests. Just a few days of permanence of the plants and controls are sufficient to significantly decrease contamination risk (Baudoin et al. 2013). Some plant pathogens are able to penetrate and survive within the seed, out of reach of surface seed treatments. They include many bacterial pathogens of vegetables as well as fungi, oomycetes, and viruses. Hot water bath at temperature that can vary from 40 to 55 °C, depending on the crop, and treatment period of 10–60 min are effective non-chemical methods to control seed-borne diseases (*Alternaria* spp., *Phoma* spp., *Septoria* spp., *Peronospora valerianellae*, *Xanthomonas* spp.) (Nega et al. 2003; Zitter 2013) on several important vegetable crops such as eggplant, pepper, tomato, cucumber, carrot, spinach, lettuce, celery, cabbage, turnip, radish, and other crucifers. It has also been reported that heat treatments on seedlings of some species inactivate plant viruses (Posnette and Cropley 1958; Grondeau et al. 1994).

Soil, Potting Soils and Substrates

Inert substrates usually utilized in hydroponic systems including perlite, expanded clay and purchased cocopeat are usually stable and sterilized products. All other substrates used in these growing systems, in particular if self-produced (rice hulls, etc.), need to be subjected to sterilization techniques, based for example on high steam temperatures (Trevors 1996).

Looking from a pest exclusion and prevention approach, also in case of cultivations on soils resting on the floor or in containers, strategies to decrease the risk of soil borne diseases, weeds and pests are recommended. This also in case of a following soil biological management with beneficial microorganisms. The most common nonchemical technique is the soil solarization, a method that uses transparent plastic sheets to mulch the soil, exploiting the solar energy for heating the first layers of soil. Solarization effectively controls a wide range of soil-borne pathogens, insects and weeds and is widely used in greenhouses and also in open field (Horowitz et al. 1983; Stapleton and DeVay 1986; Gamliel and Katan 2012). What is interesting is that the technique has been shown to be particularly effective for disinfesting small amounts of moist, containerized soil and soil in cold frames (Stapleton 2008) and it is therefore particularly suitable and easy to apply on RA contexts (Fig. 2).

Metalized plastic mulches seems to be effective to protect in particular young plants from several insects, particularly in reducing the attacks of whiteflies and thrips (Hochmuth and Sprenkel 2008). The method, indeed, has been reported as effective for tomato spotted wilt virus (TSWV) control (Paret et al. 2013).

Diagnostic and Monitoring Pests

Early detection of pests, their identification and monitoring is crucial in ecological plant protection programs. The main concept of Integrated Pest Management, indeed, is that no action should be taken against a pest unless it poses a threat to the crop.

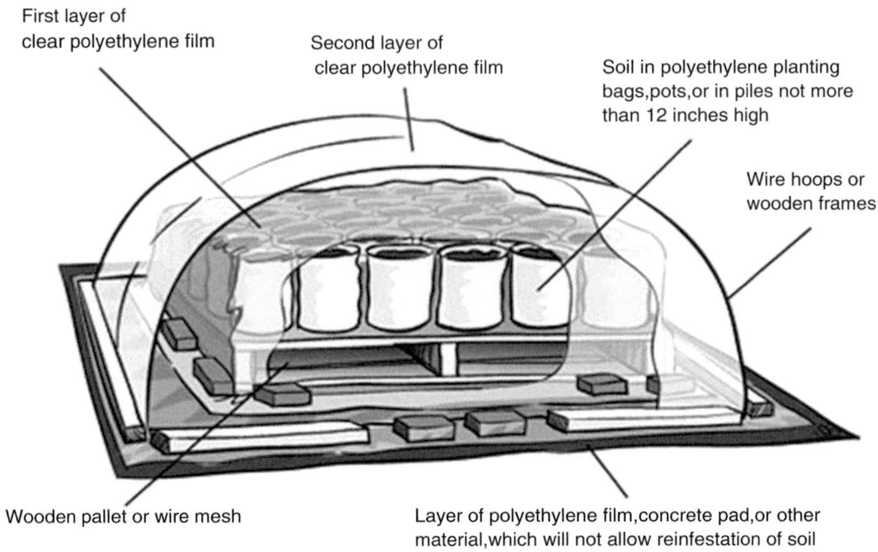

Fig. 2 Details of double-tent technique for solarizing containerized soil (From: Stapleton 2008)

Diagnostic

In recent years the diagnosis of both plant diseases and phytophagous insect has made great strides also linked to the proliferation of very comprehensive web-based knowledge exchange tools. A team of researchers at Penn State University (Pennsylvania, USA) and Swiss Federal Institute of Technology (EPFL), has recently developed "PlantVillage Images" (www.plantvillage.org), an open access collaborative and downloadable database of 50,000 images of healthy and diseased crops (Hughes and Salathé 2015). Automated systems for the recognition of insect pests (Xia et al. 2015) and plant diseases (Khirade and Patil 2015) have also been developed. Many other tools are in development and though these technologies (visual recognition) are just beginning and still not always very effective (Barbedo 2016), they will be in a short time certainly very functional for non professionals farmers.

Monitoring and Action Thresholds

A regular visual plant inspection calendar is needed to assess general plant health and to detect signs of diseases and pests at early stages. Special attention should be given to key plants (plants or cultivars with serious, persistent problems every year). In many cases "indicator plants" are a practical way to detect the presence of pests. For example, faba beans (*Vicia faba* L.) and certain petunia cultivars can detect the presence of thrips, vectors of TSWV (Tomato Spotted Wilt Virus) (Hanafi and Papasolomontos 1999).

For little flying insects (thrips, whiteflies, fungus gnats, leafminers and winged aphids), yellow and blue sticky cards of various dimensions are recommended to monitor the population. For several insects there are also available pheromone traps: they are very sensitive and can capture pest insects present in densities too low to detect using other inspection methods. The action threshold is the level of pests or disease which requires the implementation of an active treatment strategy and it is related to the number of a certain pest per plant or trap. The action (damage or economical) threshold is specific for each insect pest and plant and it is related to a number of conditions and specific considerations (e.g. if the crop production is for market or self consumption).

It is available a large mass of data on specific control and management programs based on IPM principles (monitoring methods, traps, samplings, action thresholds, etc) for different crops and pests in all the world. These data have been made available to all farmers by local, national or international authorities.

The main official resources on the web are the following. For USA: http://www.ipmcenters.org; for Europe: http://ec.europa.eu/food/plant/pesticides/sustainable_use_pesticides/ipm/index_en.htm; for Australia: http://www.naa.gov.au/records-management/agency/secure-and-store/business-continuity-planning/pest-management.aspx, for the rest of the world: http://www.fao.org/agriculture/crops/thematic-sitemap/theme/pests/ipm/en/.

There are no specific data for crops on rooftops. Given some similarities to the greenhouse, it can be applied IPM programs developed for this environment (Lenteren van et al. 1992; Baudoin et al. 2013; Badgery-Parker 2015). However, given the peculiarities of the "rooftop system", it is still necessary to study specific ecological pest management programs, sampling methods and action thresholds specific for this environment.

Pest Control

A plethora of biological, cultural and physical/mechanical means can be applied to develop integrated management of pests and diseases programs in RA without conventional chemical pesticides. Many of the following methods are, in fact, pest prevention methods, but they are usually included in IPM strategies.

Rooftop "Clean Out"

In agreement with what happens for greenhouses, in the rooftop gardens can be used the "clean out" process, not only as a generic prevention good practice (see "sanitation" above), but as a pest control strategy in strict sense. The garden clean out needs to be done as soon as the crop is finished: the old crop, the growing substrate and all plant material are completely removed from the greenhouse (Badgery-Parker 2015). The clean out activity, in fact, creates an immediate break between crops which re-establishes the "zero point" in the garden and drastically reduces the carry-over of pests and diseases (Maini and Nicoli 1990).

Weed Management

In small and soilless systems management of weeds is certainly easier than in open field, but equally important. In these conditions, such as in little vegetable gardens, hand weeding is the simplest and common way. As described before, solarisation, in particular the double-tent technique for containerized soil (Stapleton 2008), is an effective practice to avoiding the emergence of unwanted plants (Horowitz et al. 1983). In open farm conditions crop rotation and green manure cover crops help to control weeds (Teasdale 1996). However, in productive small systems, where growers work to maximize output by using intensive cropping practices, it is not always possible to design cover crops in rotation. In-season living mulches may be an effective strategy to provide the benefits of cover crops with less land commitment (Pfeiffer et al. 2015). In productive RA mulching can play a crucial role in weed management. It can be made with biodegradable black films, but also with plant materials such as different kinds of straw (Kosterna 2014). This practice provides other environmental benefits (good water management, improvement of soil biotic communities, etc.) and constitutes the basis of conservative agriculture and permaculture. Mulching with dry vegetal materials probably is the most cost-effective method of weeds prevention.

Plant Diseases Management

Because the climatic and ecological conditions, in RA plant diseases should be a less important issue than in other environments. Nevertheless careful diseases control is needed to prevent dangerous outbreaks of pathogens. The great experience possessed for greenhouses plant disease management may be exploited.

General good agricultural practice (GAP) for bacterial diseases prevention included: use certified seed from a reputable source, if you save your own seed or buy from a small producer, hot-water treatment may well be worthwhile (see preventing pest problems paragraph of this section); ensure that the seedlings grower follows a disease management programme; use sterilized potting mix; regularly check plant health and immediately remove infected and adjacent plants; minimize the period of time during which leaves are wet, carrying out irrigation early in the day and maintaining good ventilation; disinfect all benches, equipment, tools, etc. A similar set of practices should be adopted for virus diseases prevention and in addition: use metalized mulches and essential oils sprays to deter insect vectors (whiteflies, aphids, thrips, gnats flies); use organic insecticides with a rapid knock down effect (such as pyrethrins) to decrease populations of insect that can vector plant pathogens (Baudoin et al. 2013).

Plant Resistance In principle, the use of resistant cultivars has several advantages over other methods for small scale farming: cultivars resistant to the most of crop diseases exist, require little or no technology, are cost effective, etc. Many applications of this technique are, in fact, used to prevent both bacterial and virus outbreak (Nono-Womdim et al. 1991; Strobel and Kucf 1999; Nono-Womdim 2003; Russell 2013). However, complete and durable resistance is difficult to achieve, because new strains of the pathogen evolve, or because the pathogen population is a mixture of many different strains (Baudoin et al. 2013). Grafting vegetable crops is a world-wide traditional and common practice to provide resistance to soil-borne diseases. With this practice, tolerant rootstocks can bring scion of susceptible cultivars allowing productions that would be otherwise unsuccessful. The major disease problems addressed by grafting include *Fusarium* wilt, bacterial wilt, *Verticillium* wilt, *Monosporascus* root rot, and nematodes. Grafting has also been shown in some instances to increase tolerance to foliar fungal diseases, viruses, and insects (King et al. 2008). Grafting is a not easy practice to achieve and the high cost is probably the main reason of the slow adopting of this technology. Resistance, anyway, remains one of the most effective way of combating soil-borne plant diseases in IPM programs.

Agronomic Measures, Elicitors and Plant Strengtheners Although cultivated plants have partially lost during their domestication and cultivar selection history their natural defences, stimulation and activation of specific physiological pathways may improve their resistance to pathogens and insects.

General agronomic good practices such as a correct management of natural resources, soil fertility and water are essential for the plant wellness and reinforcement (Orsini et al. 2013) also because they have profound effects on soil communities and ecology. The tactics to be emphasized are those that can be actuated to reduce the likelihood of diseases problems developing. Plant density is a critical factor, poor air circulation, indeed, can encourage diseases such as *Botrytis*, *Alternaria* and Downy mildew. Planting and harvesting date adjustments can be effective pest management strategies because they lead to asynchrony of the biological cycle of pests and phenology of the plants (Anderson et al. 1996).

Plant strengtheners is a generic term for several commercially available compounds that "boost" their vigour, resilience and performance. In particular, recent studies have brought evidences that some fungal, bacterial and abiotic molecules, named elicitors, stimulate plant immunity triggering different mechanisms of resistance. Among the tested chemical elicitors that exhibit various inhibitory efficiency against bacteria, fungi, and viruses, there are: salicylic acid, methyl salicylate, benzothiadiazole, benzoic acid, chitosan (Thakur and Sohal 2013; Xing et al. 2015).

Soil Beneficial Microorganisms (Induced Systemic Resistance) As just underlined, plants are an essential and interactive component of biological control practices. Soil microorganisms include those that create symbiotic associations with plant roots (*rhizobia*, mycorrhizal fungi, *actinomycetes*, diazotrophic bacteria). Despite that they are generally evaluated based on their direct growth effects on plants, some of these organisms act as real biocontrol agents or directly acting as antagonists of soil-borne parasites and diseases (through mechanisms of parasitism, antibiosis or competition for exploitation of ecological niches) or eliciting chemical plant defences. On cucumber, for example, arbuscular mycorrhizal fungi (AMF), increase the primary defensive chemical cucurbitacin C (Barber et al. 2013). AMF are present in many crop species and enhance protection against many pathogens. Soil inoculums potential of AMF depend also by the cropping sequence: Thompson (1991) found that pre-cropping with legumes or sunflowers generated the highest results. Various non pathogenic (saprophytic) strains of fungi (*Rhizoctonia*, *Fusarium*, *Trichoderma* spp.) have been used to reduce damage caused by pathogenic fungi (e.g. *Pythium*, *Sclerotium*, *Verticillium*) (Cook 1993; Alabouvette et al. 2009). Other beneficial rhizosphere organisms have been used, mostly as seed inoculants. Generically entitled plant growth promoting bacteria (PGPB) these organisms affect plant growth through direct growth promotion (hormonal effects), and induced systemic resistance (ISR).

The use of natural inorganic products, such as cupper, sulphur, or potassium bicarbonate for plant disease control is well known and traditional. On rooftop agriculture and in particular in urban environment the use of cupper and sulphur should be limited and rational because of the risk of groundwater pollution.

Substances and products allowed in organic farming in Europe (Regulation EC No. 889/2008 (Annex II)), divided according to the different observable symptoms, are reported in Table 2 (Tonti 2013).

Insect Pest Management

Physical Control In small agricultural plots, pest exclusion nets can be effective to protected cultivation from pest insects (and birds as well). Eco-friendly nets (EFNs) have been tested in tropical and subtropical African countries, and proved effective against many pests on cabbage (Martin et al. 2006), tomato (Gogo et al. 2014) and other crops. These kinds of mechanical means are effective against several insects, including aphids, beetles (*Acalymma*, *Diabrotica*, *Leptinotarsa*, spp.), whiteflies, and their related pathogens (Licciardi et al. 2007). This method has its strength point in its intuitiveness and use easiness. It is also environmental friendly and generally well accepted by growers (Vidogbéna et al. 2015). However, there are no evidences about the relationship between the net typology (construction parameters of the net) and specific application, so that its use is generic and not always useful. The specific effect of nets on crops has to be further investigated (Castellano et al. 2008), particularly in the RA microclimatic conditions. The use of insect-proof eco-nets can be certainly recommended as covering of specific structures such as plant nurseries.

Mass trapping with pheromones or other semiochemicals traps is not a widely used methodology, because some technical and economical limits. What makes interesting this type of control for RA is that mass trapping has been demonstrated to have good potential to suppress or eradicate low-density, isolated pest populations (El-Sayed et al. 2006), which it is a highly likely condition in RA. Also the use

Table 2 Substances and products allowed in organic farming in Europe for plant diseases control [Regulation EC No. 889/2008 (Annex II)]

Symptom	Pathogen/pathogens	Product/microorganism	Use
Leaf diseases	Mildew	*Ampelomyces quisqualis* (fungus spores and mycelium)	Protectant-curative
		Potassium bicarbonate	Protectant-curative
		Spraying sulfur; wettable sulphur	Protectant-curative
	Albugo, *Alternaria* (B1, D1), *Bremia*, *Cercospora* (A1), *Cytospora*, *Colletotrichum* (E1), *Coryneum*, *Cycloconium*, *Cylindrosporium*, *Deuterophoma*, *Diplocarpon*, black mould, *Gibberella*, *Gloeosporium* (D4), *Marssonina*, *Monilia*, *Mycosphaerella*, *Nectria*, Mildew, *Penicillium*, *Peronospora*, *Phoma*, *Phyllosticta*, *Phragmidium*, *Phytophthora*, *Phomopsis*, *Plasmopara*, *Puccinia*, *Septoria*, *Sphaeropsis*, *Taphrina*, *Tilletia*, *Uromyces*, *Venturia*.	*Bacillus subtilis* (spores).	Protectant
		Trichoderma harzianum (conidia and mycelium) T-39 and T-22	Protectant
		Copper (copper hydroxide, copper oxychloride, copper sulphate, dicopper oxide, copper)	Protectant
Roots and root collars dieseases	*Fusarium* spp.(A3, B3, C3, D3, F3), *Pythium* spp., *Verticillium dahliae* (E3), *Pyrenocheta lycopersici*, *Phytophtora capsici*.	*Streptomyces griseoviridis* (bacterial cells)	Protectant
	Rhizoctonia spp., *Sclerotinia* spp., *Sclerotium rolfsii*, *Verticillium* spp. (E3), *Thielaviopsis basicola*, *Pythium* spp., *Phytophthora capsici*.	*Trichoderma asperellum* strain ICC012 (conidia)	Protectant
	Pythium spp., *Phytophthora capsici*, *Rhizoctonia solani* and *Verticillum* spp.	*Trichoderma asperellum* strain TV1 (conidia)	Protectant
	Rhizoctonia spp., *Sclerotinia* spp., *Sclerotium rolfsii*, *Verticillium* spp. (E3), *Thielaviopsis basicola*, *Pythium* spp., *Phytophthora capsici*.	*Trichoderma gamsii* strain ICC080 (conidia and mycelium) (ex *T. viride*)	Protectant
	Pythium spp., *Rhizoctonia* spp., *Fusarium* spp., (A3, B3, C3, D3, F3) *Cylindrocladium* spp., *Thielaviopsis* spp., *Myrothecium* spp. and *Armillaria mellea*.	*Trichoderma harzianum* (conidia and mycelium) T-39 and T-22	Protectant
Fruit deseases	*Slerotinia* spp. (C4)	*Coniothyrium minitans*	Protectant
	Botrytis (B4)	*Bacillus subtilis* (spores).	Protectant

Modified from Tonti (2013)

of traps commonly used for monitoring specific pests (e.g. blue sticky cards for thrips, and light and pheromone traps for *Tuta absoluta* on tomato), in small and isolated plots, probably contribute to maintain pests populations below the damage threshold and increase grower returns (Cocco et al. 2012; Sampson and Kirk 2013; Braham and Nefzaouil 2016).

Biological Control with Beneficial Insects Biocontrol with beneficial insects is a huge field of research and has many applications in organic and IPM agriculture. Over 200 species of beneficial organisms are commercially available for control of all important insect and mite pests (Lenteren van 2012). A great number of biological control programmes using insect predators and parasitoids have been developed in particular for protected crops and greenhouses. Given the similarities, this is a considerable mass of data to be drawn on in order to develop specific projects for rooftop agriculture.

In Table 1 are reported the main pests and their beneficial arthropods that can be used in RA.

The methods of release of biological agents in little systems are the following:

- *Inundative biological control*, which consist in periodically releases in large numbers in order to obtain immediate control of pests for one or two generations (the effect is similar to that of an insecticide).
- *Seasonal inoculative biological control*: natural enemies are periodically released in short-term crops (3–10 months), and the control effects are expected to last several generations. Usually a large number of natural enemies are released for an immediate control effect, plus a build-up of the natural enemy population for control later in the season.

In the second case, an almost stable relationship between the populations of prey and predator (or parasitoid and host) is generally established, as expected by the Lotka–Volterra mathematical model (Volterra 1926) and the population of the pest is expected to remain consistently below the damage threshold. Timing is the most important part of release programmes of beneficial agents. Except for generalist predators (e.g., some mite predators and preying mantis), which can survive also eating pollen or preying generic insect eggs and larvae, preys or hosts must be present on the crop before the release. On the other hand, the pest number should not be too large to achieve effective control. If pest levels are high, can be used least-toxic, short-lived natural pesticides or insecticidal soaps to break down the populations and then release beneficials to maintain control. In any case, early pest detection, identification and monitoring is an essential condition to conduct biological control strategies. Information and guidelines for specific integrated pest control programmes with beneficial insects are provided by the biofactories (commercial insectaries) whose mission is not only to sell beneficials but also provide the necessary assistance to farmers.

On the basis of experiences conducted in the last decades in greenhouses, methods to optimize the beneficial releases have been developed. "Pest in first" consist in deliberate releases of the pest in specific point of the crop because a number of hosts will be available to allow the population of natural enemies to grow so that they're in position of strength when the "natural" pest populations show up (Markkula and Tiittanen 1976; Lenteren van and Woets 1988). "Banker plants" strategy can be considered an "upgrade" of the pest in first approach. It substantially consists in a mini-rearing system that provides alternative food source and a reproduction site for a specific natural enemy (Osborne et al. 2005; Frank 2010). It is essential that the phytophagous of the banker plant is a prey of the beneficial but different by the pest of the crop that is being protected. As an example: oat or wheat grass is commonly used as a banker plant to grow aphid parasitoids for the defence of horticultural crops. The oat aphid (*Rhopalosiphon padi*), in effect, is actually a host of parasitoids (e.g. *Aphidius colemani*) of green peach and black melon aphids, but will not move to horticultural crops. Other very recent applications of beneficial insects are developing. "Predatory in first" (Kumar et al. 2015) is the use of generalist predators (such as *Amblyseius swirskii* against thrips) providing them pollen on top of the crop as alternative food until the population of their prey increases. "Living bomb" approach is, instead, the release of live insects that were pre-infected with entomopathogenic nematodes and could carry nematode infective juveniles to control insect pests living in hard-to-reach cryptic habitats (Gumus et al. 2015).

As a whole, biological control with arthropod beneficials is certainly the most "ecology based" and smart strategy of pest management in small scale agricultural sites. However, there are some important limitations to consider: it requires good production and management skills by the growers; beneficial insects are not commercialized worldwide and rare to find in developing countries; natural enemies commercial packs (number of insects, release methodology, etc) are not thought for small scale systems and not always the benefit/cost ratio is positive because of the high costs.

Functional Biodiversity for Pest Management The banker plants idea originate from the more general ecological concept of conservation biological control, that is the ways in which non crop resources and biodiversity can help natural enemies. Without entering in the huge debate regarding this topic (see Gurr et al. 2012 for a recent and exhaustive analysis), it is possible to detect some applications useful also for little and isolated agroecosystems. Shelter, nectar, alternative preys and pollen (SNAP) sources are the key factors to attract and allow the permanence of natural enemies in the system (Landis et al. 2000; Wäckers et al. 2007). In small sites, the central idea is not to make the system more complex and biodiverse in the general sense, but to use plant species and tricks that have a specific functional role in order to enhance the control of pest species (Messelink et al. 2014). Selecting specific flower plants (natural enemies are selective in their flower feeding) it is, for example, possible to attract aphidophagous hoverflies (Diptera: Syrphidae) (Colley and Luna 2000), that are good flyers and quite common also in urban environment. (i.e. alder, *Phacelia*, coriander, shy buckwheat: *Fagopyrum esculentum* and mustard: *Sinapis arvensis*) (Bazzocchi 2013). Recently, Tavares et al. (2015) showed that insectary beneficial plants can be used also in soilless systems. Furthermore, there are evidences that also a limited number of selected SNAP-plants can guarantee the constant permanence of aphid predators (Coccinellidae) in a urban garden (Bazzocchi et al. in press). The use of attractive semiochemicals and chemical and biotic elicitors and plant strengtheners enhance the attractiveness of cultivated plants to biological control agents (Sobhy et al. 2014).

Bioinseticides and Natural Insecticides Compounds and substances derived or extracted by natural materials as animals, plants, bacteria, and certain minerals can be used as bioinsecticides or natural insecticides.

Among the biological pesticides based on pathogenic microorganisms (baculovirus, fungi and bacteria) (*microbiological insecticides*) the most widely used is *Bacillus thuringiensis* (BT). It is a quite common soil bacterium producing toxins lethal to specific insects and completely harmless for humans and other animals, including invertebrates. The advantage of the BT preparations, indeed, is that they are very specific. Depending on the subspecies used, they may only attack caterpillars (*Bacillus thuringiensis kurstaki*: BTK) or beetles (*B.t. tenebrionis*: BTT) or mosquito larvae (*B.t. israeliensis*: BTI and *B. sphaericus*). Extensive is also the use of the entomopathogenic fungus, such as the *Beauveria bassiana*, that grows naturally in soils throughout the world. Products based on this fungus are very effective on several insect pests (thrips, whiteflies, aphids and different beetles.) and absolutely harmless to humans and the environment. In this case, however, the fungus attacks most of the insects, therein resulting in potential mortality also of beneficial insects, although their greater mobility compared to phytophagous insects, makes them less susceptible. Anyway, this eventuality shall be kept in mind in all cases where products, including organic ones, are not highly specific against pest insects: in order to prevent harmful effects on beneficials they should be used with special care and never while plants are in the flowering period (Bazzocchi 2013).

Recent studies are increasingly exploring the wider properties of microorganisms. New opportunities for their use in biological and integrated pest control strategies also integrating beneficial insects releases, are suggested (Gonzalez et al. 2016). Interestingly, several fungal entomopathogens can also colonize plant tissues as endophytes and affect pests systemically via the plant (Vega et al. 2009).

Among *botanical insecticides*, the extracts from *Azadirachta indica* (Neem), a tropical evergreen tree common in Asia, South America and Africa, occupy a prominent place. Their bioactivity is well known from long time and it has been amply demonstrated by a large amount of scientific researches including the factors affecting their bioefficacy (Gahukar 2014). Nowadays these products become very popular because of their biodegradability, low persistence, low toxicity to non-target organisms and easy availability. The extracts may be used both on the soil, being absorbed by the plant roots, or directly sprayed on the leaves. The use through the roots should be preferred because the buffer effect of the soil (in high doses Neem oil may have phytotoxic effects damaging plant tissues) and since the related benefits will last longer and possible side effects on beneficial insects will be avoided.

Pyrethrum, a powder extracted from *Tanacetum cinerariifolium* flowers which active compounds are named pirethrins, is probably the most widely used plant derived insecticide all over the world. It has a very powerful "knock-down" effect on the pest populations and it is also active on flying insects, but the active compounds degrade very rapidly with the sun irradiation. Pyrethrum, which exist in a many different formulations, can be used to obtain an immediate, but not durable, effect on the populations of pest. The possible side effects on beneficial insects and other arthropods should be carefully valued, spraying it only during nocturnal hours and avoiding its use during plant flowering period.

Apart from a little number of well known products, stringent regulatory requirements have prevented many other plant derived compounds from reaching the marketplace in North America and Europe. However, the awareness of economic and ecological benefits of natural insecticides is rapidly leading farmers, mainly in developing and the less-developed countries, to switch to botanical insecticides. Scientific progresses are being made in China thanks to an organized effort in using indigenous pesticidal plants and the rich tradition and knowledge of herbal medicine (Yang and Tang 1988). Similar studies, based on local traditional knowledge, are under way in India (Lal and Verma 2006), in Africa (Grzywacz et al. 2014) in Central America (Bentley 1992) and in many other part of the world (Stoll 1996), since long time. These techniques, which derive from local, traditional and scientific sources, often have to be verified, adapted or improved, but suggest that in countries where strict enforcement of pesticide regulations is impractical, and human pesticide poisonings are most prevalent, and in general, as in the most of RA situations, where human and animal health are the main asset to be preserved, botanical insecticides are a an important resource to be drawn.

Conclusions

No consolidated data are available about sustainable pest management in rooftop agriculture. New pest control strategies need to be developed starting from the ecological analysis of the "rooftop system". An ecology-based approach and in particular the ecological similarity with greenhouses seems forerunner of important applications.

Under the rooftop conditions is not recommended, and often neither permitted, the use of synthetic and broad spectrum pesticides.

Physical pest exclusion and preventing methods are fundamental for sustainable pest management in RA. Biological control with beneficial insects, in particular through specific techniques for small isolated systems such as "banker plants", is the most ecological and smart approach, but with some significant limitations in particular related to natural enemies cost and availability. The use of microbiological and botanical insecticides can be profitably integrated to the other biocontrol methods and, also used alone, probably constitute, together with physical control (heat treatments and mass trapping), the more exploitable methods both from an economic and availability standpoint. Finally, functional plant biodiversity for pest suppression is a new and promising approach in small scale and urban contexts, that should be better investigated.

Bullet Points

- The "rooftop ecosystem" is similar to the greenhouse context from the biological communities and ecological relationships standpoint, and to an island from a population dynamic perspective (habitat fragmentation). Climatic conditions are similar to a semiarid ecosystem. Any pest management strategy should start from these ecological considerations.
- One of the strengths of rooftop agriculture is the possibility of excluding potential pest from the growing area simply with preventive activities and so considerably reduce costs and product losses.
- Biological control with beneficial insects, heat treatments and other physical methods, the use of microbiological and botanical insecticides and use of functional plant biodiversity, can all contribute to integrated ecological strategy of pest control in rooftop agriculture.

References

Ackerson NO, Awuah E (2010) Urban agriculture practices and health problems among farmers operating on a university campus in Kumasi, Ghana. Field actions science reports. J Field Actions (Special Issue 1)

Alabouvette C, Olivain C, Migheli Q, Steinberg C (2009) Microbiological control of soil-borne phytopathogenic fungi with special emphasis on wilt-inducing *Fusarium oxysporum*. New Phytol 184(3):529–544

Alarcon WA, Calvert GM, Blondell JM, Mehler LN, Sievert J, Propeck M, Tibbetts DS, Becker A, Lackovic M, Soileau SB, Das R, Beckman J, Male DP, Thomsen CL, Stanbury M (2005) Acute illnesses associated with pesticide exposure at schools. JAMA 294(4):455–465

Altieri MA (1987) Agroecology: the scientific basis of alternative agriculture. Westview Press, Boulder

Anderson RL, McMullen MP, Peairs FB (1996) Integrated pest management of insects, plant pathogens, and weeds in dryland cropping systems of the Great Plains. J Prod Agric 9(2):200–208

Audenaert JA, Vangansbeke D, Verhoeven R, De Clercq P, Tirry L, Gobin B (2014). Variability in greenhouse temperature and humidity affects predation efficiency of *Phytoseiulus persimilis* and *Neoseiulus californicus* on *Tetranychus urticae*. IOBCWPRS Bull: submitted

Badgery-Parker J (2015) Keep it CLEAN: reducing costs and losses in the management of pests and diseases in the greenhouse. NSW Department of Primary Industries, Gosford

Barbedo JGA (2016) A review on the main challenges in automatic plant disease identification based on visible range images. Biosyst Eng 144:52–60

Barber NA, Kiers ET, Theis N, Hazzard RV, Adler LS (2013) Linking agricultural practices, mycorrhizal fungi, and traits mediating plant–insect interactions. Ecol Appl 23(7):1519–1530

Baudoin W, Nono-Womdim R, Lutaladio N, Hodder A, Castilla N, Leonardi C, De Pascale S, Qaryouti M, Duffy R (2013) Good agricultural practices for greenhouse vegetable crops: principles for Mediterranean climate areas. FAO plant production and protection paper (FAO)

Bazzocchi G (2013) Organic pest control in the vegetable garden. In: Orsini et al (eds) Sustainable urban garden management. Horticity Bologna: 54–64. Hortis LLP EU program: http://www.hortis-europe.net/files/documenti/inglese/final-e-books/web-e-book-2-engl-last.pdf. Last Accessed 24 Oct 2016

Bazzocchi G, Burgio G (2000) Functional response of *Lisiphlebus testaceipes* (Cresson) (Hymenoptera: Braconidae) against *Aphis gossypii* Glover (Homoptera: Aphididae) at two constant temperatures. Boll Ist Ent G Grandi Univ Bologna 54:13–21

Bazzocchi G, Pennisi G, Frabetti A, Orsini F, Gianquinto G (in press) Abundance, migration and distribution of *Coccinella septempunctata* (Coleoptera: Coccinellidae) in a highly biodiverse urban garden. Acta Horticolturae

Bennett AB, Lovell ST (2014) A comparison of arthropod abundance and arthropod mediated predation services in urban green spaces. Insect Conserv Divers 7(5):405–412

Bentley JW (1992) Alternatives to pesticides in central America: applied studies of local knowledge. Cult Agric 12(44):10–13

Berryman AA (1992) The origins and evolution of predator-prey theory. Ecology 73(5):1530–1535

Braham M, Nefzaouil F (2016) Study of the effectiveness of two methods: mass trapping and leaf and frits removal for the control of the tomato leafminer, *Tuta absoluta* Meyrick (Lepidoptera, Gelechiidae) in Tunisia. J Glob Agric Ecol 4(3):114–122

Burns CJ, McIntosh LJ, Mink PJ, Jurek AM, Li AA (2013) Pesticide exposure and neurodevelopmental outcomes: review of the epidemiologic and animal studies. J Toxicol Environ Health Part B 16(3–4):127–283

Castellano S, Mugnozza GS, Russo G, Briassoulis D, Mistriotis A, Hemming S, Waaijenberg D (2008) Plastic nets in agriculture: a general review of types and applications. Appl Eng Agric 24(6):799–808

CDPH (2014) Agricultural pesticide use near public schools in California, April 2014. California Department of Public Health, California Environmental Health Tracking Program

Celli G, Radeghieri P, Bazzocchi G (eds) (2001) Agroecology: toward a new agriculture for Europe. Brussels, GREEN/EFA

Cloyd RA (2016) Greenhouse pest management, Contemporary topics in entomology series. CRC Press, Boca Raton

Cocco A, Deliperi S, Delrio G (2012) Potential of mass trapping for *Tuta absoluta* management in greenhouse tomato crops using light and pheromone traps. IOBC-WPRS Bulletin 80:319–324

Colley MR, Luna JM (2000) Relative attractiveness of potential beneficial insectary plants to aphidophagous hoverflies (Diptera: Syrphidae). Environ Entomol 29(5):1054–1059

Cook RJ (1993) Making greater use of introduced microorganisms for biological control of plant pathogens. Annu Rev Phytopathol 31(1):53–80

De Grazio JW (1978) World bird damage problems. In: Proceedings of the 8th vertebrate Pest control conference, Davis, pp 9–24

Ecobichon DJ (2001) Pesticide use in developing countries. Toxicology 160(1):27–33

El-Sayed AM, Suckling DM, Wearing CH, Byers JA (2006) Potential of mass trapping for long-term pest management and eradication of invasive species. J Econ Entomol 99(5):1550–1564

European Parliament, European Council (2009) Establishing a framework for community action to achieve the sustainable use of pesticides. Off J Eur Union L309(52):71–87. doi:10.3000/17252555.L_2009.309.eng

FAO (2007) Profitability and sustainability of urban and peri-urban agriculture. Agricultural management, marketing and finance occasional paper No. 19 United Nations Food and Agriculture Organization, Rome

FAO (2012) Growing greener cities in Africa: first status report on urban and Peri-urban horticulture in Africa United Nations Food and Agriculture Organization, Rome

Frank SD (2010) Biological control of arthropod pests using banker plant systems: past progress and future directions. Biol Control 52(1):8–16

Furness RW, Monaghan P (1987) Seabirds as pests. In: Furness (ed) Seabird ecology. Springer, New York, pp 127–138

Gahukar RT (2014) Factors affecting content and bioefficacy of neem (*Azadirachta indica* A. Juss.) phytochemicals used in agricultural pest control: a review. Crop Prot 62:93–99

Gamliel A, Katan J (eds) (2012) Soil solarization: theory and practice. American Phytopathological Society, St. Paul

Gliessman SR (2006) Agroecology: the ecology of sustainable food systems. CRC Press, New York
Gogo EO, Saidi M, Itulya FM, Martin T, Ngouajio M (2014) Eco-friendly nets and floating row covers reduce pest infestation and improve tomato (*Solanum lycopersicum* L.) yields for smallholder farmers in Kenya. Agronomy 4(1):1–12
Gonzalez F, Tkaczuk C, Dinu MM, Fiedler Ż, Vidal S, Zchori-Fein E, Messelink GJ (2016) New opportunities for the integration of microorganisms into biological pest control systems in greenhouse crops. J Pest Sci 89(2):295–311
Grondeau C, Samson R, Sands DC (1994) A review of thermotherapy to free plant materials from pathogens, especially seeds from bacteria. Crit Rev Plant Sci 13(1):57–75
Grunert KG (2005) Food quality and safety: consumer perception and demand. Eur Rev Agric Econ 32(3):369–390. doi:10.1093/eurrag/jbi011
Grzywacz D, Stevenson PC, Mushobozi WL, Belmain S, Wilson K (2014) The use of indigenous ecological resources for pest control in Africa. Food Sec 6(1):71–86
Gumus A, Karagoz M, Shapiro-Ilan D, Hazir S (2015) A novel approach to biocontrol: release of live insect hosts pre-infected with entomopathogenic nematodes. J Invertebr Pathol 130:56–60
Gurr GM, Wratten SD, Snyder WE (eds) (2012) Biodiversity and insect pests: key issues for sustainable management. Wiley, Oxford
Hanafi A, Papasolomontos A (1999) Integrated production and protection under protected cultivation in the Mediterranean region. Biotechnol Adv 17(2):183–203
Hanski I (1998) Metapopulation dynamics. Nature 396(6706):41–49
Hochmuth RC, Sprenkel RK (2008) Exclusion methods for managing greenhouse vegetable pests. ENY-846 (IN730), Florida cooperative extension service, Institute of Food and Agricultural Sciences, University of Florida. Available at: http://edis.Ifas.Ufl.Edu/in730. Last Accessed 25 Oct 2016
Hogendorp BK, Cloyd RA (2006) Insect Management in Floriculture: how important is sanitation in avoiding insect problems? HortTechnology 16(4):633–636
Horowitz M, Roger Y, Herlinger G (1983) Solarization for weed control. Weed Sci 31:170–179
Hughes DP, Salathé M (2015) An open access repository of images on plant health to enable the development of mobile disease diagnostics. CoRR abs/1511.08060, 1
Ives AR, Settle WH (1997) Metapopulation dynamics and pest control in agricultural systems. Am Nat 149(2):220–246
Jarvis WR (1992) Managing diseases in greenhouse crops. APS Press, St. Paul
Kadas G (2006) Rare invertebrates colonizing green roofs in London. Urban habitats 4(1):66–86
Khirade SD, Patil AB (2015) Plant disease detection using image processing. In: International conference on computing communication control and automation (ICCUBEA), 2015 IEEE
King SR, Davis AR, Liu W, Levi A (2008) Grafting for disease resistance. Hortscience 43(6):1673–1676
Kosterna E (2014) The effect of different types of straw mulches on weed-control in vegetables cultivation. J Ecol Eng 15(4):109–117
Kozlova I, Singh M, Easton A, Ridland P (2005) Twospotted spider mite predator-prey model. Math Comput Model 42(11):1287–1298
Kumar V, Xiao Y, McKenzie CL, Osborne LS (2015) Early establishment of the phytoseiid mite *Amblyseius swirskii* (Acari: Phytoseiidae) on pepper seedlings in a predator-in-first approach. Exp Appl Acarol 65(4):465–481
Lal C, Verma LR (2006) Use of certain bio-products for insect-pest control. Indian J Tradit Knowl 5(1):79–82
Landis DA, Wratten SD, Gurr GM (2000) Habitat management to conserve natural enemies of arthropod pests in agriculture. Annu Rev Entomol 45(1):175–201
Lenteren van JC (2012) The state of commercial augmentative biological control: plenty of natural enemies, but a frustrating lack of uptake. BioControl 57(1):1–20
Lenteren van JC, Woets JV (1988) Biological and integrated pest control in greenhouses. Annu Rev Entomol 33(1):239–269

Lenteren van JC, Benuzzi M, Nicoli G, Maini S (1992) Biological control in protected crops in Europe. In: van Lenteren JC et al (eds) Biological control and integrated crop protection: towards environmentally safer agriculture. Wageningen, Veldhoven, pp 77–89
Licciardi S, Assogba-Komlan F, Sidick I, Chandre F, Hougard JM, Martin T (2007) A temporary tunnel screen as an eco-friendly method for small-scale farmers to protect cabbage crops in Benin. Intl J Trop Insect Sci 27:152–158
MacIvor JS, Lundholm J (2011) Insect species composition and diversity on intensive green roofs and adjacent level-ground habitats. Urban Ecosyst 14(2):225–241
Maini S, Nicoli G (1990) La serra come ecosistema. Le Scienze quaderni 53:37–43
Makris SL, Rowe JN (1998) Implementation of the food quality protection act (FQPA) as it relates to enhanced sensitivity of children. Teratology 57(4/5):246
Markkula M, Tiittanen K (1976) Pest in first and natural infestation methods in the control of *Tetranychus urticae* Koch with *Phytoseiulus persimilis* Athias-Henriot on greenhouse cucumbers. Annales Agriculturae Fenniae 15:81–85
Martin T, Assogba-Komlan F, Houndete T, Hougard JM, Chandre F (2006) Efficacy of mosquito netting for sustainable small holder's cabbage production in Africa. J Econ Entomol 99(2):450–454
Messelink GJ, Bennison J, Alomar O, Ingegno BL, Tavella L, Shipp L, Palevsky E, Wäckers FL (2014) Approaches to conserving natural enemy populations in greenhouse crops: current methods and future prospects. BioControl, 59(4):377-393
Mills NJ, Getz WM (1996) Modelling the biological control of insect pests: a review of host-parasitoid models. Ecol Model 92(2):121–143
Mittler R (2006) Abiotic stress, the field environment and stress combination. Trends Plant Sci 11(1):15–19
Nasterlack M (2006) Do pesticides cause childhood cancer? Int Arch Occup Environ Health 79(7):536–544
National Research Council (NCR) (2010) Toward sustainable agricultural systems in the 21st century. National Academies Press, Washington
Nega E, Ulrich R, Werner S, Jahn M (2003) Hot water treatment of vegetable seed–an alternative seed treatment method to control seed borne pathogens in organic farming. J Plant Dis Prot 110(3):220–234
Nono-Womdim R (2003) An overview of major virus diseases of vegetable crops in Africa and some aspects of their control. In: Plant virology in sub-Saharan Africa, proceedings conference organized by IITA. International Institute of Tropical Agriculture, Ibadan, pp 213–232
Nono-Womdim R, Marchoux G, Pochard E, Palloix A, Gebre-Selassie K (1991) Resistance of pepper lines to the movement of cucumber mosaic virus. J Phytopathol 132(1):21–32
Olkowski W, Daar S, Olkowski H (1991) Common sense – pest control. The Taunton Press, Newtown
Orsini F, Marchetti L, Magrefi F, Draghetti S Bazzocchi G (2013) Sustainable urban garden management. Horticity, Bologna. Hortis LLP EU program. http://www.hortis-europe.net/files/documenti/inglese/final-e-books/web-e-book-2-engl-last.pdf. Last Accessed 25 Oct 2016
Orsini F, Gasperi D, Marchetti L, Piovene C, Draghetti S, Ramazzotti S, Bazzocchi G, Gianquinto G (2014) Exploring the production capacity of rooftop gardens (RTGs) in urban agriculture: the potential impact on food and nutrition security, biodiversity and other ecosystem services in the city of Bologna. Food Sec 6(6):781–792
Osborne LS, Landa Z, Taylor DJ, Tyson RV (2005) Using banker plants to control insects in greenhouse vegetables. Proc Florida State Hortic Soc 118:127–128
Paret M, Pernezny K, Roberts P (2013) Disease control for Florida tomatoes. University of Florida. Cooperative Ext Serv Tech Bull EDIS PPP35. http://edis.Ifas.ufl.edu/vh056. Last Accessed 25 Oct 2016
Pfeiffer A, Silva E, Colquhoun J (2015) Living mulch cover crops for weed control in small-scale applications. Renewable Agric Food Syst 1(4):1–9
Posnette AF, Cropley R (1958) Heat treatment for the inactivation of strawberry viruses. J Hortic Sci 33(4):282–288

Racke KD, Leslie AR (1993) Pesticides in urban environments, ACS Symposium Series. Washington, DC, American Chemical Society

Rauh V, Arunajadai S, Horton M, Perera F, Hoepner L, Barr DB, Whyatt R (2015) Seven-year neurodevelopmental scores and prenatal exposure to chlorpyrifos, a common agricultural pesticide. In: Everyday environmental toxins: children's exposure risks. Boca Raton, CRC Press

Rosenzweig C, Iglesias A, Yang XB, Epstein PR, Chivian E (2001) Climate change and extreme weather events; implications for food production, plant diseases, and pests. Glob Chang Hum Health 2(2):90–104

Rotem J, Palti J (1969) Irrigation and plant diseases. Annu Rev Phytopathol 7(1):267–288

Rumei X (1991) Improvements of the plant-pest-parasitoid (PPP) model and its application on whiteily-*Encarsia* population dynamics under different release methods. J Appl Entomol 112(1–5):274–287

Russell GE (2013) Plant breeding for pest and disease resistance: studies in the agricultural and food sciences. Butterworth-Heinemann

Sabelis MW, Diekmann O, Jansen VAA (1991) Metapopulation persistence despite local extinction: predator-prey patch models of the Lotka-Volterra type. Biol J Linn Soc 42(1–2):267–283

Sampson C, Kirk WD (2013) Can mass trapping reduce thrips damage and is it economically viable? Management of the western flower thrips in strawberry. PLoS One 8(11):e80787

Sobhy IS, Erb M, Lou Y, Turlings TC (2014) The prospect of applying chemical elicitors and plant strengtheners to enhance the biological control of crop pests. Philos Trans R Soc B Biol Sci 369(1639):20120283

Stapleton JJ (2008) Soil Solarization informational web site. UC statewide IPM program; Pest notes publication 74145, Kearney agricultural center, Parlier, CA. http://ucanr.edu/sites/Solarization/files/114635.pdf. Last Accessed 26 Oct 2016

Stapleton JJ, DeVay JE (1986) Soil solarization: a non-chemical approach for management of plant pathogens and pests. Crop Prot 5(3):190–198

Steck A, Morgan S, Retzlaff W, Williams J (2015) Insect communities on green roofs that are close in proximity but vary in age and plant coverage. J Living Archit 2(1):1–11

Stenseth C (1979) Effect of temperature and humidity on the development of Phytoseiulus Persimilis and its ability to regulate populations of Tetranychus urticae [Acarina: Phytoseiidae. Tetranychidae]. Entomophaga 24(3):311–317

Stoll G (1996) Natural crop protection in the tropics. Margraf Verlag, Weikersheim

Strobel NE, Kucf JA (1999) Plant resistance to pathogens. Handb Pest Manag 175

Stuart RJ, Barbercheck ME, Grewal PS (2015) Entomopathogenic nematodes in the soil environment: distributions, interactions and the influence of biotic and abiotic factors. In: Nematode pathogenesis of insects and other pests. Springer International Publishing, Cham, pp 97–137

Suckling DM, Stringer LD, Kean JM, Lo PL, Bell V, Walker JT, Twidle AM, Jiménez-Pérez A, El-Sayed AM (2015) Spatial analysis of mass trapping: how close is close enough? Pest Manag Sci 71(10):1452–1461

Tavares J, Wang KH, Hooks CR (2015) An evaluation of insectary plants for management of insect pests in a hydroponic cropping system. Biol Control 91:1–9

Teasdale JR (1996) Contribution of cover crops to weed management in sustainable agricultural systems. J Prod Agric 9(4):475–479

Thakur M, Sohal BS (2013) Role of elicitors in inducing resistance in plants against pathogen infection: a review. ISRN Biochem 2013(2):1–10

Thompson JP (1991) Improving the mycorrhizal condition of the soil through cultural practices and effects on growth and phosphorus uptake by plants. In: Johansen C, Lee KK, Sahrawat KL (eds) Phosphorus nutrition of grain legumes in the semi-arid tropics. ICRISAT, Patancheru, pp 117–137

Timm RM, Marsh RE (1997) Vertebrate Pests. In: Mallis handbook of pest control, 8th edn. Mallis Handbook and Technical Training Company, Cleveland

Tonti S (2013) Disease management in the vegetable garden. In: Orsini et al (eds) Sustainable urban garden management. Horticity Bologna, pp 54–64. Hortis LLP EU program: http://www.hortis-europe.net/files/documenti/inglese/final-e-books/web-e-book-2-engl-last.pdf. Last Accessed 26 Oct 2016

Trevors JT (1996) Sterilization and inhibition of microbial activity in soil. J Microbiol Methods 26(1):53–59

Valcke M et al (2004) Characterization of exposure to pesticides used in average residential homes with children ages 3 to 7 in Quebec. National Institute of Public Health, Québec

Vega FE, Goettel MS, Blackwell M, Chandler D, Jackson MA, Keller S, Koike M, Maniania NK, Monzón A, Ownley BH, Pell JK, Rangell DEN, Roy HE (2009) Fungal entomopathogens: new insights on their ecology. Fungal Ecol 2(4):149–159

Vidogbéna F, Adégbidi A, Tossou R, Assogba-Komlan F, Ngouajio M, Martin T, Simon S, Parrot L, Zander KK (2015) Control of vegetable pests in Benin–Farmers' preferences for eco-friendly nets as an alternative to insecticides. J Environ Manag 147:95–107

Volterra V (1926) Variazioni e fluttuazioni del numero d'individui in specie animali conviventi. Mem Accad Lincei II 6:31–112

Wäckers FL, Romeis J, van Rijn P (2007) Nectar and pollen feeding by insect herbivores and implications for multitrophic interactions. Annu Rev Entomol 52:301–323

Wesseling C, McConnell R, Partanen T, Hogstedt C (1997) Agricultural pesticide use in developing countries: health effects and research needs. Int J Health Serv 27(2):273–308

Wezel A, Bellon S, Doré T, Francis C, Vallod D, David C (2009) Agroecology as a science, a movement and a practice. Rev Agron Sustain Dev 29(4):503–515

World Bank (2008) World development report 2008: agriculture for development. The World Bank, Washington

Wraight SP, Ugine TA, Ramos ME, Sanderson JP (2016) Efficacy of spray applications of entomopathogenic fungi against western flower thrips infesting greenhouse impatiens under variable moisture conditions. Biol Control 97:31–47

Xia C, Chon TS, Ren Z, Lee JM (2015) Automatic identification and counting of small size pests in greenhouse conditions with low computational cost. Eco Inform 29:139–146

Xing K, Zhu X, Peng X, Qin S (2015) Chitosan antimicrobial and eliciting properties for pest control in agriculture: a review. Agron Sustain Dev 35(2):569–588

Yang RZ, Tang CS (1988) Plants used for pest control in China: a literature review. Econ Bot 42(3):376–406

Zitter TA (2013) Vegetable crops. Department of Plant Pathology, Cornell University

Produce Quality and Safety

Beatrix W. Alsanius, Andrea Kosiba Held, Martine Dorais, Cecilia Moraa Onyango, and Lars Mogren

Abstract Within sustainable production, produce quality and safety are essential features. However, methods, requirements, conditions and even legislation for produce quality and safety in production in rural areas cannot always be directly transferred to production in urban areas and on rooftops. This chapter describes features of produce quality, produce safety and safety hazards in urban rooftop farming employing various technological solutions and serving various purposes in different climate zones. Sustainability is discussed in terms of product quality and safety, and requirements to resolve the principal issues are presented.

Introduction

Current considerations concerning rooftop agriculture (RA) mainly focus on city planning, demographics and food security issues or on technological solutions, urban eco-lifestyle and wellbeing. Despite their importance for sustainability, two important areas, produce quality and food safety, are only rarely considered. Given its implications for public health, food safety is a function of social sustainability. Produce quality can also affect public health, but is particularly important in terms of economic sustainability when RA produce is publicly marketed. Produce quality and safety may be viewed from the perspective of provisioning, i.e. shared responsibility for primary production, processing, storage, distribution and retail, or from the perspective of perception and demand, which is governed by consumers and

B.W. Alsanius (✉) • A.K. Held • L. Mogren
Department of Biosystems and Technology, Microbial Horticulture Unit, Swedish University of Agricultural Sciences, PO Box 103, SE-230 53 Alnarp, Sweden
e-mail: Beatrix.Alsanius@slu.se; Andrea.Held@slu.se; lars.mogren@slu.se

M. Dorais
Pacific Agri-Food Research Centre, Agriculture and Agri-Food Canada, V0M 1A0 Agassiz, BC, Canada
e-mail: Martine.Dorais@AGR.GC.CA

C.M. Onyango
Department of Plant Science and Crop protection, University of Nairobi, P.O. Box 29053-00625, Nairobi, Kenya
e-mail: cecilia.onyango@uonbi.ac.ke

F. Orsini et al. (eds.), *Rooftop Urban Agriculture*, Urban Agriculture,
https://doi.org/10.1007/978-3-319-57720-3_12

their willingness to pay (Grunert 2005). In addition, produce safety is affected by outbreaks and recalls.

The FAO/WHO (2014) can serve as a guideline to discriminate between food safety and produce quality. A food safety hazard is defined as a "biological, chemical or physical agent in, or condition of, food with the potential to cause an adverse health effect" (Codex Alimentarius 2007). Definitions of produce quality vary between different steps within the value network. Grunert (2005) developed a holistic approach to define food quality, discriminating between a "horizontal" and a "vertical" domain, reflecting attributes before and after purchase. In the present chapter, we adopt this concept to examine produce quality for RA, which includes both edible and ornamental crops (Fig. 1). As markets vary on a regional and local level, added values relevant for expected and experienced quality assessment on a societal level (e.g. sustainability properties, societal responsibility/democracy) or individual (well-being) level need to be embodied. Hence, inferior quality caused by food spoilage organisms, presence of foreign or undeclared non-hazardous items, incorrect product labelling or shelf-life date that are not adverse to human health belong to the produce quality domain, but not the safety domain. At the same time, excellent produce quality does not mean that the produce is safe.

Food safety legislation and produce quality regulations apply irrespective of production site. However, these two areas have received little awareness among all actors in the RA arena. The site, scale and purpose of urban horticultural production, as well as the logistics and distribution system, may vary, posing different demands on safety and quality. Novel sources of hazards occurring in RA also need to be considered. This chapter is restricted to interactions between site, technology, design, resources and management, produce quality and food safety, while grower safety issues are not covered in detail.

Produce Quality

Product quality is a broad term (Fig. 1) that generally includes physical characteristics, nutritive and sensory attributes and content of secondary metabolites (Rouphael et al. 2012). It has been shown that these quality attributes vary according to genetic characteristics, environmental conditions and crop management (Dorais and Alsanius 2015; Dorais et al. 2016; Dorais and Ehret 2008). Consequently, it is necessary to know the effects of both environmental factors and cultural practices prevailing for RA in order to control the product quality. In this chapter, we limit the concept of quality to measurable quality attributes and, when appropriate, differences between urban RA and conventional production in the field or greenhouse.

The heterogeneity of the environmental growing conditions in RA, e.g. building shading, wind corridors, heat island effects, air pollution, limited water resource etc., affects product quality properties. Consequently, the uniformity of quality in the produce can vary widely within this restricted cultivated area. Under greenhouse growing conditions, plants may be exposed simultaneously to more than one abiotic

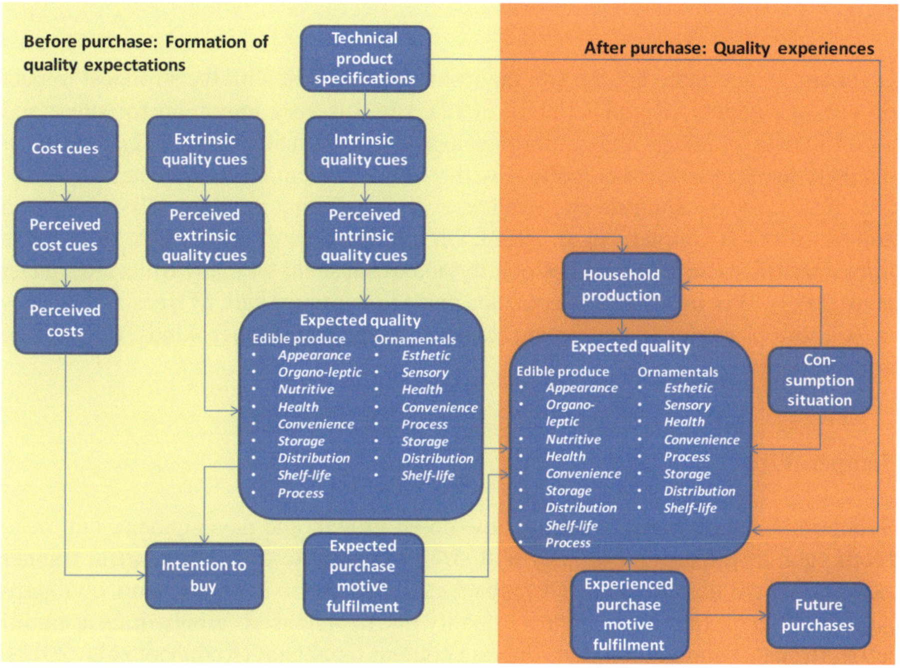

Fig. 1 A holistic model of horticultural product quality based on The Total Food Quality Model (Adapted from (Grunert 2005); reproduced with permission from Oxford University Press (license number 3946570150501)

or biotic stress at a time, and thus the effect of multiple stresses on yield and quality attributes should be considered. In fact, the presence of an abiotic stress can reduce or enhance susceptibility to a biotic stress, and vice versa (Atkinson and Urwin 2012), affecting the nutritional value of the plants.

Influence of Environmental Factors and Cultural Practices on Produce Quality

Wherever a plant is grown, the natural environment is composed of a complex mix of abiotic and biotic stresses that interact in a complex way, while the plant response is equally complex (Cramer et al. 2011; Gruda 2005).

Light

Light and temperature are the two environmental factors with the strongest impact on produce quality (Rouphael et al. 2012). Light is very important for sugar and ascorbic acid synthesis and colour development (Rouphael et al. 2012). Low light intensity leads to less sweet produce with a lower content of ascorbic acid and usually higher levels of nitrate and oxalate, compounds that are generally considered anti-nutritional (Rouphael et al. 2012). On the other hand, excess light or extreme light intensities can cause loss of quality due to sunscald on many fruits (Rouphael et al. 2012). It is important to consider that the common lack of trees and hedges in a rooftop garden environment creates a need for other kinds of shading arrangements.

Temperature

Suboptimal temperatures not only slow down growth and development, but make crops such as tomatoes less juicy and aromatic, with low acidity content, thinner skin and worse storage ability (Rouphael et al. 2012). On the other hand, cool temperatures can in some cases improve quality due to enhanced carbohydrate accumulation (tomato) and higher glucosinolate content (cabbage) (Rouphael et al. 2012). Air temperatures usually oscillate more in an uncovered rooftop environment than in more sheltered production sites on rooftops.

Wind

It should be stressed that what can be considered production in "clean air" in an urban context can be complicated to define, due to variations in sources of air pollution in the immediate vicinity and far upwind in the prevailing wind direction (Azapagic et al. 2013). The conditions are very different between a rooftop site and a lower urban street canyon site, where traffic emissions and reduced natural ventilation can result in high concentrations of pollutants (Vardoulakis et al. 2003). Wind conditions can also be very different in a rooftop environment compared with those at ground level. Apart from mechanical damage to plants, strong winds result in excessive loss of water from the plants, leading to wilting problems. Thus, in order to create optimal production conditions and assure high quality yield, the need for shelter must be taken into consideration.

Precipitation

Most fruit and vegetables contain more than 90% water. Therefore, production of high quality produce cannot exclusively rely on rainwater, but needs to be supported by irrigation. Deficit irrigation is a strategy to optimise water use, yield and produce quality (Rouphael et al. 2012), but needs constant monitoring to reduce the risk of drought and wilt problems.

Biotic Damage

There are no specific conditions regarding incidence of disease and insects in RA. If anything, pests and fungi should be less of a problem due to the non-intensive cultivated area (small area under cultivation compared with field farming) and the fact there might be more natural predators present. In a long-term perspective, problems with root-borne diseases should also be less prevalent because RA is like a soil-less growing system and the growing medium can be replaced if root disease problems occur.

Sustainability Versus Produce Quality

Species, Cultivar Selection

Plant quality attributes are mainly determined by selection of species and varieties adapted to RA, but there is no general rule of thumb to apply when selecting species or cultivars to use for RA *per se*.

Growing Medium

In soil-less cropping systems, it is important to keep track of decisive properties in the nutrient solution, such as electrical conductivity (EC), chemical forms of the elements, temperature and pH (Rouphael et al. 2012). If well managed, a soil-less cropping system gives the opportunity to improve produce quality due to precise nutrient solution practices (Rouphael et al. 2012). The source of the growing medium used in RA is critical. Apart from the major nutrients (N, P and K), it is very important to ensure that the plants get sufficient amounts of micro-nutrients, especially when growing in soil-less conditions. For example, iron (Fe) is essential for both plant productivity and product quality (Briat et al. 2014). Soil-less production in general in an urban/rooftop environment does not differ from the practices already used today in conventional production.

Plant Management

Nutrient and moisture supply are major concerns in RA, especially in extensive production. There is limited information on best management practices for RA with regard to these issues (Whittinghill et al. 2013).

Fertilizers Use of fertilisers in RA can help boost the yield and quality of the crops produced. However, excessive use of fertilisers in crops can lead to over-accumulation of dangerous or toxic substances such as nitrates (see chapter "Managing Mineral Nutrition in Soilless Culture"), oxalates and mercury, lowering the quality and safety of the produce (Shahid and Muhammad 2007). In a rooftop environment it is very important to ensure that the plants are provided with sufficient amounts of calcium, because this protects them against heat stress (mainly by induction of heat shock proteins (Hepler 2005), increases tolerance to some diseases, increases crop shelf-life and supports the accumulation of nutrients (Martin-Diana et al. 2007; Park et al. 2005). Nitrogen deficiency and abiotic and biotic stresses generally result in higher concentrations of secondary plant metabolites that are considered beneficial for human health (Dorais and Alsanius 2015; Orsini et al. 2016). Moderate salt stress affects fruit quality through comparable pathways to water deficit (see below) (De Pascale et al. 2007; De Pascale et al. 2012).

Irrigation Efficient use of irrigation water is important (Darko et al. 2016) and the concept of water footprint should be taken into consideration for all kinds of plant production (Lovarelli et al. 2016). An excess or deficit of water affects plant and produce quality. Lack of water leads to impaired plant water uptake and increases the salt concentration (including plant nutrients) in the substrate solution, which causes salt stress and hence increased leaf abscisic acid content and stomatal aperture; ethylene production; reduced transpiration rate and photosynthetic activity; decreased plant growth, development and biomass formation; increased photo-oxidative stress, involving formation of secondary metabolites; fruit abortion and negative regulation of fruit setting (Anjum et al. 2011). In contrast, waterlogging leads to inferior atmospheric conditions in the growing medium and root zone, and thus higher plant stress (epinasty, reduction of stem elongation, leaf senescence, poor root health, reduced nutrient uptake, inferior biomass formation and predisposition to plant diseases) (Barret-Lennard 2003). The crop water demand dictates the timing and volume of irrigation, with plant transpiration rate and evaporation from the growing medium being decisive factors. These differ depending on environmental conditions (temperature, light intensity, relative humidity, wind speed/ventilation), but also on the design of the rooftop surface (tunnel, greenhouse, outdoor).

The quality of irrigation water varies between different sources, as has recently been comprehensively reviewed by Alsanius (2014a) and Dorais et al. (2016). In urban contexts the use of rainwater, contained stormwater and reclaimed wastewater is often suggested by city planners in terms of sustainable city development. Therefore, this chapter mainly focuses on these sources. Safety aspects considering water are discussed in a later section of this chapter. Although their quality is undisputed, these sources of water may import acids, nutrients and undesired substances, which need to be considered with respect to amount of fertilisers supplied. Furthermore, rainwater and stormwater collected close to the sea may contain elevated concentrations of sodium chloride (NaCl). Likewise, municipal water may contain high concentrations of sulphate, sodium and chloride. For both rainwater and contained stormwater, mode of collection and storage are important features for quality.

Low electrical conductivity in the growing medium promotes plant water uptake, which causes inferior fruit and vegetable nutrient content and taste (Dorais and Ehret 2008; Dorais et al. 2001). However, moderately restricted water supply is an effective tool to improve quality attributes in RA, as it results in accumulation of critical attributes for taste, such as fruit-soluble sugars, organic acids and aroma compounds (Ripoll et al. 2014).

Pest Management

All farmers, regardless of where they grow their crops, have to cope with pests. RA involves growing crops where people live. Therefore use of synthetic pesticides which have been reported to cause serious environmental and health problems may not be sustainable. Use of biological control and integrated pest management may be the most sustainable way of managing pests without compromising plant and produce quality (National Research Council -NCR 2010).

Produce Safety

Legislation

With changing food production patterns, trade between different regions, changes in technology and increased public expectations regarding health protection, there are continuous demands on the environment in which food safety systems operate (FAO/WHO 2006). The main challenge facing most governments today is to secure food systems both in the present and the future. Moreover, great pressure is being placed on agricultural land due to increasing population, increased urbanisation and predicted global warming factors, which in combination are likely to lead to changes in crop production practices, for instance bringing humans, animals and crop production closer together (Berger et al. 2010b), a likely scenario in RA. Hence, there is an urgent need to address issues associated with the supply of safe and healthy food. This calls for governments to develop legislation governing RA systems, especially if the crops produced are made available for sale. All actors in the production and sale of these crops need keen attention by the food safety agencies to uphold the application of HACCP and fair trade. This can be done through enhancing the capacity of all stakeholders involved in the value chain to deal with food safety concerns and their implications for human health.

A major proportion of urban food production is currently carried out in home gardens or in private initiatives (Eigenbrod and Gruda 2015), but urban food production is an increasing societal movement and is encouraged by various authorities

(Lohrberg et al. 2015). In contrast to professional growers, amateurs lack knowledge of the correct use and handling of plant protection products (PPPs). This poses a risk of misuse or over-dosage and consequently pesticide residues in produce. The European legislation on PPPs considers this fact and establishes categories for consumer and professional use when products are registered (Sustainable Use Directive 2009/128/EC) (European Parliament and European Council 2009a). PPPs can also contribute to achieving high yields in RA. However, the areas of use that are covered in the application process today (*e.g.* EU Regulation 1107/2009) (European Parliament and European Council 2009b) do not include the diverse features of RA systems. Stormwater systems, microclimate, variation in the soils used and many other factors are not considered in the traditional models used to calculate the environmental impact of PPPs.

Hazards of Interest for Roof Top Farming

Physical, biological and chemical hazards can compromise food safety, including that of ***edible plant food*** grown on rooftops. Food safety is the responsibility of all, regardless of production system. Studies have shown that qualitative safety hazards differ between production systems (Bourn and Prescott 2002; Maghos et al. 2006; Winter and Davis 2006). Due to the current lack of information on outbreak statistics, it is not possible to generalise that food safety risks are greater in one system than another. However, apart from the hazards listed in the Codex Alimentarius, the lack of regulation for the specific conditions prevailing in RA poses a hazard *per se*, i.e. a regulatory hazard.

Physical Hazards

Foreign bodies, such as soil/growing medium, stones, wood and bone chips, metal or glass plastic items, and injuries caused by foreign bodies constitute physical food hazards. Growing system design (e.g. recycled pallet rim), choice and source of the growing medium (e.g. wood, bone and shell chips), soil management strategy (e.g. mulching) and distance between soil/growing medium and the harvested produce are decisive factors. Foreign bodies may cause internal lacerations upon ingestion. The risk for chips from broken glass or plastic poses special demands on the stability of the covering material in roof greenhouse constructions. In this context it also needs to be considered that public greenhouse areas are governed by different legal requirements than commercial greenhouses. Inferior waste disposal methods and low quality construction material in low tech systems favor the occurrence of such hazards. Also air quality, in terms of load of coarse and fine particles, as well as nano-particles are important factors in RA and crops grown in urban areas can become contaminated by airborne particles (Vittori Antisari et al. 2015; Kim et al. 2015; Säumel et al. 2012). Inhalation of coarse, fine and nano-particles may inflict with the urban growers' health (Jayawardena et al. 2009), but not necessarily affect produce safety. Moreover, varieties in the toxicity profile for urban air has been reported for cold climate (Salonen et al. 2004). However, fine particles including nano-particles may invade the body tissue causing harmful effects in different organs (Chen et al. 2015; Imrich et al. 2000; Mattsson et al. 2015).

Biological Hazards

Biological hazards may be provoked by helminths, protozoa, fungi (yeasts and molds), bacteria or viruses and may cause either food infections (gastroenteritis) as a result of organisms proliferating in the human intestinal tract or food intoxications due to toxin formation before ingestion or during passage through the gastro-intestinal tract. The various biological hazards differ in infectious dose, incubation time as well as manifestation and severity of illness. Outbreaks associated to horticultural produce including sprouted seeds have increasingly been reported globally (Fett 2006; Warriner and Smal 2015) and several explanations have been indicated (Tauxe et al. 1997). Among these, two factors are of particular interest for RA, namely an increasing number (i) of biological agents with very low infectious doses and (ii) of immunodepressed persons. Prominent microbial agents involved in food borne illnesses in fruit, berries and vegetables are enterotoxigenic as well as shiga-toxigenic *E. coli*, *Salmonella*, *Yersinia enterocolitica*, *Listeria monocytogenes*, *Bacillus cereus*, *Cryptospora*, *Cryptosporidium parvum*, as well as Norovirus and Hepatitis A.

Chemical Hazards

Chemical hazards arise in the form of heavy metals, undesired organic contaminants; in particular persistent organic products, the presence of mycotoxins as well as antimicrobial compounds. These contaminants can occur either naturally or may be introduced during primary production or postharvest handling. Due to knowledge gaps, food safety hazards are difficult to assess and may be under- or overestimated (Gallaher et al. 2013). Crops grown in urban areas can become contaminated by heavy metals in soils, quality of reused water and aerial depositions (Binns et al. 2004). Produce contamination is therefore dependent on which part of the plant is consumed and crops should thus be chosen carefully depending on the conditions of the growing system and its environment.

Routes of Transmission

All plant surfaces as well as the inner of plants are colonized. However, the majority of organisms associated to plants are not harmful to humans. Nonetheless, among the ones inferior to human health, many follow the fecal-oral route of transmission and are closely related to the animal-human bond where animals may display an asymptomatic reservoir. Others are also ubiquitous in the environment (*Listeria monocytogenes*, *Bacillus cereus*). Most of them have not been grouped as plant associated organisms. However, they can colonize the plant surfaces (*epiphytic*) in microbial assemblages (biofilms) or be internalized and colonize the inner of plants (*endophytic*) (Dorais and Alsanius 2015). General routes for the transmission of human pathogens in horticultural production chains are plant-related (seeds, small plants), input-related (soil and growing medium, water, organic fertilisers, composts), management-related (rotation, sprays, tools, management practices, harvesting and post-harvest practices, anthropogenic activities) and site-specific (surroundings of the growing site, practices within the cropping system or rotation, plant cultures close to the growing site, wild animals). Routes of transmission have been surveyed comprehensively by Bihn and Gravani (2006), Gerba (2009) and Matthews et al. (2014) and are therefore not described in detail here. Several routes of transmission result in microbial prevalence in RA, which cannot merely be counteracted by covering (greenhouse, tunnel) (see below). As contamination at an earlier stage within the farm-to-fork chain cannot be counteracted at a later stage (e.g. by washing), actions to prevent biological hazards need to be taken within the entire production and consumption chain. In this context, RA growers need to be aware that their products may be consumed in an unconventional way (e.g. eaten raw instead of cooked). Figure 2 shows routes of transmission for biological hazards. In RA systems, insects are a possible source of contamination, especially in congested urban environments with poor waste disposal. In most developing countries, heaps of garbage are found in majority of urban centres, attracting huge numbers of flies belonging to the Muscidae and Calliphoridae families. Studies have shown that contaminated flies and vectors can transfer bacteria and *E. coli* O157:H7 to plant leaves or fruit on farms (Beuchat 2006; Iwasa et al. 1999; Sela et al. 2005; Talley et al. 2009).

Inputs such as soil, growing medium, soil improvers, fertilisers, water and products to maintain plant health are the main route for transmission of chemical hazards to RA systems and crops. Likewise, chemical and physical hazards may be imported to RA systems through production inputs. However, the environment surrounding the rooftop garden is also crucial, e.g. air-borne physical contaminants may settle on the foliage. Due to the novelty of RA as a production form, available scientific information on hazard incidence and health impacts is scarce.

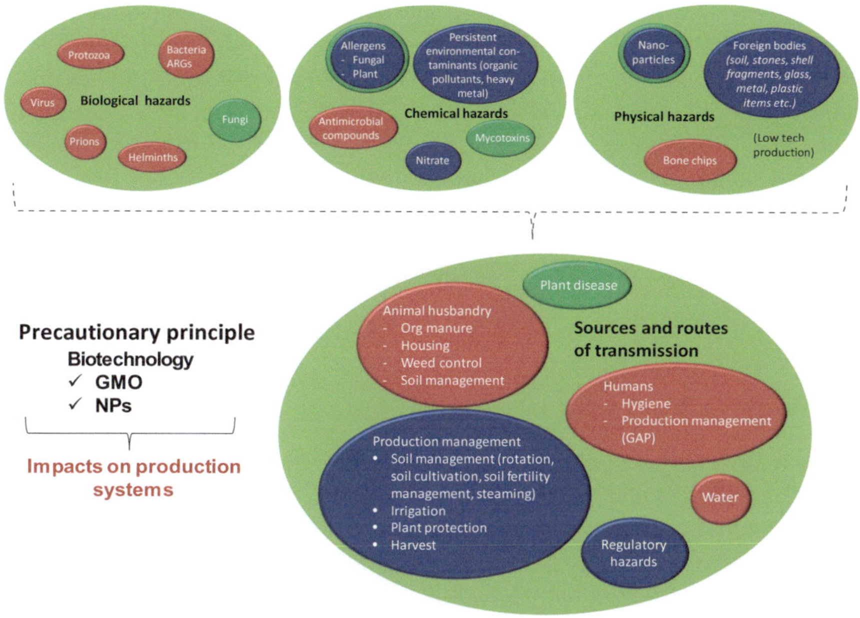

Fig. 2 Food safety hazard interactions with sources and routes of transmission in rooftop agriculture (RA) (Modified after (Alsanius et al. 2016); ARGs = antibiotic resistance genes; GAP = good agricultural practice). Source: BioGreenhouse Factsheet Food Safety. The figure is reprinted with permission of the editor/project chair

Water

All plant production is dependent on water. Irrigation water sources differ in terms of their predisposition to biological hazards. Contaminated irrigation water and handling of produce by infected individuals are the major mechanisms of contamination of fresh produce (Berger et al. 2010a). From a biological point of view, municipal water, rainwater and groundwater are perceived as safe, whereas surface water and different types of sewage water can act as vectors for food-borne pathogens (Alsanius 2014a). With respect to water sources considered safe, the microbial quality at collection and the hygiene conditions during water storage are equally important. Therefore, precautions to prevent contamination of water collection tanks and pipelines must be considered (e.g. covering, cleaning) (Alsanius 2014a; Dorais et al. 2016). Use of dual water tanks for animal husbandry (e.g. fish, crustaceans) and irrigation water storage must also be reconsidered. Furthermore, the reuse of reclaimed, but not reconditioned, water (brown/black water) and wastewater loops within buildings or cities to irrigate food crops does not comply with food safety.

Water also acts as a carrier of chemical hazards. In this regard, the origin of the water and inputs from different discharge systems (e.g. industry) play an important role. In the presence of discharges by industry to water sources, persistent environmental pollutants (e.g. heavy metals, persistent organic compounds) can be expected. Challenges by pharmaceutical and personal care products (PPCPs) and PPP residues in water are novel problems. The degradation of some PPP residues in aqueous solutions is very slow (Abbate et al. 2009; Abbate et al. 2007; Alsanius and Bergstrand 2014; Alsanius et al. 2013; Kreuger et al. 2010). Little is known about plant root uptake or invasion and translocation of such agents, especially when water and other resources are reused within production systems (closed systems). However, some reports indicate plant uptake of PPCPs (Herklotz et al. 2010; Karnjanapiboonwong et al. 2011; Redshaw et al. 2008; Shenker et al. 2011; Winker et al. 2010).

Growing Medium

The source of the growing media used in RA is critical for produce quality and safety. Great care should be taken to ensure that the soil used is not contaminated, especially with heavy metals. This could be the case for soils obtained from old garage sites or dumping sites. While most governments in developing countries advocate the use of unleaded fuel, poor handling of garage waste continues to add heavy metals to the soils in the vicinity (Mutuku 2013). If such soils are subsequently used in RA, the risk of contamination of the crops produced is high and this poses a potential health risk to consumers. If plants are grown in contaminated substrate, the concentrations of lead (Pb), cadmium (Cd), nickel (Ni), cobalt (Co), and chromium (Cr) in produce usually increase with plant age (Naser et al. 2012), indicating that special attention is needed to limit heavy metal concentrations in growing medium used for crops with a long growing season.

Nutrients

Plant nutrients are supplied from either inorganic or organic sources. In particular, leafy vegetables require considerable amounts of nitrogen for biomass formation. Supplying excess nitrogen leads to high nitrate concentrations in some leafy vegetables and may cause adverse health effects. Within the European Union, the threshold levels for leaf nitrate content are regulated in the EU Nitrates Directive (European Union 1991). Due to changes in the transmittance light spectrum caused by greenhouse and tunnel covering material, nitrate accumulates to a higher extent in covered cropping systems than in open fields (Alsanius et al. 2016).

Inorganic and organic fertilisers may also be carriers of other chemical agents than the intended macro- and micronutrients. Within the EU, the cadmium concentration in phosphorus fertilisers is limited to 114.5 mg kg^{-1} P. Use of organic fertilisers can result in import of heavy metals to the cropping system (Table 1). A comprehensive review of heavy metal concentrations has been performed by Möller and Schultheiss (2014). Determination of the load of heavy metals in proportion to the content of macronutrients, such as nitrogen or phosphorus, is a suitable measure

Table 1 Specific heavy metal content per 100 kg of nitrogen (N), heavy metal load (HML) and potential microbial hazards of selected organic fertilisers

Fertiliser	Specific heavy metal load, per 100 kg N							HML	Potential microbial hazard
	Cu	Zn	Pb	Cd	Cr	Ni	Hg		
Dry chicken manure[a]	229	417	5.67	1.21	13.5	9.53	–	5.9	*Campylobacter*, *Salmonella*
Cattle manure	15.8	91.5	2.71	0.35	6.90	3.53	0.05	0.10	*Salmonella*, EHEC[b], *Listeria monocytogenes*, *Campylobacter jejuni*
Pig manure	33.1	207	1.49	0.23	23.7	5.13	0.04	0.13	*Salmonella*, ESBL[c], *Yersinia enterocolitica*, *Listeria monocytogenes*, *Campylobacter*
Blood meal	11.8	38.7	1.49	0.08	2.66	0.37	0.01	0.02	
Bone meal	11.4	109	2.97	0.21	13.7	3.31	0.02	0.04	BSE[d]
Feather meal	9.26	111	0.15	0.03	1.20	0.61	<0.05	0.03	*Salmonella*
Hair meal	16.2	192	<1.0	<0.2	7.50	<5.2	<0.05	0.07	None expected
Mushroom compost	60.4	159	7.01	0.29	7.22	4.47	0.06	0.17	*Bacillus cereus*

Alsanius (2014b) and Möller and Schultheiss (2014)
[a]Dry matter content: 60.1%
[b]*EHEC* enterohaemorrhagic *Escherichia coli*
[c]*ESBL* extended spectrum betalactamase-producing bacteria
[d]*BSE* bovine spongiform encephalopathy

to avoid their accumulation in RA produce. As a result of animal husbandry, organic manure may also be a source of residual antibiotic compounds and a reservoir for antibiotic resistance genes (Udikovic-Kolic et al. 2014). Therefore, knowledge of the husbandry methods used on the farm of origin is an important parameter for predicting the health risks.

From the perspective of microbial contaminants, organic nutrient sources, including nutrients extracted from urban resource cycles, are of great significance. Due to the importance of the faecal-oral transmission route of human pathogens, animal manure needs to be sanitised before being applied to horticultural crops. Proper composting of organic manure is an efficient method to destroy non-sporulating human pathogens. However, uniform heat distribution within the compost heap is necessary to prevent survival and recontamination by undesired organisms (Bollen and Volker 1996; Franke-Whittle and Insam 2013). Not all biological agents are sensitive to the temperatures generated during composting. For instance, safe destruction of bacterial spores of *Clostridium* ssp. requires a temperature of 133 °C and 3 bar of pressure (Franke-Whittle and Insam 2013). For successful removal of prions, the material has to be exposed to 850 °C or alkaline hydrolysis (Franke-Whittle and Insam 2013), limiting the use of bone meal from BSE-infected cattle. Undesired human pathogens may also occur in connection with composting. Indeed, several serotypes of *Legionella* have been detected in the compost production chain, among these *L. pneumophila*, *L. bozemanii*, *L. micdadei*, *L. oakridgensis*, *L. jamestowniensis* and *L. cincinnatiensis* (Casati et al. 2010). *Legionella longbeachae* isolated from the growing medium of pot plants in the immediate vicinity of diseased individuals has been shown to survive for 7 months in growing medium kept at room temperature (Steele et al. 1990).

Digestates generated during anaerobic digestion (AD) and biogas production are viewed as hygienically safer than different types of non-treated farmyard manure. However, the source of substrate used for the fermentation process and its level of contamination dictate the level of hazard and need for sanitation before or after AD. Enteric bacteria, such as *E. coli* and *Salmonella*, are efficiently reduced during mesophilic digestion, whereas *Campylobacter*, *Clostridium* and *Bacillus* survive and elimination cannot be guaranteed during thermophilic digestion (Möller and Schultheiss 2014).

Safe storage conditions for organic fertilisers are essential to prevent their recontamination, irrespective of previous treatment.

Wildlife and Domestic Animals

Wildlife important for RA include birds, small rodents (mice, rats) and insects. Birds are well-known shedders of some microbial contaminants (Heddema et al. 2005; Nielsen et al. 2004; Wallace et al. 1997). Covered rooftop gardens (greenhouses, tunnels) are partly shielded from contamination by birds. Rodents and insects acting as vectors for biological contaminants are more difficult to control, even in RA. From a hygiene point of view, it is clearly questionable whether domestic animals, e.g. fowl, should be part of RA. The same applies to the presence of other domestic animals, such as dogs or cats (Gröndalen et al. 2008), which also act as carriers and shedders of multiresistant bacteria (Schaufler et al. 2015).

Cross Contamination

To avoid cross contamination, the cropping area on rooftops needs be well structured and compartmented, separating the zone of input means (e.g. compost, organic manure) from the zone of crop production, harvest and storage of harvested produce, as well as wastes, sanitary installations and food consumption. The transport streams for harvested produce should not cross those for input means. Furthermore, leakage from the storage area of organic manure or composts must be prevented.

Humans, both rooftop growers and visitors, pose a substantial route for cross contamination. The health status of all individuals visiting an RA system is essential. Hygiene awareness and cultural differences in perceptions of hygiene and health are crucial. A simple, but important and effective, approach is the maintenance of good hand hygiene (and glove hygiene). Special attention needs to be paid to compartmentalisation of rooftops in cases of mixed uses, for example RA and social contexts.

Sustainability Versus Produce Safety

Water

Water is a fundamental factor for sustainable urban development and sustainable urban horticulture. The intersection of these two domains is an arena for environmental, social and economic sustainability. The main focus in this section is on agricultural water, i.e. water used in RA. This water can be used for irrigation, fertiliser and pesticide application, frost protection and reduction of transpiration after harvest (Alsanius 2014a). The water can be obtained from rivers, streams, ditches, open canals, wells, municipal supplies, wastewater from kitchens and sewerage (Pimentel et al. 2004). Collected stormwater from heavy rain events is of particular interest for urban environments. The source of water dictates the safety of produce. If the water used during production and post-harvest handling of produce is contaminated, the produce may be contaminated and hence pose a health risk to consumers. The transmission of biological agents can be moderated through choice of irrigation method (canopy overhead irrigation, drip irrigation, subirrigation). However, contamination may also occur by splash from contaminated growing medium or plant parts to non-contaminated plants (Monaghan and Hutchinson 2012).

There is a tendency for urban agriculture practitioners to use sewerage water that is of questionable quality. This is mainly the case in developing countries that have fast-growing urban centres with limited access to clean water (Onyango et al. 2008). To prevent contamination of crops produced in RA, especially in urban centres, it is important to be aware of the microbial quality of water used during crop production. The hygiene quality of water sources should be tested regularly, at a frequency depending on the source of the water and the chances that the water may be contaminated. Critical limits for any type of contamination will depend on the kind and point of application and intended use (pre- or post-harvest) (CFSAN 1998). Very high safety standards, comparable to those for drinking water, are necessary for any post-harvest use. Growers should always ensure that the water used is free from contamination. If the quality of the water is uncertain, the producers and the handlers of the crops should apply good agricultural practices (GAPs) that minimise the risk of contamination arising from use of the water (CFSAN 1998).

Nutrients

As in the case of water, closed resource flow in urban settings is a vital part of the sustainable city concept and the reuse of waste material rich in plant nutrients is often mentioned as an incentive for urban plant food production. However, it is not only the presence of nutrients that is vital to plant biomass formation, but also the timing of nutrient supply in relation to crop requirements. This means that not all sorts of wastes are suitable for plant production. The requirements may vary between different crops, different developmental stages of the crop and quality criteria dictated by consumers. Public health is an important dimension of social and economic sustainability. To date, the linkage between public health and urban crop production as related to urban resource flows has been insufficiently studied on a scientific basis.

Plant Disease Control

Plant diseases and pests reach urban growing sites via the growing medium, on seeds and seedlings or with humans or pets as vectors. However, known disease and pest development patterns may be altered by differences in microclimate, lack of natural enemies or other factors caused by urban infrastructure. The diversity in urban and peri-urban areas with regard to cropping systems in and on buildings differs greatly in different parts of the world and investigations on plant disease control specifically focusing on urban horticulture are therefore still scarce. Overuse and misuse of pesticides due to lack of knowledge among non-professional growers is critical in edible crops (Lagerkvist et al. 2012; Ngowi et al. 2007).

Air

Pollutants in the air are difficult to control. Previous studies have considered airborne pollutants on a short distance basis from roads with heavy traffic to rooftop gardens (Binns et al. 2004), as well as pollutants translocated between countries (Kim et al. 2015). Residues on the leaves can be removed by washing, but have to be perceived as a health risk (Lagerkvist et al. 2012) when setting mitigation measures.

Needs for Maintained Produce Safety and Quality

Produce safety and quality need to be ensured throughout the production and supply chain. To maintain high quality and long shelf-life, produce needs to be stored in a continuous cooling chain from harvest to kitchen. Depending on the social framework of RA and the distance between production site and consumer, the cooling facilities on site and the cooling chain may be designed in different ways. Cooling slows down the proliferation of most microorganisms acting as foodborne pathogens, but notable exceptions are *Listeria monocytogenes* and *Yersinia enterocolitica*, which can tolerate and grow at refrigerator temperature. To ensure not only attractive, but also safe produce, transport of the harvested produce to the storage and packaging unit needs to meet operative standards for safety, e.g. individual food transport systems which are not used by the public. Furthermore, sanitary equipment to ensure personal hygiene has to be considered. All these requirements should be considered early during the planning process for the building.

Leadership and commitment are fundamental elements for produce safety and quality. It is a critical task for the owner or steering board to develop and implement a produce safety and quality framework and continuously monitor compliance. Furthermore, all RA participants or staff need to assign themselves to that framework. Well-designed safety and quality guidelines based on verbal and non-verbal communications need to be developed and distributed to all participants. Repeated training sessions for safety and quality need to be arranged. The framework should be based on identification, impact assessment of different critical hazard points and risk ranking. The existing GAP standards (FAO 2003) may be used for guidance. This includes the development and allocation of responsibility for a risk communication plan within the organisation or RA collective.

Conclusions

Although urban farming practices, including RA, are promoted as popular measures in sustainable city development, very little attention is given to produce quality and safety aspects. Ultimately, these are important for (i) competition (produce quality) of products originating from RA, especially when aiming at commercialisation of the produce and (ii) public health (produce safety and produce quality). Irrespective of the ambition level and size of the target consumer group, the risk of cross contamination needs to be acknowledged. There is a considerable need to integrate these aspects into urban farming concepts and awareness of them is required among stakeholders engaged in RA on all levels (politicians, municipalities, authorities, city planners, construction industry, advisory services, practitioners, consumers, researchers). This is ultimately a question of leadership. Knowledge gaps concerning produce quality and safety within RA must be met by relevant research and provision of guidelines, manuals and adequate advisory services.

Bullet Points

- Produce quality and food safety are key aspects with respect to environmental, economic and social sustainability. Methods, requirements and even legislation need to be adopted to the preconditions for urban production systems, especially rooftop agriculture.
- To meet quality experiences after purchase, quality expectations of horticultural product qualitys (food plant produce and ornamental plants) need to be formed before production and influence the primary production value network. Management and environmental factors Cultural management and environmental factors, such as light, temperature, wind, precipitation, biological damage, as well as postharvest management affect product quality and are thus decisive.

- To guarantee the health of RA product consumers, produce safety is an important feature and the urban food system must be secured. Legislation does not consider the specific challenges with RA and, in contrast to professional growers, RA amateurs often lack knowledge for the correct use of hazardous compounds and critical biological, chemical or physical hazards. Concepts for produce safety must be addressed in terms of legislation, construction, production and capacity building.

Acknowledgment This chapter was partly supported by the project "*African urban agriculture: Environmental, social and economic challenges and prospects under changing global and demographic realities*", funded by the Swedish research council Formas, which is gratefully acknowledged.

References

Abbate C, Borzi D, Caboni P, Baglieri A (2007) Behavior of fenhexamid in soil and water. J Environ Sci Health B 42(7):843–849. doi:10.1080/03601230701555088

Abbate C, Borzi D, Baglieri A, Gennari M (2009) Degradation of the fungicide pyrimethanil in water and soil. Agrochimica 53(6):1–7

Alsanius BW (2014a) Hygien och bevattning (trans: Technology DoBa), vol 2014:10. Swedish University of Agricultural Sciences, Alnarp

Alsanius BW (2014b) Mikrobiologiska faror i grönsakskedjan under primärproduktion, vol 2014:12. SLU, Alnarp

Alsanius BW, Bergstrand K-J (2014) Degradation of pesticides in nutrient solution from closed hydroponic systems. Acta Hortic 1034(2014):33–40

Alsanius BW, Bergstrand K-J, Burleigh S, Gruyer N, Rosberg AK (2013) Persistence of fenhexamid in the nutrient solution of a closed cropping system. Agric Water Manag 127:25–30

Alsanius BW, M Dorais, O. Doyle, F. Oancea, D. Spaddaro (2016) Potential food hazards from organic greenhouse horticulture. BioGreenhouse Fact sheet 17:3. COST project FA 1105 "BioGreenhouse". Wageningen. http://edepot.wur.nl/375668

Anjum SA, Xie X-Y, Wang L-C, Saleem MF, Man C, Lei W (2011) Morphological, physiological and biochemical responses of plants to drought stress. Afr J Agric Res 6(9):2026–2032

Antisari LV, Orsini F, Marchetti L, Vianello G, Gianquinto G (2015) Heavy metal accumulation in vegetables grown in urban gardens. Agron Sustain Dev 35(3):1139–1147

Atkinson NJ, Urwin PE (2012) The interaction of plant biotic and abiotic stresses: from genes to the field. J Exp Bot 63:3523–3543

Azapagic A, Chalabi Z, Fletcher T et al (2013) An integrated approach to assessing the environmental and health impacts of pollution in the urban environment: methodology and a case study. Process Saf Environ Prot 91(6):508–520. doi:10.1016/j.psep.2012.11.004

Barrett-Lennard EG (2003) The interaction between waterlogging and salinity in higher plants: causes, consequences and implications. Plant Soil 253(1):35–54

Berger CN, Sodha SV, Shaw RK et al (2010a) Fresh fruit and vegetables as vehicles for the transmission of human pathogens. Environ Microbiol 12(9):2385–2397. doi:10.1111/j.1462-2920.2010.02297.x

Beuchat LR (2006) Vectors and conditions for pre-harvest contamination of fruits and vegetables with pathogens capable of causing enteric diseases. Br Food J 108(1):38–53

Bihn EA, Gravani RB (2006) Role of good agricultural practices in fruit and vegetable safety. In: Matthews KR (ed) Microbiology of fresh produce. ASM Press, Washington, DC, pp 21–53

Binns HJ, Gray KA, Chen T et al (2004) Evaluation of landscape coverings to reduce soil lead hazards in urban residential yards: the safer yards project. Environ Res 96(2):127–138
Bollen GJ, Volker D (1996) Phytohygienic aspects of composting. In: Bartoldi M, Sequi P, Lemmes B, Papi T (eds) Science of composting. Blackie Academic and Professional, London, pp 233–246
Bourn D, Prescott J (2002) A comparison of the nutritional value, sensory qualities, and food safety of organically and conventionally produced foods. Rev in Food Sci Nutr 42(1):1–34
Briat J-F, Dubos C, Gaymard F (2014) Iron nutrition, biomass production, and plant product quality. Cell Press 20(1):33–40. doi:10.1016/j.tplants.2014.07.005
Casati S, Conza L, Bruin J, Gaia V (2010) Compost facilities as a reservoir of Legionella pneumophila and other Legionella species. Clin Microbiol Infect 16:945–947. doi:10.1111/j.1469-0691.2009.03009.x
CFSAN (1998) Guide to minimize microbial food safety hazards for fresh fruits and vegetables. In: http://wwwfoodsaftetygov/~acrobat/prodguidpdf. Accessed 27 Apr 2007
Chen LQ, Fang L, Ling J, Ding CZ, Kang B, Huang CH (2015) Nanotoxicity of silver nanoparticles to red blood cells: size dependent adsorption, uptake, and hemolytic activity. Chem Res Toxicol 28(3):501–509. doi:10.1021/tx500479m
Codex Alimentarius (2007) Principles and guidelines for the conduct of microbiological risk management (MRM), vol CAC/GL 63-2007. FAO, Rome, pp 1–15
Cramer GR, Urano K, Delrot S, Pezzotti M, Shinozaki K (2011) Effects of abiotic stress on plants: a systems biology perspective. BMC Plant Biol 11. doi:10.1186/1471-2229-11-163
Darko RO, Yuan S, Hong L, Liu J, Yan H (2016) Irrigation, a productive tool for food security - a review. Acta Agric Scand Sect B Soil Plant Sci 66(3):191–206. doi:10.1080/09064710.2015.1093654
De Pascale S, Martino A, Raimondi G, Maggio A (2007) Comparative analysis of water and salt stress-induced modifications of quality parameters in cherry tomatoes. J Hortic Sci Biotechnol 82:283–289
De Pascale S, Orsini F, Caputo R, Palermo MA, Barbieri G, Maggio A (2012) Seasonal and multiannual effects of salinisation on tomato yield and fruit quality. Funct Plant Biol 39:689–698
Dorais M, Alsanius BW (2015) Advances and trends in organic fruit and vegetable farming research. Hortic Rev 43:185–267
Dorais M, Ehret DL (2008) Agronomy and the nutritional quality of fruit. In: Tomás-Barberán FA, Gil MI (eds) Improving the health-promoting properties of fruit and vegetable products, 1st edn. Woodhead Publishing, Cambridge, pp 346–391
Dorais M, Papadopoulos AP, Gosselin A (2001) Influence of EC management on greenhouse tomato yield and fruit quality. Agronomie 21:367–384
Dorais M, Alsanius BW, Voogt W, Pépin S, Tüzel H, Tüzel Y, Möller K (2016) Impact of water quality and irrigation management on organic greenhouse horticulture. Wageningen, www.biogreenhouse.org. 98 pages. ISBN: 978-94-6257-538-7. doi:10.18174/373585
Eigenbrod C, Gruda N (2015) Urban vegetable for food security in cities. Agron Sustain Dev 35:483–498
European Parliament, European Council (2009a) Establishing a framework for community action to achieve the sustainable use of pesticides. Off J Eur Union 52:71–87
European Parliament, European Council (2009b) Legislation concerning the placing of plant protection products on the market and repealing council directives 79/117/EEC and 91/414/EEC. Off J Eur Union 52(309):1–50
European Union (1991) Council directive of 12 December 1991 concerning the protection of waters against pollution caused by nitrates from agricultural sources (91/676/EEC). In: European Council (ed) Official journal of the European communities, vol L 375. European Union, Brussels, p 8
FAO (2003) Development of a framework for good agricultural practices. In: Committee on Agriculture (ed). FAO, Rome, p 10

FAO/WHO (2006) Food safety risk analysis A guide for national food safety authorities. In: FAO/WHO (ed). FAO/WHO, Rome

FAO/WHO (2014) Codex Alimentarius: principles and guidelines for the conduct of microbiological risk assessment. In: Codex Alimentarius (ed) CAC/GL 30-1999. FAO/WHO, Rome, pp 1–5

Fett W (2006) Seed sprouts: the state of microbiological safety. In: Matthews KR (ed) Microbiology of fresh produce. Emerging issues in food safety. ASM, Washington, DC, pp 167–220

Franke-Whittle I, Insam H (2013) Treatment alternatives of slaughterhouse wastes, and their effect on the inactivation of different pathogens: a review. Crit Rev Microbiol 39(2):139–151. doi:10.3109/1040841X.2012.694410

Gallaher CM, Mwaniki D, Njenga M, Karanja NK, Winkler Prins AM (2013) Real or perceived: the environmental health risks of urban sack gardening in Kibera slums of Nairobi, Kenya. EcoHealth 10(1):9–20. doi:10.1007/s10393-013-0827-5

Gerba CP (2009) The role of water and water testing in produce safety. In: Fan BA, Niemira CJ, Doona FE, Feeherry, Gravani RB (eds) Microbial safety of fresh produce. X. IFT Press/Wiley-Blackwell, Ames, pp 129–142. ISBN: 978-0-8138-0416-3

Gröndalen J, Saevik B, Sörum H (2008) Companion animals as reservoir for zoonotic diseases. Eur J Campanion Anim Pract 18(3):213–222

Gruda N (2005) Impact of environmental factors on product quality of greenhouse vegetables for fresh consumption. Crit Rev Plant Sci 24(3):227–247. doi:10.1080/07352680591008628

Grunert KG (2005) Food quality and safety: consumer perception and demand. Eur Rev Agric Econ 32(3):369–390. doi:10.1093/eurrag/jbi011

Heddema ER, ter Sluis S, Buys JA, Vandenbroucke-Grauls CMJE, van Wijnen JH, Visser CE (2005) Prevalence of Chlamydophila psittaci in fecal droppings from feral pigeons in Amsterdam, The Netherlands. Appl Environ Microbiol 72:4423–4425

Hepler PK (2005) Calcium: a central regulator of plant growth and development. Plant Cell 17(8):2142–2155. doi:10.1105/tpc.105.032508

Herklotz PA, Gurung P, Vanden Heuvel B, Kinney CA (2010) Uptake of human pharmaceuticals by plants grown under hydroponic conditions. Chemosphere 78:1416–1421

Imrich A, Ning YY, Kobzik L (2000) Insoluble components of concentrated air particles mediate alveolar macrophage responses in vitro. Toxicol Appl Pharmacol 167(2):140–150

Iwasa M, Makino S, Asakura H, Kobori H, Morimoto Y (1999) Detection of Escherichia coli O157:H7 from Musca domestica (Diptera: Muscidae) at a cattle farm in Japan. J Med Entomol 36:108–112

Jayawardena U, Tollemark L, Tagesson C, Leanderson P (2009) Pyrogenic effect of respirable road dust particles. J Phys Conf Ser 151(012015):1–9. doi:10.1088/1742-6596/151/1/012015

Karnjanapiboonwong A, Chase DA, Cañas JE et al (2011) Uptake of 17a-ethynylestradiol and triclosan in pinto bean, Phaseolus vulgaris. Ecotoxicol Environ Saf 74:1336–1342

Kim HS, Kim R-K, Lim G-H, Kim J-W, Kim K-H (2015) Influence of airborne dust on the metal concentrations in crop plants cultivated in a rooftop garden in Seoul. Soil Sci Plant Nutr 61:1–10. doi:10.1080/00380768.2015.1028873

Kreuger J, Graaf S, Patring J, Adielsson S (2010) Pesticides in surface water in areas with open ground and greenhouse horticultural crops in Sweden 2008 Ekohydrologi, vol 117. Division of Water Quality Management, Uppsala, p 49

Lagerkvist CJ, Ngigi M, Okello JJ, Karanja N (2012) Means-end chain approach to understanding farmers' motivations for pesticide use in leafy vegetables: the case of kale in peri-urban Nairobi, Kenya. Crop Prot 39:72–80. doi:10.1016/j.cropro.2012.03.018

Lohrberg F, Licka L, Scazzosi L, Timpe A (2015) Urban agriculture Europe. JOVIS, Berlin

Lovarelli D, Bacenetti J, Fiala M (2016) Water footprint of crop productions: a review. Sci Total Environ 548:236–251. doi:10.1016/j.scitotenv.2016.01.022

Maghos F, Fotini A, Aampelas A (2006) Organic food: buying more safety or just peace of mind? A critical review of the literature. Crit Rev Food Sci Nutr 46:23–56

Martin-Diana AB, Rico D, Frias JM, Barat JM, Henehan GTM, Barry-Ryan C (2007) Calcium for extending the shelf life of fresh whole and minimally processed fruits and vegetables: a review. Trends Food Sci Technol 18(4):210–218. doi:10.1016/j.tifs.2006.11.027

Matthews KR, Sapers GM, Gerba CP (2014) The produce contamination problem, 2nd edn. Academic, Amsterdam
Mattsson K, Ekvall MT, Hansson A, Linse S, Malmendal A, Cedervall T (2015) Altered behavior, physiology, and metabolism in fish exposed to polystyrene nanoparticles. Environ Sci Technol 49(1):553–561
Möller K, Schultheiss U (2014) Organische Handelsdüngemittel im ökologischen Landbau, vol 499. KTBL, Darmstadt
Monaghan JM, Hutchinson ML (2012) Distribution and decline of human pathogenic bacteria in soil after application in irrigation water and the potential for soil-splash-mediated dispersal onto fresh produce. J Appl Microbiol 112:1007–1019. doi:10.1111/j.1365-2672.2012.05269.x
Mutuku MS (2013) Lead, cadmium and zinc speciation in garage soils, their levels in kales and water along Katothyani stream. Kenyatta University, Machakos
Naser HM, Sultana S, Mahmud NU, Gomes R, Noor S (2012) Heavy metal levels in vegetables with growth stage and plant species variations. Bangladesh. J Agric Res 36(4):563–574. doi:http://dx.doi.org/10.3329/bjar.v36i4.11743
National Research Council (NCR) (2010) Toward sustainable agricultural systems in the 21st century. National Academies Press, Washington, DC
Ngowi AV, Mbise FTJ, Ijani ASM, London L, Ajayi OC (2007) Smallholder vegetable farmers in northern Tanzania: pesticides use practices, perceptions, cost and health effects. Crop Prot 26(11):1617–1624
Nielsen EM, Skov MN, Madsen JJ, Lodal J, Jespersen JB, Baggesen DL (2004) Verocytotoxin-producing Escherichia coli in wild birds and rodents in close proximity to farms. Appl Environ Microbiol 70(11):6944–6947. doi:10.1128/AEM.70.11.6944-6947.2004
Onyango MC, Shibairo SI, Imungi JK, Harbinson J (2008) The physico-chemical characteristics and some nutritional values of vegetable amaranth sold in Nairobi-Kenya. Ecol Food Nutr 37(4):382–398
Orsini F, Maggio A, Rouphael Y, De Pascale S (2016) Physiological quality of organically grown vegetables. Sci Hortic 208:131–139
Park S, Cheng NH, Pittman JK et al (2005) Increased calcium levels and prolonged shelf life in tomatoes expressing Arabidopsis H+/Ca-2 transporters. Plant Physiol 139(3):1194–1206. doi:10.1104/pp.105.066266
Pimentel D, Berger B, Filiberto D et al (2004) Water resources: agricultural and environmental issues. Bioscience 54(10):909–918
Redshaw CH, Wootton WG, Rowland SJ (2008) Uptake of the pharmaceutical fluoxetine hydrochloride from growth medium by Brassicaceae. Phytochemistry 69:2510–2516
Ripoll J, Urban L, Staudt M, Lopez-Lauri F, Bidel LPR, Bertin N (2014) Water shortage and quality of fleshy fruits-making the most of the unavoidable. J Exp Bot 65(15):4097–4117. doi:10.1093/jxb/eru197
Rouphael Y, Cardarelli M, Bassal A, Leonardi C, Giuffrida F, Colla G (2012) Vegetable quality as affected by genetic, agronomic and environmental factors. J Food Agric Environ 10(3–4):680–688
Salonen RO, Hälinen AI, Penanen AS et al (2004) Chemical and in vitro toxicologic characterization of wintertime and springtime urban-air particles with an aerodynamic diameter below 10 microm in Helsinki. Scand J Work Environ Health 30:80–90
Säumel I, Kotsyuk I, Hölscher M, Lenkereit C, Weber F, Kowarik I (2012) How healthy is urban horticulture in high traffic areas? Trace metal concentrations in vegetable crops from plantings within inner city neighbourhoods in Berlin, Germany. Environ Pollut 165:124–132
Schaufler K, Bethe A, Lübke-Becker A et al (2015) Putative connection between zoonotic multiresistant extended-spectrum beta-lactamase (ESBL)-producing Escherichia coli in dog feces from a veterinary campus and clinical isolates from dogs. Infect Ecol Epidemiol 5:25334
Sela S, Nestel D, Pinto R, Nemny-Lavy E, Bar-Joseph M (2005) Mediterranean fruit fly as a potential vector of bacterial pathogens. Appl Environ Microbiol 71:4052–4056
Shahid U, Muhammad I (2007) Nitrate accumulation in plants, factors affecting the process, and human health implications. A review. Agron Sustain Dev 27(1):47–57

Shenker M, Harush D, Ben-Ari J, Chefetz B (2011) Uptake of carbamazepine by cucumber plants – a case study related to irrigation with reclaimed wastewater. Chemosphere 82:905–910
Steele TW, Lanser J, Sangster N (1990) Isolation of Legionella longbeachae serogroup 1 from potting mixes. Appl Environ Microbiol 56(1):49–53
Talley JL, Wayadande AC, Wasala LP et al (2009) Association of Escherichia coli O157:H7 with filth flies (Muscidae and Calliphoridae) captured in leafy greens fields and experimental transmission of E. coli O157:H7 to spinach leaves by house flies (Diptera: Muscidae). J Food Prot 72:1547–1552
Tauxe R, Kruse H, Hedberg C, Potter M, Madden J, Wachsmuth K (1997) Microbial hazards and emerging issues associated with produce: a preliminary report to the National Advisory Committee on microbiologic criteria for foods. J Food Prot 60(11):1400–1408
Udikovic-Kolic N, Wiechmann F, Broderick NA, Handelsman J (2014) Bloom of resident antibiotic-resistant bacteria in soil following manure fertilization. Proc Natl Acad Sci 111(42):15202–15207. doi:10.1073/pnas.1409836111
Vardoulakis S, Fisher BEA, Pericleous K, Gonzalez-Flesca N (2003) Modelling air quality in street canyons: a review. Atmos Environ 37(2):155–182. doi:10.1016/s1352-2310(02)00857-9
Wallace JS, Cheasty T, Jones K (1997) Isolation of vero cytotoxin-producing Escherichia coli O157 from wild birds. J Appl Microbiol 82:399–404
Warriner K, Smal B (2015) Microbiological safety of sprouted seeds: interventions and regulations. In: Matthews KR, Sapers GM, Gerba CP (eds) The produce contamination problem: causes and solutions, Food Science and Technology, International Series, 2nd edn. Elsevier/Academic, Amsterdam, pp 237–268
Whittinghill LJ, Rowe DB, Cregg BM (2013) Evaluation of vegetable production on extensive green roofs. Agroecology Sustain Food Syst 37(4):465–484. doi:10.1080/21683565.2012.756847
Winker M, Clemens J, Reich M, Gulyas H, Otterpohl R (2010) Ryegrass uptake of carbamazepine and ibuprofen applied by urine fertilization. Sci Total Environ 408:1902–1908
Winter CK, Davis SF (2006) Organic foods. Scientific status review. J Food Sci 7(9):R117–R124

Part IV
Multifunctional Rooftop Agriculture

Francesco Orsini

Rooftop Gardening for Improved Food and Nutrition Security in the Urban Environment

W. Baudoin, Y. Desjardins, M. Dorais, U.R. Charrondière, L. Herzigova, U. El-Behairy, N. Metwaly, C. Marulanda, and N. Ba

Abstract With the expanding population and the shrinking area for conventional horticulture and home gardens within cities, the idea of roof gardening in towns has evolved. New technologies have been developed to grow a series of horticultural crops in different types of containers. Information is provided on the productivity of different technologies used for roof-top gardening. A review is made of the diversity of species and cultivars which can be grown on the roofs of buildings, and of how they can contribute to supplying a variety of nutrients, which help to meet the requirements for a healthy diet and contribute to the food and nutrition security of the increasing urban population.

Disclaimer The views expressed in this publication are those of the author(s) and do not necessarily reflect the views or policies of the Food and Agriculture Organization of the United Nations.

W. Baudoin (✉) • U.R. Charrondière • L. Herzigova
Food and Agriculture Organisation of the United Nations (FAO), Rome, Italy
e-mail: wilfried.baudoin@gmail.com; ruth.charrondiere@fao.org; lucie.herzigova@gmail.com

Y. Desjardins • M. Dorais
Laval University, Quebec City, QC, Canada
e-mail: yves.desjardins@fsaa.ulaval.ca; martine.Dorais@AGR.GC.CA

U. El-Behairy
Ain Shams University, Cairo, Egypt
e-mail: el_behairy2003@hotmail.com

N. Metwaly
Agriculture Research Center, Giza, Egypt
e-mail: neveen2000eg@yahoo.com

C. Marulanda
Food and Agriculture Organization of the United Nations (FAO), Bogotà, Colombia
e-mail: cesarmarulanda51@hotmail.com

N. Ba (deceased)
Ministry of Agriculture, Dakar, Senegal

F. Orsini et al. (eds.), *Rooftop Urban Agriculture*, Urban Agriculture,
https://doi.org/10.1007/978-3-319-57720-3_13

Introduction

As described in **Part 2**, roof-top gardens are grown in different shapes and sizes. They are essentially container-based production systems using different technologies ranging from plants grown in pots filled with soil or substrates, to more sophisticated and sometimes automated hydroponic or aeroponic systems. Organic roof-top gardening is also possible, in compliance with guidelines relating to the characteristics of growing media and compost; seed and seedlings as well as non-chemical pest and disease control; fertilizers; biostimulants; irrigation water management; hygiene and sanitation. (Dorais 2016).

Roof gardening is a component of a global endeavor to "green" the cities developed in the "Growing Greener Cities" programme framework (FAO 2015). Besides their primary role of facilitating access to nutritious food, roof-top gardens play a special role in lowering the temperature inside the building and have a potential social benefit by bringing people together (El Behairy 2012). From a food and nutrition point of view, roof gardens help to increase the availability of, and facilitate the access to fresh fruits and vegetables, which is perceived as a contribution to balanced diets for all and to the decrease of malnutrition affecting especially the urban poor.

Productivity and Efficiency

Roof-top gardens can be highly productive and therefore most efficient in terms of amounts harvested and water used per cultivated area and per year. Studies conducted by the Food and Agriculture Organization of the United Nations (FAO) have shown that a roof garden of one square meter could produce a seizable amount of fruits and vegetables, as shown in Table 1.

Crops can be grown in monoculture systems or in mixed cropping, associating short-cycle crops with medium- to long-cycle crops, e.g. lettuce and tomato, or lettuce and zucchini; also associating perennial herbs and condimental species like chive, thymus, laurus and rosemary, with any annual vegetable crop. Border areas can be planted with short growth habit plants, like strawberry or cherry radishes, while the central part is grown with taller species, like eggplant or okra (Fig. 1).

Table 1 Examples of commodities produced per year within 1 m^2 of roof garden in Senegal (project data GDCP/SEN/002/ITA)

Commodities	Number of unit or kg	Average time requested
Lettuce	36 units	Every 60 days
Cabbage	10 units	Every 90 days
Potato	10 kg	Within 100 days
Tomato	100 units	Within 180 days
Leafy condiments (e.g. mint)	2 bunches	Every day

Fig. 1 Mixed cropping (Photo: W. Baudoin)

Productivity can be further increased and improved by covering the roof garden with a net that will create a protected environment, with the multiple functions of providing shade, while protecting against insects, wind, dust and bird droppings (Fig. 2). The physical protection of the crop can be further improved according to the need to protect against low/high temperatures or heavy rains. Different types of simple lightweight shelter structures can be established on a roof using polyethylene, polycarbonate or similar transparent materials (Fig. 3). Simple greenhouse structures with a saddle roof or round-arched roof can be equipped with gutters. This will allow to harvest the rainwater (Fig. 4). On part of the roof, photovoltaic panels can be placed to generate the needed power for an automated drip irrigation system, sourcing the water from a reservoir filled with rainwater and tapped up with water from the city distribution network as needed.

In terms of productivity and efficiency, roof garden systems are a living example of what can be achieved more with less in line with the save of grow principles (FAO 2011).

- ***More yield*** *per unit of cultivated surface, water and time*
- ***More diversity*** *in a small space*
- ***More people*** *in the family can grow a roof garden, young and elderly, as well as disabled*

- ***Less space*** *occupied*
- ***Less soil*** *or no soil*
- ***Less water***
- ***Less*** *soil-borne* ***disease***
- ***Less pesticide***

Fig. 2 Simple and lightweight net-house for rooftop gardens (Photo: W. Baudoin)

Fig. 3 Simple lightweight shelter structures for roof gardens (Photo: W. Baudoin)

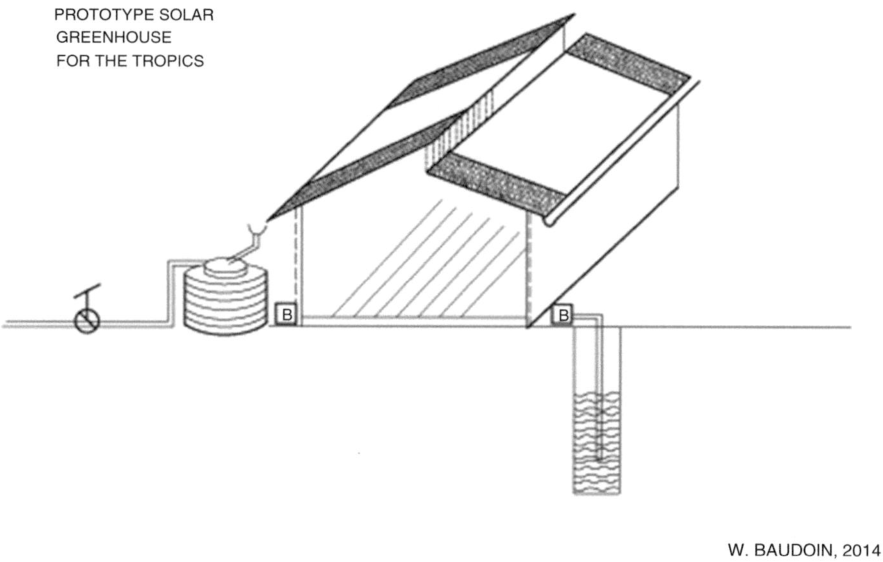

Fig. 4 Rainwater harvesting in small, simple greenhouse for roof gardens in the tropics

- ***Less fertilizer*** *or no mineral fertilizers*
- ***Less physical effort***
- ***Less transport and packaging***
- ***Less food waste***

Smart innovations have made roof gardens still more productive, like the **"cubic-garden"**, system, which takes advantage of the volume available and not only of the area (FAO 2016b). This is possible with climbing or hanging plants like various mint cultivars, Ceylon spinach and water spinach (Figs. 5 and 6).

From field observations, it is considered that a 1-m^2 roof garden can provide a series of diverse condiments to daily enrich and flavor the meals of the family. A 10-m^2 roof garden will allow growing a variety of crops to enhance the consumption of fruit and vegetables as recommended by WHO and FAO (WHO, 2003 http://apps.who.int/iris/bitstream/10665/42665/1/WHO_TRS_916.pdf). A 40-m^2 will offer surplus for sale and generate income to meet modest expenses and improve the livelihood of the family (Baudoin 2014).

Selected data on the field performances of horticulture cultivars, extracted from HORTIVAR (www.fao.org/hortivar) illustrate the productivity presented in Table 2. In South America, the yields obtained are shown in Table 3.

Food that is grown and consumed in cities has other advantages: During times of abundance, it may cost less than supermarket fare that has come long distances, and during times of emergency – when transportation and distribution channels break down – it can contribute filling a vegetable supply void.

Fig. 5 A cubicgarden with Ceylon spinach in Dakar, Senegal (Photo: W. Baudoin)

Fig. 6 A cubicgarden with Spanish mint in Dakar, Senegal (Photo: W. Baudoin)

Table 2 Productivity of rooftop agriculture systems

Species	Cultivar	Site	Density (plants/$m^{2)}$	Cycle length (days)	Yield (kg/ m^2/cycle)
Head cabbage	Africa cross F1	Senegal, Dakar	8	78	5.1
Eggplant	AB-1 AB-2	DR Congo, Kisangani	3.33	130	1.75
Garlic	Ajo-local	Bolivia, El Alto	36	258	4.2
Cucumber	Beregovoi	Russian Federation	2.1	134	0.9
Corn salad	Big Holland	Bolivia, El Alto	400	90	3.6
Swiss chard	Greenwave	Namibia, Katutura	16	128	10
Radish	Supun	Mapalana, Sri Lanka	1,5	150	0,9
Cabbage	Ditmar	Vinica, FYR Macedonia	5	90	5
Eggplant	Black Beauty	El Alto, Bolivia	12	121	5,6
Water cress	Amarilla	El Alto Bolivia	16	210	4.31

FAO Hortivar (2016a)

Table 3 Yields in South America

Species	Species	Density (plants / m^2)	Cycle length (days)	Yield (kg/ m^2/cyle)
Common name	Botanical name			
Chinese cabbage (F.S)	*B oleracea* var. *chinensis*	10	36	9
Swiss chard (S.S.)	*B. vulgaris* var. *cycla*	10	52	5
Basil (F.S.)	*O. bassilicum*	8	55	6
Leaf celery (F.S.)	*A. graveolens*	17	65	8
Sweet potato (S.S.)	*I. batata*	3	85	4.5
Water cress (F.S.)	*N. officinalis*	7	50	3.5
Welsh onion (S.S.)	*A. fistulosum*	17	62	7.5
Head lettuce abierta (F.S.)	*L. sativa*	21	42	10
Cherry tomato (S.S.)	*L. pimpinelifolium*	2	85	2
Salad tomato (S.S.)	*L. esculentum*	2	90	4.5

Field notes by César H. Marulanda Tabares – FAO, in 12 Latin American countries
F.S. floating system (hydroponic cultivation), *S.S.* solid substrate cultivation

Vegetable Production on Extensive Green Roofs

Rooftop vegetable gardening is a production system in urban agriculture, based on green roof technology. Three growing systems – a 10-cm deep green roof, raised green roof platforms with 10 cm of substrate, and in-ground – were evaluated for vegetable and herb production over three growing seasons (2009–11). Tomatoes (*Solanum lycopersicum*), green beans (*Phaseolus vulgaris*), cucumbers (*Cucumis sativus*), peppers (*Capsicum annuum*), basil (*Ocimum basilicum*) and chives (*Allium schoenoprasum*) were studied because of their common use in home gardens. Results suggest that with proper management, vegetable and herb production in an extensive green roof system is possible and productive (Whittinghill and Rowe 2012).

Recent reports on *The Ecologist* suggested that urban agriculture could supply all of London with substantial amounts of fruit and vegetables if part of the 1650 hectares of roof space would be to be converted to growing space. The city has 20,000 ha of roof space, most of which is pitched and residential. By converting half this area to commercial grade horticultural greenhouses, a very conservative annual productivity of 40 kg per m^2 could supply all of the 8.2 million people with 1.3 kg of fruits and vegetables per day – far more than the average UK daily consumption per person of 346 g (Dring 2014). A case study from Bologna, a city in Italy with some 375,000 inhabitants, suggests that if all available and suitable flat roof space (0.82 km^2) were utilized for urban agriculture, roof-top gardens in the city could supply around 12,500 tons of vegetables a year. This means that, based on actual consumption data for the city, roof-top gardens could meet 77% of residents' needs for vegetables, while also providing a range of ecosystem services, according to the researchers (European Commission 2015).

Species and Cultivars

Roof-garden systems allow growing a broad range of short-cycle crops, and reaching high levels of yields per unit of area, time and water, which make them attractive as a model for small-scale crop diversification and intensification. There are a wide range of vegetables and fruit crops as well as root and tuber crops, ornamentals, condiments and medicinal plants that can be grown on a roof providing nutritious food or for ornamental purposes (Table 4 and Fig. 7). Species and cultivars will vary according to the agroecological zone and the season as well as the consumer and market requirements.

Of special interest for roof gardens and exiguous places, are plants and cultivars with a dwarf vegetative development (McLaughlin 2011). They are cultivated either for their commercial or for their nutritional value. They enable taking advantage of little spaces or adopting higher planting densities to increase yields per unit of area and time. Examples of dwarf plant species and cultivars can be found in specialized literature and in commercial seed catalogues. Dwarf cultivars do exist for a series of species like: cauliflower, broad bean, chili pepper, sweet pepper, golden berry, snap beans, sweet pea and dill.

Table 4 Common species grown in roof-top gardens

Common name	*Latin name*
Leafy vegetables	
Broccoli	*Brassica oleracea* var. *italica*
Brussel sprouts	*Brassica oleracea* var.
Cabbage (green)	*Brassica oleracea* convar. *capitata* var. *alba*
Chicory	*Cichorium intybus* L.
Chicory var. Treviso	*Cichorium intybus* L.
Curly endive	*Cichorium endivia* var. *crispum*
Endive	*Cichorium endivia* var. *latifolium*
Fennel	*Foeniculum vulgare* Mill.
Kale	*Brassica oleracea* var. *sabellica*
Lettuce (butter head)	*Lactuca sativa* L.
Lettuce (crispy)	*Lactuca sativa* L.
Lettuce (roman)	*Lactuca sativa* L. var. *longifolia*
Malabar spinach; Ceylon spinach	*Basella alba*
New Zealand Spinach, Tetragon	*Tetragonia tetragonioides*
Rocket salad	*Eruca sativa* Miller
Savoy cabbage	*Brassica oleracea* var. *sabauda* L.
Silverbeet, Swiss chard	*Beta vulgaris* L. var. *cicla*
Spinach	*Spinacia oleracea*
Water spinach	*Ipomoea aquatica*
Water cress	*Nasturtium*
Garden cress	*Lepidium sativum*
Water cress	*Nasturtium officinale*
Amaranth	*Amaranthus* spp.
Black night shade	*Solanum nigrum*
Fruits and fruiting vegetables	
Cape gooseberry	*Physalis peruviana*
Cucumber	*Cucumis sativus* L.
Eggplant	*Solanum melongena*
Green bean (pod climbing)	*Phaseolus officinalis*
Green bean (pod dwarf)	*Phaseolus officinalis*
Pea (sweet, mangetout)	*Pisum sativum*
Pumpkin	*Cucurbita maxima*
Spaghetti squash	*Cucurbita pepo* subsp. *pepo*
Strawberry	*Fragaria* x *annanassa* Duch.
Sweet pepper	*Capsicum annuum* L.
Tomato	*Solanum lycopersicum*
African eggplant (scarlet eggplant)	*Solanum aethiopicum*
Zucchini	*Cucurbita pepo*
Medicinal plants (neutroceuticals)	
Aloe	*Aloe vera*
Artemisia	*Artemisia annua*

(continued)

Table 4 (continued)

Common name	*Latin name*
Camomilla	*Matricaria chamomilla* L.
Mélisse	*Melissa officinalis*
Condiments (herbs and spices)	
Basil	*Ocimum basilicum* L.
Bay	*Laurus nobilis* L.
Celery (branch)	*Apium graveolens* var. *dulce* (Mill.) Pers.
Chervil	*Anthriscus cerefolium*
Chive	*Allium schoenoprasum* L.
Cicely	*Myrrhis odorata*
Coriander	*Coriandrum sativum* L.
Dill	*Anethum graveolens*
Fenugreek	*Trigonella foenum-graecum*
Hot pepper	*Capsicum frutescens* L.
Lemon grass	*Cymbopogon citratus*
Marjoram	*Origanum majorana*
Mint	*Mentha* L.
Oregano	*Origanum vulgare* L.
Parsley	*Petroselinum crispum*
Rosemary	*Rosmarinus officinalis* L.
Sage	*Salvia officinalis*
Satureja	*Satureja* L.
Stevia	*Stevia rebaudiana*
Tarragon	*Artemisia dracunculus*
Thyme	*Thymus vulgaris* L.
Bulb, root and tuber vegetables	
Carrot	*Daucus carota* L.
Celery (root)	*Apium graveolens* var. *dulce* (Mill.) Pers.
Garlic	*Allium sativum* L.
Onion	*Allium cepa* L. var. *cepa*
Potato	*Solanum tuberosum* L.
Radish (cherry)	*Raphanus sativus*
Red beet	*Beta vulgaris* L. var. *crassa*
Scallion	*Allium cepa*
Shallot	*Allium cepa* L. var. *aggregatum* G. Don.
Sweet potato	*Ipomoea batatas* (L.) Lam.
Turnip	*Brassica rapa* subsp. *rapa*
Fruit trees	
Moringa	*Moringa oleifera L.*
Peach	*Prunus persica L. Batsch*
Olive	*Olea europea L.*
Mandarin, tangerine	*Citrus reticulata Blanco x*
Orange, sweet	*Citrus sinensis L.Osbeck*

Table 4 (continued)

Common name	*Latin name*
Lime, tahiti	*Citrus latifolia Tan.*
Passion fruit	*Passiflora edulis*
Kiwi	*Actinidia chinensis Planch.*
Grape	*Vitis vinifera*
Pomegrenate	*Punica granatum L.*
Banana, dwarf	*Musa acuminata (dwarf cavendish)*

Data compiled by Wilfried Baudoin and Lucie Herzigova. Information about the characteristics and field performances of horticulture cultivars can be retrieved from the FAO website: www.fao.org/hortivar

Fig. 7 A broad range of different species can be grown on rooftops (Photo: W. Baudoin)

Contribution to Food and Nutrition Security

Horticulture varieties are renowned for their content in vitamins, minerals and micronutrients, and bioactive compounds including polyphenol, and carotenoids with antioxidant action, which make them a unique "wealth" of "health" ingredients. As such, horticulture crops are the ideal "companion" crops to meet a family's needs for a well-balanced diet. With roof-garden systems, families can grow their own vegetables and pick frequently some fresh produce to enrich the menu.

As a result of the broad range of horticulture plants which can be grown in roof gardens, they can play a key role in improving the diet of urban dwellers and can reduce the risk of micronutrient deficiencies, when combined with nutrition education leading to increased consumption of a diversified range of vegetables and condiments grown on the roof.

The reduced time between production and consumption in an urban garden can lead to the nutrient content of produce being higher. For example, in conventional production, the loss in nutrients can be as high as 30–50% in the 5–10 days it takes to travel from farm to table (Bellows et al. 2013).

Fruit and vegetables are an important component of a healthy diet and, if consumed daily in sufficient amounts, could help prevent major diseases such as Cardio Vascular Diseases (CVDs), and certain cancers. According to The World Health Report 2002, low fruit and vegetable intake is estimated to cause about 31% of ischemic heart disease and 11% of strokes worldwide (WHO 2002). Overall it is estimated that up to 2.7 million lives could potentially be saved each year if fruit and vegetable consumption was sufficiently increased. Recommendations in this direction tend to complement and reinforce other valid messages based on the long-known health benefits of consuming vegetables and fruit as dietary sources of fiber, vegetable proteins and protective micronutrients. The joint FAO/WHO Expert Consultation on diet, nutrition and the prevention of chronic diseases (WHO, 2003 http://apps.who.int/iris/bitstream/10665/42665/1/WHO_TRS_916.pdf), recommended the intake of a minimum of 400 g of fruit and vegetables per day (excluding potatoes and other starchy tubers) for the prevention of chronic diseases such as heart disease, cancer, diabetes and obesity, as well as for the prevention and alleviation of malnutrition, especially micronutrient deficiencies, especially in less developed countries. The recommendation thus adds to the already strong case for the health benefits to be gained from the consumption of fruit and vegetables and paves the way for concrete action advocating increased consumption of these commodities (WHO-FAO 2004). The production in roof top gardens of 0.7 to 12 kg of vegetable, spices or fruits per m2 and per year, could increase substantially the supply of these healthy foods.

Crop production could be expressed in "Nutrient Productivity", which is proposed by FAO to assess the real potential of vegetables, herbs or fruits to contribute to a sustainable diet and reduce the risk of micronutrient deficiencies. The nutrient productivity is expressed as the percentage of DRI (Daily Reference Intakes), to be met by 10 adults per year from an agricultural product produced in 1 ha per year, either for one or for all selected nutrients, which are namely energy, protein, dietary fiber, Iron, Zinc, Calcium, Vitamin A, Vitamin C and folate. (FAO 2017 in preparation)

Nutritional Value

Vegetables have a great importance to human health for the following reasons:

1. ***Vegetables fruits and herbs are a good source of many nutrients, especially micronutrients such as minerals and vitamins***, as illustrated in Table 5.
2. ***Vegetables, fruits and herbs – particularly rich in fiber – stimulate the small intestine movement*** to reduce constipation and feed a healthy microbiota. The most important vegetables: leafy vegetables, such as cabbage, spinach, and lettuce are appreciated for their satietogenic properties. Generally, all vegetables can be considered as filling materials, especially leafy vegetables and root vegetables.

Table 5 Nutrient composition of selected plants

Nutrient	Name of plants and their richness in certain nutrients.Compositional data is expressed per 100 g edible portion on fresh weight basis
Carbohydrates	Potato: 16,9 g* – Taro: 19g/20g (raw/boil)*
Calcium	Turnip leaves: 41 mg* – Parsley: 217 mg* $$
Iron	Parsley: 4.9 mg$$ – Spinach: 2.5–3.1 mg (boil/raw) $ – Kale: 1.47 mg – Peas: 2 mg
Vitamin A	Orange and dark green colored ones: Carrots: 835–852 mcg (raw/boil)* $$ – Spinach: 387–409 (boil/raw) mcg*$$ – Turnip leaves: 579 mcg*$$ – Parsley: 583 mcg*$$ – Orange sweet potato: 377–397 (boil/raw) mcg*$ – Pumpkin: 100–104 mcg (raw/boil)* – Broccoli: 31–77 mcg (raw/cooked)*
Vitamin B_6	Turnip leaves: 0.263 mg* – Mushroom 0.04–0.293 mg – Parsley: 0.22 mg* – Okra – Okra: 0.17–0.22 (boil/raw) mg* Spinach: 0.13–0.19 mg*
Niacin	Mushrooms: high* – Sweet corn: 1.7 mg* – Potato: 0.9–1.2 mg (boil/raw) – Taro: 0.8–0.6 g (raw/boil) – Okra 0.5–0.7 mg (boil/raw) – Asparagus: 0.978–1.08 mg (raw/boil)*.
Vitamin C	Parsley: 175 mg*$$ – Turnip leaf: 60 mg$$ – Broccoli: 65–8 9 mg (boil/raw)* $$ – Brussels sprout: 62–85mg (boil/raw)* $$ – Cauliflo wer: 48 mg (raw) $$ – Spinach: 15–36 mg$$ – Cabbage: 22–54 mg (boil/raw) $$ – Green beans: 12 mg$ – Okra: fruit: 19–28 (boil/raw) mg$$ –Tomato: 23–30 mg* Sweet pepper (red, green or yellow): 80–110(boil) 110–160 (raw) mg *$$ –Sweetpepper(red,greenoryellow): 80 – 110(boil)110 – 160(raw)mg*

*a good source according to FAO (2013 Eating well for good health. Lessons on nutrition and healthy diets http://www.fao.org/do crep/017/i3261e/i3261e00.htm) p. 122–123 and 129–131 source according to FAO (2005). Codex nutritional labelling ftp://ftp.fao.org/docrep/fao/005/y2770E00.pdf $ high source according to FAO (2005) Codex nutritional labelling ftp://ftp.fao.org/docrep/fao/005/y2770E/y2770E00.pdf
no standard exists for this nutrient in FAO (2005) Codex nutritional labelling ftp://ftp.fao.org/docrep/fao/005/y2770E/y2770E00.pdf. Sources of composition data: FAO (2012) West African Food Composition Table (http://www.fao.org/infoods/infoods/tables-and-databases/faoinfoods-datab ases/en/) and USDA SR28 (2015) USDA National Nutrient Database for Standard Reference (https://www.ars.usda.gov/northeast-area/beltsvillemd/beltsville-human-nutrition-research-center/nutrient-data-laboratory/docs/usda-national-nutrient-database-for-standard-reference/)

3. ***Vegetables, fruits and herbs are generally nutrient dense and energy poor*** with low levels of fats, thus increasing their consumption will not lead to obesity. On the other hand, tubers such as potatoes, are rich in carbohydrates and are often fried (French fries) and can thus contribute to overweight – contrary to vegetables.

Conclusions

In the present chapter the potential contribution of rooftop agriculture to food and nutrition security in the urban environment has been described, including the direct and indirect effect on food supply, diet diversification and resilience to food crisis. Accordingly, rooftop agriculture has a great potential to improve city liveability and health of the citizens, reduce the urban environmental impact and overall provide a great range of ecosystem services. Rooftop gardens in order to be viable over time, some initial investments is needed in the material and knowledge, but also market access if vending of surplus products are envisaged and a continuous supply of seeds. Any promotion of rooftop gardens should be based on a careful cost-benefit analysis of inputs (funds, time, material) versus potential benefits in terms of income or nutrients produced.

The sustainability of rooftop gardens as a source of fruits, herbs and vegetables for health is linked to the technology used, its economic viability, the social acceptance and the environmental impact. All these aspects have to be taken into consideration when planning future expansion in the context of the growing urban population and establishing new housing schemes.

The successful management of urban rooftop gardens will rely on the level of inputs required and the technical skills of the growers.

Ultimately, the overall impact of the rooftop gardens on food and nutrition security, will depend on the increased levels of consumption, which requires advocacy and nutrition education initiatives.

Bullet Points

- Rooftop Agriculture can improve city food and nutrition security by facilitating access to a number of different nutritional vegetables, herbs and fruits.
- The adoption of intercropping and species selection (including adoption of dwarf cultivars) may enable to improve space and water use efficiency.
- By shortening the food chain it is possible to improve the sustainability of the system, reduce the environmental impact and ensure food security in times of crisis and prevent nutritional depletion of the produce.

References

Baudoin W (2014) Integrated microgarden concept. FAO

Bellows AC, Brown K, Smit J (2013) Health benefits of urban agriculture. Paper and research conducted by members of the community food security coalition's North American Initiative on Urban Agriculture. Available at Foodsecurity.org. Accessed 1 Apr 2013

Dorais M (2016) Chapter 36: Sustainable and organic greenhouse tomato. In: Mattoo A, Handa AK (eds) Achieving sustainable tomato cultivation. Burleigh Dodds Science Publishing, Cambridge

Dring R (2014) Coming to a rooftop near you – the urban growing revolution. Available at: http://www.theecologist.org/green_green_living/2533583/coming_to_a_rooftop_near_you_the_urban_growing_revolution.html

European Commission (2015) Science for Environment Policy – rooftop gardens could grow three quarters of city's vegetables. Available at: http://ec.europa.eu/environment/integration/research/newsalert/pdf/rooftop_gardens_could_grow_three_quarters_of_citys_vegetables_409na2_en.pdf

FAO (2011) Save and grow. Food and Agricultural Organization of the United Nations, Rome. Available at: http://www.fao.org/ag/save-and-grow/

FAO (2013) Eating well for good health. Lessons on nutrition and healthy diets. http://www.fao.org/docrep/017/i3261e/i3261e00.htm

FAO (2015) Growing greener cities. Available at: www.fao.org/ag/agp/greenercities/

FAO (2016a) HORTIVAR Horticulture cultivars performance database. Available at: www.fao.org/hortivar

FAO (2016b) Microgarden technologies for nutrition and food security (in press)

FAO (2017) Nutrient productivity for nutrition and agriculture (in preparation)

FAO, El Behairy (2012) Green food from green roof manual, FAO TCP/EGY/1066

McLaughlin (2011) Have a small-space garden? Try dwarf vegetable varieties. Available at: http://www.vegetablegardener.com/item/10658/have-a-small-space-garden-try-dwarf-vegetable-varieties

Whittinghill LJ, Rowe DB (2012) The role of green roof technology in urban agriculture. Renew Agric Food Syst 27(4):314–322

WHO (2002) The World Health Report 2002. Available at: http://www.who.int/whr/2002/en/

WHO-FAO (2004) Kobe Joint FAO-WHO Workshop ("Kobe Framework"). Available at: www.who.int/dietphysicalactivity/

Biodiversity of Flora and Fauna

Francesca Bretzel, Francesca Vannucchi, Stefano Benvenuti, and Heather Rumble

Abstract Rooftop farming can contribute to urban biodiversity in terms of providing habitats and creating an urban green network. In addition, the cultivation of plants on roofs can provide places where wild animals and plants can survive and reproduce. Chosen cultivation practices and plant species can improve habitats and present more opportunities for wildlife and if flowering plants are grown together with vegetables, wild pollinators and domestic bees are attracted, providing pollination for edible species too.

The presence of pollinators also contributes to a trophic web, attracting other species, such as predators (spiders and birds) and parasites (e.g. wasps). Thus, green roofs are an opportunity to create greenways in anthropized areas, combating the habitat fragmentation caused by urban expansion. This chapter highlights the differences between different levels of green roof management in relation to their contribution to urban biodiversity and considers agrobiodiversity in relation to cultivated species and local cultivars.

Finally, rooftop fauna has been considered in order to assess the attraction of rooftop habitats to animal species, especially wild and domestic pollinators.

F. Bretzel (✉)
CNR-Institute for Ecosystem Studies, Pisa, Italy
e-mail: francesca.bretzel@ise.cnr.it

F. Vannucchi
Institute of Life Sciences, Scuola Superiore Sant'Anna, Pisa, Italy
e-mail: f.vannucchi@santannapisa.it

S. Benvenuti
Department of Agriculture, Food and Environment, Pisa University, Pisa, Italy
e-mail: stefano.benvenuti@unipi.it

H. Rumble
Department of Geography, University of Portsmouth, Portsmouth, UK
e-mail: heather.rumble@port.ac.uk

F. Orsini et al. (eds.), *Rooftop Urban Agriculture*, Urban Agriculture,
https://doi.org/10.1007/978-3-319-57720-3_14

Introduction

Rooftop agriculture provides ecological benefits and services in urban areas, creating novel habitats with multifunctional properties (Carter and Butler 2008). Cultivation on roofs is a strategy for contributing to environmental sustainability and for intensifying urban agriculture (Hui 2011). The cultivation of vegetables, edible plants and flowering species increases the habitat available for plants and animals, giving more opportunities for wildlife: if flowering plants are grown together with vegetables, wild pollinators are attracted, contributing to the persistence of plant communities and to agricultural production (Colla et al. 2009). The presence of pollinators also supports trophic webs and attracts other species, thus roofs have the potential to create greenways in urban areas, combating habitat fragmentation due to urban expansion (Goddard et al. 2010; MacIvor and Lundholm 2011). Moreover, the use of different substrate types in rooftop agriculture creates a structural complexity leading to different capabilities to host wild flora and fauna (Madre et al. 2013). Rooftop agriculture also provides a valuable way to produce and propagate local cultivars and vegetables, locally adapted, that perform well in terms of yield and nutritional qualities and provide a great opportunity for agrobiodiversity conservation (Havaligi 2011).This chapter highlights the role of rooftop agriculture in providing ecosystem benefits, in terms of increasing wild plant and animal diversity and agrobiodiversity as well as providing services, such as pollination and agricultural production.

Urban Biodiversity: Contribution of Rooftop Agriculture to the Network of Ecological Corridors

Creation of New Habitat to Flying Insects and Wind Transported Seeds

Rooftop agriculture provides an opportunity to create new habitats, enhancing the biodiversity in urban areas: invertebrates and plants can spontaneously colonize and spread on vegetated roofs (Kadas 2006; Dunnett et al. 2008). Rooftops are potentially valuable sites for pollinators if planted with diverse native forbs to provide foraging resources (Tonietto et al. 2011). Ksiazek et al. (2012) states that although a lower number and diversity of pollinators are observed on rooftops, compared to ground-level, the ability of native plant species to produce seed is not hindered. A plants ability to disperse to rooftops is related to seed size and habitat condition. Colonizing species that produce smaller seeds are more easily dispersed by wind, whereas larger seeds typically rely on zoochory (Fenner 1984). The rooftops condition, in terms of age, surface area, height, substrate depth and maintenance intensity, are filters for colonizing species, structuring the plant communities (Madre et al. 2014).

Use of Herbaceous Plants to Improve Biodiversity

The increase in plant diversity using edible, ornamental and wild plant species contributes to improving the long-term functioning of cultivated rooftops (Cook-Patton and Bauerle 2012). Specifically, the use of flowering herbaceous plants, including those used for vegetable cultivation, increases the structural complexity of rooftops, providing food resources for many beneficial invertebrates, especially wild bees (Fenster et al. 2004; Goddard et al. 2010). The selection of species (Table 1) with morphological and physiological adaptations to stress conditions and with different

Table 1 Plant functional types useful to select herbaceous species for enhancing wildlife and the properties in terms of sustainability and ecosystem services

Functional types	Features	Properties	References
Life form	Terophytes, emicriptophytes, geophytes, chamaephytes	Habitat heterogeneity	Cornelissen et al. (2003)
Life cycle	Annual, biennial; perennials	Production of flowers and seeds in the short and long period; soil nitrogen retention	Van Mechelen et al. (2014) and Maron and Jefferies (2001)
Position in the trophic web	Production of nectar (insects) or seeds (granivorous birds)	Creation of forage habitat for pollinators and birds	Braman et al. (2002), Matteson and Langellotto (2011) and Blaauw and Isaacs (2014)
Photosynthetic pathways	C3, C4, CAM, nitrogen-fixing plants	Self-sustaining vegetation	Tilman (2001), Tilman and Downing (1994), De Deyn et al. (2009) and Lambers et al. (2011)
Type of pollination	Entomophilous	Pollinator attraction	Benvenuti et al. (2007), Ollerton et al. (2011) and Haaland and Gyllin (2011)
Flower morphology	Attractive flowers, flower abundance	Increase in insect diversity	Haaland and Gyllin (2011)
Plant morphology	Small plants (<1 m), hairy leaves, needle-like leaves	Drought-resistant	Van Mechelen et al. (2014)
Flowering length	Wide flowering period	Continuous forage for pollinators	Blaauw and Isaacs (2014)
Ecological strategies	Stress tolerant and ruderals	Vegetation resistance to rooftop environment	Hodgson et al. (1999), Grime (2001) and Pierce et al. (2013)

flowering times provides economic and ecological advantages, as well as requiring lower maintenance (Oberndorfer et al. 2007; Menz et al. 2011).

The establishment of season-long floral resources is an important strategy in order to enhance the crop yield, as well as conserve wild pollinators by increasing forage habitat for bees (Blaauw and Isaacs 2014). Flower abundance, seed mixture, vegetation structure and management are factors affecting insect abundance and diversity (Haaland and Gyllin 2011). The presence of pollinators contributes to the persistence of plant communities, which attract other species that are dependent on bee-pollinated plants for food or refuge, as well as to increase agricultural production. The presence of species belonging to different functional groups (forbs, graminoids and succulents) in a plant community confer to the vegetation resilience after stress and disturbance as well as to environmental change (Lavorel et al. 1998). Moreover, rooftop ecosystem services can be improved by increasing the diversity of plant life forms (Lundholm et al. 2010). The vegetation performance in terms of survival, diversity, size and flowering is influenced by the substrate depth and an advanced soil formation makes the substrate environment stable, thus improving nutrient content and soil biota activity (Dunnett et al. 2008; Schrader and Böning 2006).

Opportunities for Biodiversity Conservation

Urban development has affected ecosystem processes and functions through fragmentation and degradation of natural habitats, as well as homogenising biota. This can be remediated by enhancing urban biodiversity (Alberti 2005; McKinney 2006) and the creation of new habitats on roofs could contribute to this. Species that may benefit from green roofs are primarily small organisms, such as insects, which have low resource requirements and are able to disperse, or large organisms, such as birds, that are very mobile. These species may complete their life cycle on the roof, or use the roof as a habitat as part of their wider range. In terms of planning green roofs to optimise biodiversity, providing good foraging sites has been shown to sustain populations of rare and endangered species (Kowarik 2011; Kadas 2006). Thus, planning novel habitats on roofs contributes to landscape scale urban and rural ecological networks, remediating the habitat fragmentation caused by urban expansion (Ignatieva et al. 2011; Williams et al. 2014). Flower provision is a major factor in supporting this species richness, with some species supported regardless of plant origin (Matteson and Langellotto 2011) and other species affected by specific mixes of native and introduced species (e.g. moths, see Tallamy and Shropshire 2009). Plants also provide architecture and structure, creating hiding places and nesting sites for fauna. Biodiverse roofs, because they are built with wildlife in mind, also often include bare areas, logs and stones to provide shelter for specific organisms and sometimes include deeper areas of substrate that stay moist in summer. There are a number of studies being undertaken currently to assess the impact of these interventions (Ishimatsu and Ito 2013).

Specificities of Rooftop Agriculture Elements and Their Effect on Biodiversity

Compared to ground-level environments, rooftops can provide equivalent habitat for many urban insects. Thus, there is an opportunity to increase and manage the associated ecosystem services these species provide, such as decomposition, pollination and biological control (Matteson and Langellotto 2011). Particularly, the creation of structural complexity, using different substrate types and depth, creates microhabitats suitable for plants and insects, leading to an increase in the biotic component (Madre et al. 2013). The microclimate on a green roof results in earlier cropping compared to that of ground-level soils, with spring crops emerging earlier in the season. This is due partly to the more rapid heating of building materials, with respect to common agricultural soil. Consequently, this potential ecological niche, as part of a mosaic within the surrounding agricultural landscape, could increase the period during which food for wildlife is available (early and late season). This availability of food can be of crucial importance in terms of both fruits, in the case of fruit-eating birds, and pollen and nectar in the case of pollinating insects. In other words, the particular micro-environment of rooftops increases ecological complexity, thus increasing biodiversity connected with provided habitats. Substrates are a particular area where green roofs may present an opportunity for enhanced and unique habitat in terms of the broader landscape. This has been demonstrated for ground-nesting bumblebees, for which a typical green roof substrate (lightweight volcanic material with coarse texture) may present the ideal environment (Svensson et al. 2000). Ground nesting bumblebees are often rare (Goulson et al. 2005) and their conservation is therefore critical (Goulson et al. 2008) in both natural and anthropized ecosystems. The limiting factor for these insects is the availability of dry micro-environments, which are frequently rare or even absent in a common agricultural soil, which is often subject to periodic waterlogging. A further diversification on green roofs could be in the form of and facilitated by weed communities. Indeed, green roofs are primarily invaded by pioneer flora via anemochory. This implies that there is an increase in the total (conventional and roof agriculture) weed biodiversity of a landscape, since in common agroecosystems other dispersal strategies prevail (Benvenuti 2007) and this could, in turn, increase the biodiversity of species reliant upon them.

Comparative Analysis of Beneficial Effects on Biodiversity in Rooftop Agriculture vs Biodiverse Roofs

Agricultural rooftops and ornamental green roofs are both important in terms of biodiversity with both contributing to urban ecology (Petchey and Gaston 2002). However, the simultaneous presence of both kinds in cities of course presents the highest opportunity for ecological services. This is true from both a biological perspective (greater number of useful species) and a psychological one, with greater wellbeing afforded by a complex, living cityscape. Most of the ecological characteristics of agro and green roofs overlap. However, the higher productivity of the agro-roof, due to the use of improved crops, irrigation and fertilizers, implies greater intensity of management. Consequently, both organisms: undesired for this kind of gardening (i.e. aphids, scales, lepidopteran larvae) and their predators tend to increase (Cardinale et al. 2003), as it occurs in the food pyramid of any agro-ecosystem. However, the presence of beneficial insects in urban areas, such as predatory ladybugs, can occur very soon after conversion of a rooftop to agriculture, even in the most cemented environment. Thus, the increased spread of phytophagous that agriculture encourages could be counterbalanced by the increased presence of their predators, supported by the diversity of habitats presented by rooftop gardening and other urban green spaces. This biological complexity will improve proportionally to the greater number of plants genetic variety (crops, varieties, landraces) adopted in space and time (Hajjar et al. 2008).

Agrobiodiversity

Use of Local Cultivars of Vegetables

As cities grow, so do jobs and services, adding complexity to the environment (Nugent 2000). The use of local horticultural varieties can play a role in creating an alternative to the "globalized" food market arising from conventional agriculture. Indeed, traditional crop varieties, commonly referred to as landraces, are severely threatened by genetic extinction, primarily due to their replacement by modern genetically uniform varieties obtained to optimize agronomic performance (high productivity, simultaneous ripening, improved storage properties, etc.). Although it is not easy to define a landrace (Zeven 1998), it can be understood as a variety characterized by historical origin, high genetic diversity, local genetic adaptation, recognizable identity, lack of formal genetic improvement and an association with traditional farming systems (Villa et al. 2005). Cultivating landraces represents not only an opportunity for producing different food varieties to those commonly found in traditional agricultural systems, but also rekindles links with rural traditions (Altieri and Merrick 1987) and even with human history and culture. Moreover, this taxonomic crop complexity has been considered to have useful effects for the social integrity and ecological health of people (Johns et al. 2013). Whilst modern horticultural varieties perfectly meet the agronomic requirements of modern agriculture, the old landraces appear to fit better with urban horticulture. A clear example is relative to the time of fruit ripening in tomatoes: in modern agriculture this must be simultaneous due to mechanized harvesting methods (Sim et al. 2011), while the de-synchronized ripening of landraces is absolutely preferable to expand the calendar of productivity for the urban self-producing gardener. Local varieties have a germplasm of particular attitude for urban agriculture: its nutraceutical (overall in terms of antioxidant power), organoleptic (in terms of sensorial profile) and agricultural (prolonged ripening and harvest time) performances are more appreciable by the amateur self-producer gardeners (Table 2).

Table 2 Botanical, agronomic and nutraceutical characteristics of some vegetable landraces suitable for the urban agro-environment

Botanical family	Crop species	Landrace	Properties	References
Alliaceae	*Allium cepa*	Vermelha da	Rich of quercetin	Rodrigues et al. (2011)
Apiaceae	*Apium graveolens* var. *dulce*	Nero di Trevi	Sensory profile	Torricelli et al. (2013)
	Daucus carota	Carota di Polignano	Rich of anthocyanin	Cefola et al. (2012)
Asparagaceae	*Asparagus officinalis*	Violetto di Albenga	Rich of anthocyanin	Castro et al. (2014)
Brassicaceae	*Brassica oleracea* var. *botrytis*	Violetto di Catania	Rich of anthocyanin and ornamental	Scalzo et al. (2008)
	Brassica oleracea var. *Italica*	Broccolo fiolaro	Rich of sulforaphane	Vischi et al. (2008)
	Brassica oleracea var. *viridis*	Cavolo nero fiorentino	Rich of carotenoids and sensory profile	Bacchetti et al. (2014)
		Galega kale	High and prolonged productivity	Dias et al. (1995)
Cucurbitaceae	*Cucumis melo* var. *Cantalupensis*	Mediterranean landraces	Sensory profile	Aubert and Bourger (2004)
	Cucumis melo var. *chate*	Carosello pugliese	Sensory profile	Laghetti et al. (2008)
	Cucurbita maxima	Turkish germplasm	Ornamental and sensory profile	Balkaya et al. (2010)
	Cucurbita moschata	Trombetta di Albenga	Ornamental and prolonged production	Gaetano et al. (2012)
	Lagenaria siceraria var. *longissima*	Zucca serpente	Ornamental and sensory profile	Branca and La Malfa (2008)
Fabaceae	*Vigna unguiculata*	Fagiolina del Lago Trasimeno	Low water needs and sensory profile	Negri (2003)
	Phaseolus coccineus	European landraces	Pollinators attractivity	Rodriguez et al. (2013)
	Phaseolus vulgaris	Zolfino	Sensory profile and flavonoid quality	Romani et al. (2004)
		Piattella pisana	Sensory profile	Baldanzi and Pardini (2003)

(continued)

Table 2 (continued)

Botanical family	Crop species	Landrace	Properties	References
Lamiaceae	*Ocimum basilicum*	Genovese	Sensory profile	Miele et al. (2001)
Malvaceae	*Abelmoschus esculentus*	Lemnos germplasm	Functional and ornamental	Roy et al. (2014) and Thomas et al. (2012)
Solanaceae	*Solanum lycopersicum*	Pera abruzzese and Canestrino	Sensory profile	Mazzuccato et al. (2010)
		Pisanello	Low allergens	Bencivenni et al. (2012)
	Solanum melanogena	Andalusian germplasm	Rich of phenolics	Raigón et al. (2008)

Urban Agro-Landscaping

The views of natural landscapes have positive influences on emotional and psychological health (Ulrich 1986; Fleischer and Tsur 2000), reducing stress by providing restorative landscapes (Grahn and Stigsdotter 2003). Agricultural landscape is a key element (Ode et al. 2008) of the environmental perception of an aesthetically pleasant natural setting and, as such, some agro-indicators could be utilized in the planning of future green scenarios (Fry et al. 2009). Crop heterogeneity enriches the rural landscape (Arriaza et al. 2004) in terms of biodiversity (Hietala-Koivu et al. 2004) and functionality (Berkel and Verburg 2014) and provides visual quality. These characteristics are well represented in rural landscapes of the past and could inform planning of future landscapes by providing knowledge about human interactions with their environment (Antrop 2005). Since modern citizens almost exclusively spend their daily life within the city, it follows that some aesthetic elements of the agricultural landscape, which includes plants and animals, can have a substantial benefit even in the urban environment.

Wild Herb Cropping

Many spontaneous species common in urban ecosystems are important food plants of ethnobotanical tradition (Benvenuti 2004). They are not edible in cities as they grow in soil and sediments polluted by traffic and are potentially contaminated (Bretzel et al. 2014). Nevertheless, growing these species on rooftops is conceivable and safer from the majority of urban traffic pollution. Due to their weight, heavy metals tend to concentrate in the lower layers of the urban atmosphere. Aside from the well-known health properties of many edible wild herbs (Pardo de Santayana et al. 2013), cultivating these plants represents an opportunity to satisfy the

Fig. 1 *Urospermum dalechampii* a common wild herb of urban ecosystem: could we host it on our roof? (Photo Stefano Benvenuti)

psychological pleasure of finding ecological space in the city for those species called "vagabond plants" (Clément 2002) and that are widely part of the common imaginary representation of the countryside. In this perspective, species that have evolved to be tolerant to water stress, such as wild herbs, appear to be ideal for rooftop farming, since their growth is self-sufficient in terms of water need. The maximum degree of suitability is shown by those species already present in urban buildings, especially in the case of ancient monuments (Caneva et al. 2003). Some species adapted to a Mediterranean and temperate climate are: *Bunias erucago* L., *Diplotaxis erucoides* (L.) DC., *Picris hieracioides* L., *Plantago lanceolata* L., *Reichardia picroides* (L.) Roth., *Sonchus tenerrimus* L. and *Urospermum dalechampii* (L.) F. W. Schmidt (Fig. 1). Their cultivation on green roofs could provide a source of nutraceutical foods in the context of urban biodiversity and agricultural sustainability.

Educational Involvement

In the past, agricultural landscapes represented a sort of "calendar", producing tangible evidence for changes in season. This could be revived in the grey urban landscape, where biophilic emotions are often absent (Conradson 2005) and could be especially important for children in their developmental age. The shapes and colours of crop plants, along with their flowers and fragrances are missing in the urban landscapes, consequently, children benefit from therapeutic encounters with agro-biodiversity in the rural environment (Bagdonis et al. 2009). Being able to observe the flight of a butterfly or dragonfly (Fig. 2) that lands on a sunflower can be an effective experience with nature.

Placing crops on rooftops can create, in miniature, an urban counterpart to these unfamiliar agro-landscapes. It is also clear that this landscape-mediated well-being

Fig. 2 Dragonfly on a green roof (Photo Heather Rumble)

has a neglected but beneficial effect on the psyche not only of children but on citizens of any age (Abraham et al. 2010). If traditional rural landscapes are now scarce or have disappeared (Plieninger et al. 2006), agro-roofs may not be a total fiction or utopia (Antrop 2006), but rather an opportunity to improve the habitability of urban ecosystems in the future. Crop biodiversity, threatened by agricultural globalization, could be the agronomic challenge of this new millennium. The urban cropping of ancient wheat cultivars (*Triticum* spp.), emmer wheat (*Triticum dicoccon*) and buckwheat (*Fagopyrum esculentum*) could be a clear example of this new urban concept, inspired by the agricultural scenarios of the past. The presence of pollinating insects on buckwheat (Patten et al. 1993), enhance the perception of a "living" agricultural landscape, due to the buzzing and the movement of bees and bumble bees during the flowering periods. In the case of the ancient cereals, the post-harvest disposition in sheaves during the summer could be an additional tool for memorialising forgotten rural landscapes. Their arrangement on the rooftop, especially with a geometric and/or artistic shape, could be an important element of de-globalization both in space (crops geographically climate-defined) and time (dynamically season-dependent landscapes scenarios).

Wildlife

Rooftop Fauna

Green roofs have the potential to provide valuable habitat for species in impoverished urban environments. In this section, the current state of knowledge in terms of habitat provision for all types of green roofs will be outlined, with a view to advising those wishing to design agricultural green roofs to also benefit wildlife. As an emerging field, there are few studies specifically addressing the wildlife provision of agricultural green roofs, but factors affecting wildlife provision on green roofs can be inferred from general green roof studies, as well as knowledge of ground-level urban agriculture provision.

An animals ability to utilise a green roof is affected by plant choice, substrate type and depth, provision of open ground patches, regularity of disturbance by humans, accessibility and the landscape surrounding the roof. Animals require three major elements from their habitat: shelter/breeding sites, food and water. A single habitat may provide all of these elements or species may move between habitats that provide these different functions in different spaces or times. For immobile or low mobility species, such as soil organisms, habitat must provide all three. For more mobile species, such as flying insects and birds, a habitat may just provide one of these services, or may contribute to services provided by the wider landscape. Braaker et al. (2014) note that for mobile species, green roofs can be a stepping stone between habitats, but this is not the case for less mobile species. It is also not yet understood if the quality of a green roof as a habitat affects this. Thus, the suitability of green roofs in the wider landscape will be species and roof specific. When thinking about green roofs to support biodiversity it is key to understand the ecology of the species concerned.

The majority of wildlife found on green roofs have colonised of their own accord, often producing unique species assemblages. Soil organisms, such as mites and springtails, vital for nutrient cycling, have been found on green roofs (Schrader and Böning 2006), though sometimes in unstable, drought limited populations (Rumble and Gange 2013); however in agricultural systems one would expect higher moisture contents, perhaps alleviating this problem. Surprisingly, snail communities have also been found on green roofs, presumably transported there via phoresy on birds (Kadas 2006; Rumble 2014). Green roofs have been found to be a significant contributor to urban insect and spider diversity, remediating the loss of rare species from brownfield sites (Kadas 2006). Encouraging these species onto agricultural green roofs requires plants that are structurally complex, in addition to providing food species such as flowering plants.

Green roofs can even support larger organisms, such as birds and bats. In London, many green roofs have been built to support the black redstart (*Phoenicurus ochruros*) population, dwindling after it's favoured rubble strewn habitat began being built on in London's rapid urban expansion (Gedge 2003; Grant 2006). Studies into the use of green roofs by bats have also been conducted and biodiverse roofs in particular may contribute to a wider landscape suitable for bats (Pearce and Walters 2012), particularly if insect species can be encouraged to provide a food source.

Whilst these are examples of specific species that may benefit from green roofs, there is much research needed to determine if these communities are sustainable and how green roofs contribute to the wider ecology of urban environments (Fig. 3). As a man-made environment, it is important to consider the aims of a roof in terms of wildlife. Some animals and plant species may be rare in nature for a reason; others may be common but have simply been overlooked. As landscape planners, there is the opportunity to decide if rare species can become common through the employment of green roofs and other green infrastructure, or if naturally occurring

Fig. 3 Trials on a green roof with native herbaceous species to improve the urban biodiversity (Photo Francesca Bretzel)

fauna is sensitively represented. To date, most colonisation of green roofs by animals has been secondary to an engineering design, with little consideration for the species that will colonise these spaces. This passive approach to design could be positive, creating unique, urban specific communities, but it could represent a missed opportunity to use green roofs as a tool to combat habitat loss through development.

Pollination

Pollinators, such as bees and flies, have been known to use green roofs as commuting spaces and for feeding. Colla et al. (2009) found that green roofs in Toronto harboured as diverse species assemblages of bees as ground-level urban habitats, though they did not specifically observe feeding. In Chicago, Tonietto et al. (2011) found that bees were visiting rooftop flowers, though less often than those feeding in parks. However, they noted that green roofs were a particularly good habitat for ground-dwelling bees and cavity nesters, which were otherwise rare in urban environments. Tonietto et al. (2011) also found that bee diversity was correlated with the diversity of flowering plants on rooftops, suggesting that mixed rooftop agriculture may be best for supporting these pollinating species. Thus, the maintenance of bloom and habitat heterogeneity on rooftops helps to attract wild bees, providing potential reservoirs in urban habitats (Tommasi et al. 2004).

Urban Beekeeping

Urban beekeeping can be practiced on rooftops as hives only require a small amount of space, minimal maintenance and low equipment cost (Petts 2000). Urban apiculture on rooftops has economic and conservational values, promoting sustainable agriculture and the conservation of honey bee diversity. *Rooftop honey* in Melbourne (Australia) includes beekeepers that help to save the honeybee from the various threats of disease and human activities, bringing bees back to the city. The *Chicago Honey Co-op* (US) has over 100 rooftop beehives, selling products (honey and candles) at local farmers' markets and online (Broadway 2009). In addition, honeybees provide ecosystem services, acting as bioindicators of urban environmental quality (Badiou-Bénéteau et al. 2013). The *Urbees* project promotes the beekeeping in the city of Torino (Italy), in order to produce honey and wax, as well as the biomonitoring through honey analysis.

Conclusions

Rooftop agriculture can contribute to urban biodiversity, if specific aspects are taken into account, such as the use of pollinating plants, the ecological complexity of vegetation and organic maintenance. Increasing urban biodiversity has important consequences in terms of ecosystem services, for example by enhancing environmental resilience, combatting habitat fragmentation, producing food locally and reducing of the use of pesticides. Rooftop agriculture can also be a great opportunity to spread landraces, local cultivars and to preserve traditional agricultural heritage. In an urban context, small spaces of a usually neglected roof surface can have significant value, not only due to the biodiversity outlined in the chapter and its complexity and contribution as a food source, but even in terms of psychological wellness. As a man-made environment, green roofs present a perfect opportunity to cultivate and design biodiverse spaces that are both pleasing and helpful to urban inhabitants. Indeed, cultivation on green roofs can be a clear example of therapeutic agriculture aimed to elicit the "lost" ecological perception of "Gaia" (Lovelock, 2000), even in an urban ecosystem. In conclusion, the urban affirmation of agroroofs will transform grey and emotionally inhospitable cities to the green and biodiverse environments in which man slowly evolved.

Bullet Points

- Rooftops are valuable sites for pollinators if planted with diverse vegetation able to provide foraging resources;
- Rooftop agriculture combats habitat fragmentation caused by urban expansion and can provide space to conserve rare urban species;
- Rooftops can become a kind of "open house" to support any biodiversity useful for food and/or medicine;
- The perception of biodiversity on a roof can have psychologically therapeutic value.

References

Abraham A, Sommerhalder K, Abel T (2010) Landscape and well-being: a scoping study on the health-promoting impact of outdoor environments. Int J Public Health 55:59–69

Alberti M (2005) The effects of urban patterns on ecosystem function. Int Reg sci rev 28(2):168–192

Altieri MA, Merrick L (1987) In situ conservation of crop genetic resources through maintenance of traditional farming systems. Econ Bot 41(1):86–96

Antrop M (2005) Why landscapes of the past are important for the future. Landsc Urban Plan 70(1):21–34

Antrop M (2006) Sustainable landscapes: contradiction, fiction or utopia? Landsc Urban Plan 75:187–197

Arriaza M, Canas-Ortega JF, Canas-Madueno JA, Ruiz-Aviles P (2004) Assessing the visual quality of rural landscapes. Landsc Urban Plan 69(1):115–125

Aubert C, Bourger N (2004) Investigation of volatiles in Charentais cantaloupe melons (Cucumis melo var.cantalupensis). Characterization of aroma constituents in some cultivars. J Agric Food Chem 52:4522–4528

Bacchetti T, Tullii D, Masciangelo S, Gesuita R, Skrami E, Brugè F, Silvestri S, Orlando P, Tiano L, Ferretti, G (2014) Effect of black and red cabbage on plasma carotenoid levels, lipid profile and oxidized low density lipoprotein. J Funct Food 8:128–137

Badiou-Bénéteau A, Benneveau A, Géretc F, Delatte H, Becker N, Brunet JL, Reynaud B, Belzunces LP (2013) Honey bee biomarkers as promising tools to monitor environmental quality. Environ Int 60:31–41

Bagdonis JM, Hinrichs CC, Schafft KA (2009) The emergence and framing of farm-to-school initiatives: civic engagement, health and local agriculture. Agric Hum Values 26:107–119

Baldanzi M, Pardini G (2003) Evaluation of lines selected in the common bean folk variety Piattella Pisana or San Michele (Phaseolus vulgaris L. – Tuscany). Sementi Elette 49:50–56

Balkaya A, Özbakir M, Kurtar ES (2010) The phenotypic diversity and fruit characterization of winter squash (Cucurbita maxima) populations from the Black Sea Region of Turkey. Afr J Biotechnol 9:152–162

Bencivenni M, Faccini A, Bottesini C, Rao R, Detoraki A, Ridolo E, Marone G, Dall'Aglio PP, Dossena A, Marchelli R, Sforza S (2012) Assessing allergenicity of different tomato ecotypes by using pooled sera of allergic subjects: identification of the main allergens. Eur Food Res Technol 234:405–414

Benvenuti S (2004) Weed dynamics in the Mediterranean urban ecosystem: ecology, biodiversity and management. Weed Res 44:341–354

Benvenuti S (2007) Weed seed movement and dispersal strategies in the agricultural environment. Weed Biol Manag 7:141–157

Benvenuti S, Loddo D, Basteri G, Russo A (2007) Insect-pollinated weeds as indicator of the agroecosystem biodiversity. Agr Med 137(3/4):132–137
Berkel DB, Verburg PH (2014) Spatial quantification and valuation of cultural ecosystem services in an agricultural landscape. Ecol Indic 37:163–174
Blaauw BR, Isaacs R (2014) Flower plantings increase wild bee abundance and the pollination services provided to a pollination-dependent crop. J Appl Ecol 51:890–898
Braaker S, Ghazoul J, Obrist MK, Moretti M (2014) Habitat connectivity shapes urban arthropod communities: the key role of green roofs. Ecology 95:2010–2021
Braman SK, Pendely AF, Corley W (2002) Influence of commercially available wildflower mixes on beneficial arthropod abundance and predation in turfgrass. Environ Entomol 31:564–572
Branca F, La Malfa G (2008) Traditional vegetables of Sicily. Chronica horticulturae 48(1):20–25
Bretzel F, Benvenuti S, Pistelli L (2014) Metal contamination in urban street sediment in Pisa (Italy) can affect the production of antioxidant metabolites in *Taraxacum officinale* Weber. Environ Sci Pollut Res 21:2325–2333
Broadway M (2009) Growing urban agriculture in north American cities: the example of Milwaukee. Focus Geogr 52(3):23–30
Caneva G, Pacini A, Grapow LC, Ceschin S (2003) The Colosseum's use and state of abandonment as analysed through its flora. Int Biodeterior Biodegrad 51:211–219
Cardinale BJ, Harvey CT, Gross K, Ives AR (2003) Biodiversity and biocontrol: emergent impacts of a multi-enemy assemblage on pest suppression and crop yield in an agroecosystem. Ecol Lett 6:857–865
Carter T, Butler C (2008) Ecological impacts of replacing traditional roofs with green roofs in two urban areas. CATE 1:1–17
Castro P, Rubio J, Gil J, Moreno R (2014) Introgression of new germplasm in current diploid cultivars of garden asparagus from a tetraploid spanish landrace "Morado de Huétor". Scientia Horticulturae 168:157–160
Cefola M, Pace B, Renna M, Santamaria P, Signore A, Serio F (2012) Compositional analysis and antioxidant profile of yellow, orange and purple Polignano carrots. Ital J Food Sci 24(3):284
Clément G (ed) (2002) Eloge des vagabondes. Herbes, arbres et fleurs à la conquête du monde. Nil Éditions, Paris
Colla SR, Willis E, Packer L (2009) Can green roofs provide habitat for urban bees (Hymenoptera: Apidae)? Cities Environ 2(1):1–12
Conradson D (2005) Landscape, care and the relational self: therapeutic encounters in rural England. Health Place 11:337–348
Cook-Patton SC, Bauerle TL (2012) Potential benefits of plant diversity on vegetated roofs: a literature review. J Environ Manag 106:85–92
Cornelissen JHC, Lavorel S, Garnier E, Díaz S, Buchmann N, Gurvich DE, Reich PB, ter Steege H, Morgan HD, van der Heijden MGA, Pausas JG, Poorter H (2003) Handbook of protocols for standardised and easy measurements of plant functional traits worldwide. Austr J Bot 51:335–380
De Deyn GB, Quirk H, Yo Z, Oakley S, Ostle N, Bardgett RD (2009) Vegetation composition promotes carbon and nitrogen storage in model grassland communities of contrasting soil fertility. J Ecol 97:864–875
Dias JS (1995) The Portuguese tronchuda cabbage and galega kale landraces: a historical review. Genet Resour Crop Evol 42:179–194
Dunnett N, Nagase A, Hallam A (2008) The dynamics of planted and colonizing species on a green roof over six growing seasons 2001–2006: influence of substrate depth. Urban Ecosyst 11:373–384
Fenner M (1984) Seed characteristics in relation to succession. In: Gray AJ, Crawley MJ, Edwards PJ (eds) Colonization, succession and stability. Blackwell Scientific Publications, London, pp 103–114
Fenster CB, Armbruster WS, Wilson P, Dudash MR, Thomson JD (2004) Pollination syndromes and floral specialization. Annu Rev Ecol Evol Syst 35:375–403

Fleischer A, Tsur Y (2000) Measuring the recreational value of agricultural landscape. Eur Rev Agric Econ 27(3):385–398

Fry G, Tveit MS, Ode Å, Velarde MD (2009) The ecology of visual landscapes: exploring the conceptual common ground of visual and ecological landscape indicators. Ecol Indic 9(5):933–947

Gaetano L, Giovanni G, Fabiano M, Salvatore C, Domenico P, Karl H (2012) Collecting crop genetic resources in two Italian linguistic (Occitan and Ladin) islands and West Liguria with historical and ethnobotanical notes. Int J Biodivers Conserv 4:54–70

Gedge D (2003) From rubble to redstarts. In: Proceedings of the first annual greening rooftops for sustainable communities conference. Awards and Trade Show, Chicago

Goddard MA, Dougill AJH, Benton TG (2010) Scaling up from gardens: biodiversity conservation in urban environments. Trends Ecol Evol 25:90–98

Goulson D, Hanley ME, Darvill B, Ellis JS, Knight ME (2005) Causes of rarity in bumblebees. Biol Conserv 122:1–8

Goulson D, Lye GC, Darvill B (2008) Decline and conservation of bumble bees. Annu Rev Entomol 53:191–208

Grahn P, Stigsdotter UA (2003) Landscape planning and stress. Urban For Urban Green 2(1):1–18

Grant G (2006) Extensive green roofs in London. Urban Habitats 4:51–65

Grime JP (ed) (2001) Plant strategies, vegetation progress and ecosystem properties. Wiley, New York

Haaland C, Gyllin M (2011) Sown wildflower strips – a strategy to enhance biodiversity and amenity in intensively used agricultural areas. In: Jordi LA (ed) The importance of biological interactions in the study of biodiversity. InTech, Rijeka, pp 155–172

Hajjar R, Jarvis DI, Gemmill-Herren B (2008) The utility of crop genetic diversity in maintaining ecosystem services. Agric Ecosyst Environ 123:261–270

Havaligi N (2011) Contribution of urban agriculture to food security, biodiversity conservation and reducing agricultural carbon footprint. In: Filho WL (ed) The economic, social and political elements of climate change. Springer, Berlin, pp 99–112

Hietala-Koivu R, Lankoski J, Tarmi S (2004) Loss of biodiversity and its social cost in an agricultural landscape. Agric Ecosyst Environ 103(1):75–83

Hodgson JG, Wilson PJ, Hunt R, Grime JP, Thompson K (1999) Allocating C-S-R plant functional types: a soft approach to a hard problem. Oikos 85:282–294

Hui SCM (2011) Green roof urban farming for buildings in high-density urban cities. Hainan China World Green Roof conference, 18–21 March 2011, Hainan (Haikuo, Boao and Sanya), China

Ignatieva M, Stewart GH, Meurk C (2011) Planning and design of ecological networks in urban areas. Landsc Ecol Eng 7(1):17–25

Ishimatsu K, Ito K (2013) Brown/biodiverse roofs: a conservation action for threatened brownfields to support urban biodiversity. Landsc Ecol Eng 9:299–304

Johns T, Powell B, Maundu P, Eyzaguirre PB (2013). Agricultural biodiversity as a link between traditional food systems and contemporary development, social integrity and ecological health. J Sci Food Agric 93(14):3433–3442

Kadas G (2006) Rare invertebrates colonizing green roofs in London. Urban Habitats 4:66–86

Kowarik I (2011) Novel urban ecosystems, biodiversity, and conservation. Environ Pollut 159(8):1974–1983

Ksiazek K, Fant J, Skogen K (2012) An assessment of pollen limitation on Chicago green roofs. Landsc Urban Plan 107:401–408

Laghetti G, Accogli R, Hammer K (2008) Different cucumber melon (Cucumis melo L.) races cultivated in Salento (Italy). Genet Resour Crop Evol 55(4):619–623

Lambers H, Brundrett MC, Raven JA, Hopper SD (2011) Plant mineral nutrition in ancient landscapes: high plant species diversity on infertile soils is linked to functional diversity for nutritional strategies. Plant Soil 334:11–31

Lavorel S, Touzard B, Lebreton J, Clement B (1998) Identifying functional groups for response to disturbance in an abandoned pasture. Acta Oecol 19:227–240

Lovelock J (2000) Gaia: a new look at life on earth. Oxford University Press, Oxford. Paperbacks
Lundholm J, MacIvor JS, MacDougall Z, Ranalli M (2010) Plant species and functional group combinations affect green roof ecosystem functions. PloS one 5(3):e9677
MacIvor JS, Lundholm J (2011) Insect species composition and diversity on intensive green roofs and adjacent level-ground habitats. Urban Ecosyst 14:225–241
Madre F, Vergnes A, Machon N, Clergeau P (2013) A comparison of 3 types of green roof as habitats for arthropods. Ecol Eng 57:109–117
Madre F, Vergnes A, Machon N, Clergeau P (2014) Green roofs as habitats for wild plant species in urban landscapes: first insights from a large-scale sampling. Landsc Urban Plan 122:100–107
Maron JL, Jefferies RL (2001) Restoring enriched grasslands: effects of mowing on species richness, productivity and nitrogen retention. Ecol Appl 11(4):1080–1100
Matteson KC, Langellotto GA (2011) Determinates of inner city butterfly and bee species richness. Urban Ecosyst 13:333–347
Mazzuccato A, Ficcadenti N, Caioni M, Mosconi P, Piccinini E, Sanampudi VR, Sestili S, Ferrari V (2010) Genetic diversity and distinctiveness in tomato (Solanum lycopersicum L.) landraces: the Italian case study of 'A pera Abruzzese'. Scientia horticulturae 125:55–62
McKinney ML (2006) Urbanization as a major cause of biotic homogenization. Biol Conserv 127:247–260
Menz MHM, Phillips RD, Winfree R, Kremen C, Aizen MA, Johnson SD, Dixon KW (2011) Reconnecting plants and pollinators: challenges in the restoration of pollination mutualisms. Trends Plant Sci 16:4–12
Miele M, Dondero R, Ciarallo G, Mazzei M (2001) Methyleugenol in Ocimum basilicum L. Cv. genovese gigante. J Agric Food Chem 49:517–521
Negri V (2003) Landraces in central Italy: where and why they are conserved and perspectives for their on-farm conservation. Genet Resour Crop Evol 50:871–885
Nugent R (2000) The impact of urban agriculture on the household and local economies. In: Bakker N, Dubbeling M, Gündel S, Sabel-Koshella U, de Zeeuw H (eds) Growing cities, growing food. Urban agriculture on the policy agenda. Zentralstelle für Ernährung und Landwirtschaft (ZEL), Feldafing, pp 67–95
Oberndorfer EC, Lundholm JT, Bass B, Coffman R, Doshi H, Dunnett N, Gaffin S, Kohler M, Liu K, Rowe B (2007) Greenroofs as urban ecosystems: ecological structures, functions and services. Bio Sci 57(10):823–833
Ode Å, Tveit MS, Fry G (2008) Capturing landscape visual character using indicators: touching base with landscape aesthetic theory. Landsc Res 33(1):89–117
Ollerton J, Winfree R, Tarrant S (2011) How many flowering plants are pollinated by animals? Oikos 120:321–326
Pardo-de-Santayana M, Pieroni A, Puri RK (eds) (2013) Ethnobotany in the new Europe: people, health and wild plant resources. Berghahn Books, Oxford/New York
Patten KD, Shanks CH, Mayer DF (1993) Evaluation of herbaceous plants for attractiveness to bumble bees for use near cranberry farms. J Apic Res 32:73–79
Pearce H, Walters CL (2012) Do green roofs provide habitat for bats in urban areas? Acta Chiropterologica 14:469–478
Petchey OL, Gaston KJ (2002) Functional diversity (FD), species richness and community composition. Ecol Lett 5:402–411
Petts J (ed) (2000) Edible buildings-benefits, challenges and limitations. In: Edible Cities-A report of visits to urban agriculture projects in the USA. www.sustainweb.org
Pierce S, Brusa G, Vagge I, Cerabolini BEL (2013) Allocating CSR plant functional types: the use of leaf economics and size traits to classify woody and herbaceous vascular plants. Funct Ecol 27:1002–1010
Plieninger T, Höchtl F, Spek T (2006) Traditional land-use and nature conservation in European rural landscapes. Environ Sci Pol 9:317–321
Raigón MD, Prohens J, Muñoz-Falcón JE, Nuez F (2008) Comparison of eggplant landraces and commercial varieties for fruit content of phenolics, minerals, dry matter and protein. J Food Compos Anal 21:370–376

Rodrigues AS, Pérez-Gregorio MR, García-Falcón MS, Simal-Gándara J, Almeida DPF (2011) Effect of meteorological conditions on antioxidant flavonoids in Portuguese cultivars of white and red onions. Food Chem 124(1):303–308
Rodriguez M, Rau D, Angioi SA, Bellucci E, Bitocchi E, Nanni L, Knüpffer H, Negri V, Papa R, Attene G (2013) European Phaseolus coccineus L. landraces: population structure and adaptation, as revealed by cpSSRs and phenotypic analyses. PloS one 8(2):e57337
Romani A, Vignolini P, Galardi C, Mulinacci N, Benedettelli S, Heimler D (2004) Germplasm characterization of Zolfino landraces (Phaseolus vulgaris L.) by flavonoid content. J Agric Food Chem 52:3838
Roy A, Shrivastava SL, Mandal SM (2014) Functional properties of Okra Abelmoschus esculentus L.(Moench): traditional claims and scientific evidences. Plant Sci Today 1:121–130
Rumble H (2014) Quantifying the soil community on green roofs. Doctoral dissertation, University of London
Rumble H, Gange AC (2013) Soil microarthropod community dynamics in extensive green roofs. Ecol Eng 57:197–204
Scalzo RL, Genna A, Branca F, Chedin M, Chassaigne H (2008) Anthocyanin composition of cauliflower (Brassica oleracea L. var. botrytis) and cabbage (B. oleracea L. var. capitata) and its stability in relation to thermal treatments. Food Chem 107:136–144
Schrader S, Böning M (2006) Soil formation on green roofs and its contribution to urban biodiversity with emphasis on Collembolans. Pedobiologia 50:347–356
Sim SC, Robbins MD, Van Deynze A, Michel AP, Francis DM (2011) Population structure and genetic differentiation associated with breeding history and selection in tomato (Solanum lycopersicum L.). Heredity 106(6):927–935
Svensson B, Lagerlöf J, Svensson BG (2000) Habitat preferences of nest-seeking bumble bees (Hymenoptera: Apidae) in an agricultural landscape. Agric Ecosyst Environ 77:247–255
Tallamy DW, Shropshire KJ (2009) Ranking lepidopteran use of native versus introduced plants. Conserv Biol 23(4):941–947
Thomas K, Thanopoulos R, Knüpffer H, Bebeli PJ (2012) Plant genetic resources of Lemnos (Greece), an isolated island in the Northern Aegean Sea, with emphasis on landraces. Genet Resour Crop Evol 59:1417–1440
Tilman D (2001) Functional diversity. In: Levin SA (ed) Encyclopedia of Biodiversity 3:109–120, Academic
Tilman D, Downing JA (1994) Biodiversity and stability in grasslands. Nature 367:363–365
Tommasi D, Miro A, Higo HA, Winston ML (2004) Bee diversity and abundance in an urban setting. Can Entomol 136:851–869
Tonietto R, Fant J, Ascher J, Ellis K, Larkin D (2011) A comparison of bee communities of Chicago green roofs, parks and prairies. Landsc Urban Plan 103:102–108
Torricelli R, Tiranti B, Spataro G, Castellini G, Albertini E, Falcinelli M, Negri V (2013) Differentiation and structure of an Italian landrace of celery (Apium graveolens L.): inferences for on farm conservation. Genet Resour Crop Evol 1–12
Ulrich RS (1986) Human responses to vegetation and landscapes. Landsc Urban Plan 13:29–44
Van Mechelen C, Dutoit T, Hermy M (2014) Mediterranean open habitat vegetation offers great potential for extensive green roof design. Landsc Urban Plan 121:81–91
Villa TCC, Maxted N, Scholten M, Ford-Lloyd B (2005) Defining and identifying crop landraces. Plant Genet Resourc 3(3):373–384
Vischi M, Fiori M, De Paoli E, Padovan S, Guarda M, Olivieri A (2008) Broccoli 'Fiolaro'(Brassica oleracea) an endangered typical Italian cultivar. A genetic analysis by SSR markers. Plant Genet Resour Charact Util 6:215–221
Williams NSG, Lundholm J, MacIvor J (2014) Do green roofs help urban biodiversity conservation? J Appl Ecol 51(6):1643–1649
Zeven AC (1998) Landraces: a review of definitions and classifications. Euphytica 104(2):127–139

City Resilience to Climate Change

Teodoro Georgiadis, Ana Iglesias, and Pedro Iglesias

Abstract This chapter addresses the problem of city resilience to climate change and to the applicable methodologies to improve the capacity to ameliorate the population wellbeing. Future scenarios of the physiological equivalent temperature indicate the magnitude of the phenomenon. The best available technologies to mitigate such problem seem to be the utilization of vegetation in open spaces, including roof top farming. The modelling of real cases demonstrate the effectiveness of such green improvements, encouraging policies to greenness.

Resilient Cities

In Sicily (Italy) an old proverb states "*càliti juncu ca passi la china*". That means '*bend reed that goes the full*'.This is the deep substance of the term *resilience,* the capacity of a system to survive, or to adapt, to changes. The 5th Assessment Report of IPCC (2013) clearly addressed the urgency for the city systems to adapt to climate changes by including into the urban plans technical solutions able to increase the resilience capacity in all the compartments of the so-called "urban metabolism".

A city can be depicted as a complex organism, such as a human body, in which multiple processes strictly inter-correlated take place, and where the failure of one of the compartments reflects causing impacts on the entire city body, and thus compromising the survival of the organism itself. The complexity and importance of such organism it is multiplied by the fact that nowadays almost 50% of the world population lives in cities. This number is continuously increasing and the economic and social activities, extensive infrastructures and government operations, are concentrated within the city body implementing and creating unprecedented relationships between the different activities. The structure of a city, arrangement of buildings and streets is obviously strictly related to urban transportation. Number,

T. Georgiadis (✉)
Institute of Biometeorology IBIMET-CNR, Bologna, Italy
e-mail: t.georgiadis@ibimet.cnr.it

A. Iglesias • P. Iglesias
Universidad Politécnica de Madrid, Architecture, Madrid, Spain

F. Orsini et al. (eds.), *Rooftop Urban Agriculture*, Urban Agriculture,
https://doi.org/10.1007/978-3-319-57720-3_15

distribution and circulation of vehicles reflects immediately on pollution levels and consequently on human health and wellness. Surface sealing, produced by the necessity of urban movements, on the other way gives rise to flooding episodes which can cause strong damages to the structures, as well to energy systems impacting again on population health and security. All the city systems are strictly interconnected and interdependent and the vulnerability of one can have major impacts on others (Bugliarello 2003). These considerations assure about how the metabolic approach is reasonable and points the need that planners address the solution of city problems in a holistic view.

Because of such complex approach, it is necessary to incorporate in the study of city aspects of many different disciplines, including meteorology and atmospheric sciences, architecture and engineering, urban and landscape planning, as well economics and social sciences (Mills 2014). A real problem is then to merge knowledge coming from the different sciences in a unique comprehensive language. It is also important to focus on a common target to converge on, and to center a new scientific paradigm, stating that urbanization is an unstoppable process which will render in few decades any person on the planet a citizen. In presence of a growing concern for the human condition of such citizens, where *dignity* of life will be a primary requirement not assured at present, the 'quest' of citizenship, should be centralized in all the physiological, social, individual aspects and expectations we may think about.

The basic question is *how to make cities resilient*? We know that the technologies we will decide to apply will necessarily embed three main requisites: they should be cheap enough to ensure their applicability, suitable to small and distributed application and compatible with human's needs (The urban technologist 2014). Many technologies directly deal with accessibility, increasing the capacity of the individuals to an active participation to the city life. Others are tailored to contribute to the concept of *smart urbanism*, where distributed improvements of the urban architecture and texture may ameliorate the population's wellness (Dickson et al. 2009; Grimm et al. 2008; Helliwell et al. 2013), promote a sustainable development, and guarantee an inclusive growth. Such techniques directly reflect in increasing the capacity of the urban system in terms of resilience trough the mitigation and the adaptation to adverse effects caused by climate variability. A wide spectrum of strategies is currently available to meet this target as for example cool roofs, urban greeneries (i.e. public and private green, green roofs and rooftop farming), de-sealing of surfaces, rainwater caption and sustainable drainage systems.

In order to fight the consequences of the heat waves and urban heat island (the first ones produced by synoptic conditions on large atmospheric scales and the second caused by the physical properties of urban materials), the most promising strategies between those previously mentioned are the cool roofs, the urban greenery and the de-sealing of surfaces. These strategies, relatively cheap and easily to be distributed in the urban texture represent a good line of defense to protect cities and citizens from hazards when other assets, like energy and transportation systems, fail.

Impacts on Population and Living Organisms

In a city, the microclimate significantly differs from the surrounding areas because of the different partition of the solar radiation, in which a marked heat sensible flux is prevalent. This leads to the occurrence of exacerbated events of discomfort and may cause direct impacts on human health such as respiratory difficulties, fatal and non-fatal strokes, alteration of the sleep cycle (Georgiadis 2015).

Kalkstein et al. (2011) estimated, for major U.S. cities, that excessive heat events induce 1′300 excess mortality per year. The historic summer heatwave in 2003 accounted for a heat death-related mortality to exceed 15′000 cases only in France (Koppe et al. 2004; Poumadère et al. 2005). Conti et al. (2005) strongly highlights the role of UHI (urban heat island) during the occurrence of heat-waves and demonstrates, on epidemiological base, that the higher increase in mortality occur in the elderly population. Such episodes represent not only a health but also an economic problem, because of the consequent increase in the hospital admissions. Analysis of co-morbidity revealed excess admissions for cardiovascular diseases, diabetes, renal diseases and nervous system disorders that were significantly rising during the heat wave episodes (Semenza et al. 1999).

The peculiar susceptibility of elderly population to temperature extremes have been highlighted also in a different epidemiological study conducted in Stockholm County (Roclöv et al. 2014), reporting the effects on human health due to high summer and low winter temperatures. While for winter hazards the residential population is sufficiently advised and prepared, temperature extremes occurring during summer represent an increasing danger, specifically in the urban open spaces, and adaptation plans are being foreseen worldwide (Carmin et al. 2012).

A specific heat exposure index (HEI) described by Rey et al. (2009) for the extreme heat event of 2003 clearly indicated the most remarkable impacts in the most urbanized areas, where heat waves and urban heat island effects synergistically operate and negatively affect the health of the populations.

The exacerbation of the urban heat island impact during the heat wave episodes is not the only influence reflecting on population. Such episodes are often related to synoptic high pressure system which can cause air stagnation and consequently induce the increase of air pollution, because of the almost entire suppression of air masses exchange processes (Fortezza et al. 1993). Stagnation of pollutants over a city can produce excess values, when compared to the health standards recommendations, for a wide variety of compounds and particles. Trapped within the urban boundary layer, pollutants strongly increase in concentration during the night when the height of the mixed layer, sustained by the solar energy exchange during the daytime, collapses.

The effects of both heat fluxes and pollutants powerfully affect population and, in particular, vulnerable population: but citizens are not the only living organisms affected by such processes: spores, fungi, bacteria, insects, and more generally biodiversity, are directly impacted. Along with direct effect of pollutants on such organisms (biochemical processes) we should also consider the impacts produced by the physical processes induced by the presence of the city.

The heat fluxes that are at the basis of the formation of heat island can be represented as follows: air masses in close contact to the paved surfaces gain energy and consequently expand decreasing their density; the consequence is a less dense bubble of air floating in a heavier atmosphere thus creating thermals.

When the entire city contribute to such processes a 'thermal wall' will form, that prevent small organism to enter in the city itself, causing a detriment of the urban biodiversity. In this case all the ecological chains are directly impacted and, sometimes, definitely broken (it is easy to think to the bird-insect relationships) or moved to others territories (Levizzani et al. 1998).

A possible mitigation of the previous hazard is to make the city "thermally permeable", inducing continuity solutions in the thermal wall by creating cold pools in the interior of the urban environment. This makes the air to be advected horizontally, creating anemologic flows and, consequently, makes the transport of small organisms possible.

The less expensive and environmental sustainable technique is the introduction of green areas in the city texture. Vegetated areas because of their structural properties and physiological processes, evapotranspiration *in primis*, are commonly much cooler than the surrounding environment. The effects of urban greenery can actually be optimized in their performance and management by landscape planners, using specific computational tools.

Modeling the Urban Environment

The meteorological parameters are not sufficient alone in evaluating the bioclimatic performances of a specific location. This description requires a complex evaluation of thermo-physiological values in order to properly address the effects of the environment on human beings. In literature, some methodologies and indices are provided to define the wellbeing of an individual placed within an open space of an urban architecture, as well as to describe the indoor behavior (Oke 1987, 1988, 2006; Peng et al. 2012).The most common used indices in current models are the PMV (Predicted Mean Vote), and the PET (Physiological Equivalent Temperature), capable to evaluate the thermal heat stress basing their computation on the wider concept of the energy balance equation of a living body (Bethea and Parsons 2002; Souch and Grimmond 2006).

The PMV is an index that predicts the mean value of the votes of a large group of persons on a thermal sensation scale (from cold to hot), and it is based on the heat balance of the human body. To determine the index a combination of measurements of air temperature, mean radiant temperature, relative humidity, air speed, metabolic rate, and clothing insulation is required (Fanger 1972).

The PET index (Matzarakis et al. 1999), which units are reported in °C, makes results easy understandable for potential users. PET defines the equilibrium

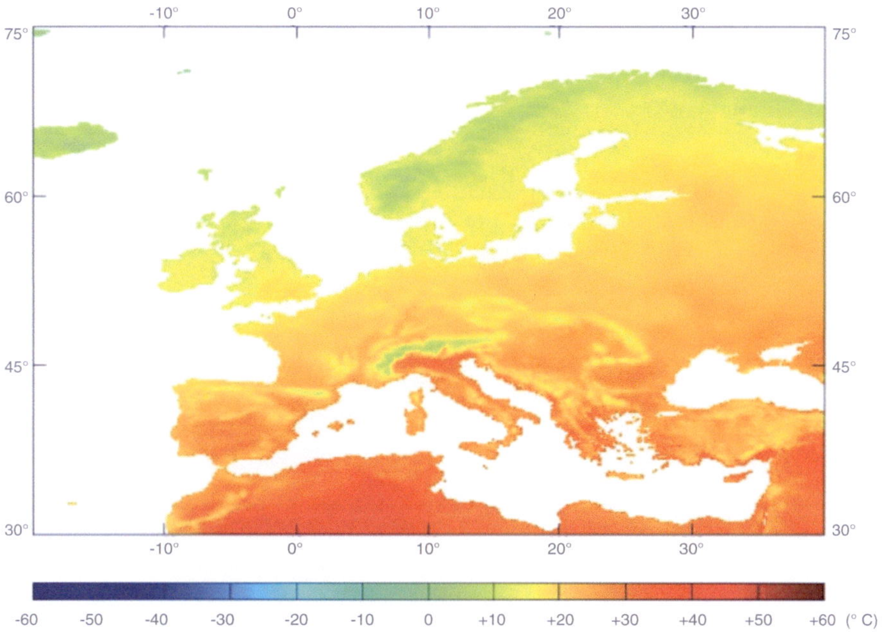

Fig. 1 PET estimation for July (period 1961–1990) (Matzarakis et al. 2007, with kind permission of Società Italiana di Fisica)

temperature of a human body, and enables planners to compare the integral effects of a complex thermal outdoor environment utilizing different parameters such: air temperature, air humidity, wind speed and short- and long-wave radiation measured or modelled in an outdoor environment on small and large scale representations.

In Fig. 1, the PET's for July, as averages over the period 1961–1990, are reported. It is particularly interesting to draw a baseline of wellness on a large scale, capable to compare conditions within Europe as depicted for the climatic period considered in the IPCC Assessment Reports (Matzarakis et al. 2007). Thus, trends can be derived for the different scenarios projected up to year 2020 for temperatures which indicate strong impact over European population. It is also possible to demonstrate the utility to apply such models to describe the bioclimatic behavior in the interior of a city (Oke et al. 1991; Sailor and Lu 2004; Taha 1997; Georgiadis et al. 2013). Particularly, impacts on vulnerable population, such as elderly and children, can be obtained by overlapping bioclimatic data with population-age density maps resulting in a direct visualization of critical occurrence within the city structure (Morabito et al. 2015). In the last study proposed, long time-series of remote sensing MODIS data have been utilized along with elderly population data extracted from the JRC population grid (100 m) from the 2001 census (Eurostat Census Hub (2011)

Nowadays, some models based on Computational Fluid Dynamics (CFD) (Wesseling 2001), or directly on the solution of Navier-Stokes equations (Cuvelier et al. 1986) which regulate the fluid motions, offer robust support in designing the

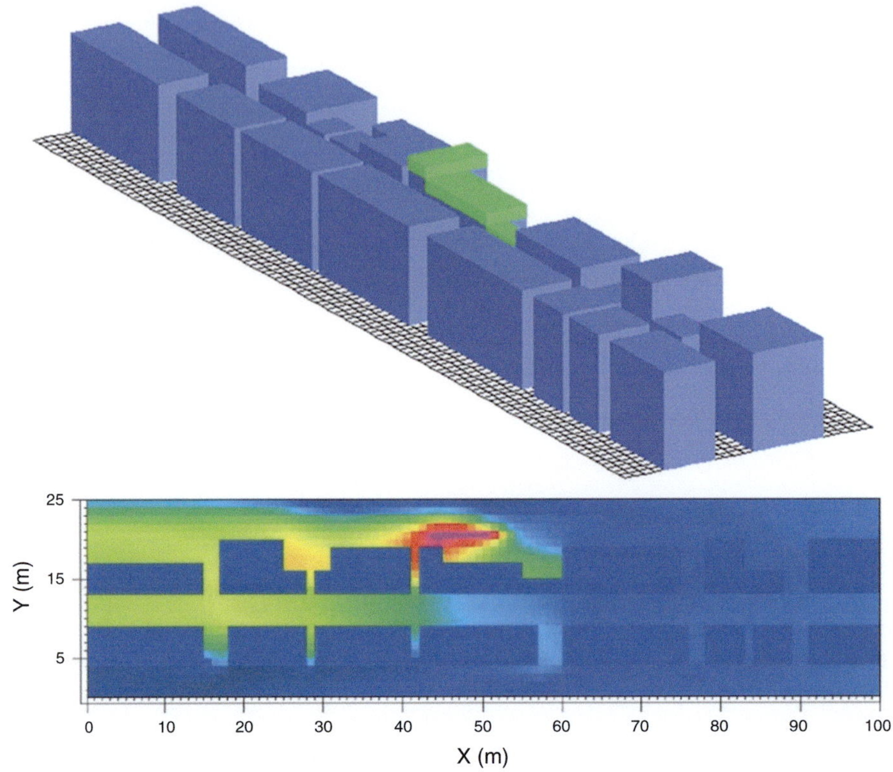

Fig. 2 Top, a sketch of the street canyon with the green roof and bottom, the air temperature differences (Image. T.Georgiadis)

urban architecture taking into account the micro-physics of the interactions/feedbacks between environmental parameters and the built environment. These models are based on the representation of the urban architecture along with some properties of its materials (i.e. albedo of the surfaces and thermal characteristics) and on their interactions between with the environmental parameters such as air temperature and humidity, wind speed and direction.

The simulations obtained as output of these models enable planners to depict *ex-ante* and *ex-post* scenarios and to properly choose the arrangements of the urban texture most optimized to obtain the effects looked-for. Not only the properties of abiotic materials can be represented by models: actually, new generation of models include more and more parameters linked to biotic elements, such as trees of different species, bushes and grasses. One of the most utilized microphysical model is the ENVI-Met software (Bruse and Fleer 1998) which solves the Navier-Stokes equation set and allows to directly calculate the surface fields of atmospheric parameters as well as the PMV. In Fig. 2 a simulation of the effects of a small green roof in a little urban canyon on air temperature is represented. On the top, the architecture of the canyon with the building around it is shown, with, in green color, a roof garden with an approximate surface of 12 m^2. It is then evident that when a roof top grass coverage is utilized, and even more when the presence of vegetation is marked such as in roof gardening, a mitigation on air temperature occurs, along with a better insulation of buildings. Mitigation does not only regards the building itself, but reflects also on the surrounding urban spaces, representing an adjunct value for all the city's environment.

Mitigation and Adaptation Strategies

A performance-based approach to resilience may provide a more holistic perspective of the city's ability to fulfill its essential functions (Rockefeller Foundation ARUP 2014). The necessity of a holistic perspective is widely recognized to face adaptation and to increase the city resilience. To act in a holistic way means to abandon the centralistic approach (systems governed by central units) much vulnerable, and to follow a distributed strategy of mitigation and adaptation. As far as mitigation methodologies applied are self-regenerative and self-sufficient, risks and hazards could be better faced. Some functions of the city result strategic for fighting risks: the safeguarding of human life and human health, and the delivery of basic needs.

The distribution of urban greenery is crucial to some of these functions be properly performed, and its design is not merely ornamental but should be carefully studied to assure the necessities of life and health by the utilization of specialized vegetation species.

Urban agriculture and rooftop farming result as the most promising strategies to face food scarcity. A remarkable example of urban agriculture is offered by the city of Yokohama (Japan) which population is about 3.7 millions, where 6′500 are engaged with agriculture. This last is not a large number, but the farmlands, covering a total area of about 3′000 hectares, account for the 7.3% of the total city area (Niitsu and Tokura 2015).

Rooftop farming results the most promising strategy because of recent technological advancements: hydroponic and other alternative methodologies solve the need of soil and consequently reduce the impact of soil weight on the structures of the buildings.

The characteristics of the urban environment largely determine the opportunities for including agriculture as an option to reduce the urban heat island effect (Table 1). In general rooftop agriculture is less effective than "on the ground" agriculture or green spaces. However, rooftop gardens or farms have large co-benefits to the local communities, such as reduction of noise, cultural and aesthetic change. And demonstration of the possibilities of sustainable cities.

The described techniques can be considered self-regenerative (being based on biotic productions), and assure, in normal conditions, the city metabolism and consequently do not aggravate the city assets during crisis periods. The entire range of such strategies becomes part of what are often called 'green infrastructures' (Foster et al. 2011), considered as a cost-effective adaptation and mitigation strategy for

Table 1 Characteristics of the rooftop gardens or farms and the open air gardens or farms in urban areas

	Types of gardens or farms in urban areas	
Characteristics	Roof top	On the ground, open air
Level of reduction of urban heat	Implemented in small areas with local benefits extending the building	Implemented in larger scales with benefits to the urban climate. Larger effects for mitigation of climate change
Co-benefits to the urban environment	Cultural change and aesthetic change, reduction of noise, examples towards sustainable buildings	Positive social interaction, individual health, environmental restoration, landscape change
Opportunities	Implementation in large, overpopulated cities	More suited in peri-urban areas
Challenges	Technical problems due to water leakage and increased insect populations	Market competition with the traditional rural producers
Examples of successful demonstration farm projects	In Cairo, rooftop agriculture developed by the Ain Shams University since the 1990s	Small urban farms in Amsterdam supported by NGOs since the 1980s
	In New York City, supported by private initiatives and incentives from The Green Roof Tax Abatement Program and Green Infrastructure Grant Program since the 2000s	Urban farms in Melbourne, Australia since the 2000s.
	In The Hague, the UF002 De Schilde rooftop farm is the largest in the world, launched in 2016	In Cape Town, South Africa, more than 400 micro farms are supported by community programs since the 2000s

solving urban and climatic challenges by building with nature. They imply standards and economics considerations exactly as the others, giving support to land value, quality of life, public health, and city economy. The advantage of such infrastructures is the possibility to apply them both in a public view of the city planning and in the private property parts of the city contributing to economic benefits. It is important to highlight their multiple beneficial effects, which can account not only for adaptation, but also mitigation of climate change. The amelioration to the local climate the green infrastructures can furnish reflects in benefits to both private and public properties, i.e. increasing the value of the building insisting on the mitigated area. In addition, a further element to consider is that, influencing wide spaces of the urban environment, their effects are not limited to a single class of citizenship but contribute to social equity.

Even if cities are commonly described as systems ‘out of equilibrium’ (Carter et al. 2015), a proper planning, in which resilience is considered an integrated concept, a central pillar, of the plan itself, can satisfy the increasing demand of security needs and sustainability.

Bullet Points

- City resilience to climate change may improve the capacity to ameliorate the population wellbeing.
- Future scenarios of the physiological equivalent temperature indicate the magnitude of the potential climate mitigation provided by urban green infrastructures.
- The modelling of real cases demonstrate the effectiveness of such green improvements, encouraging policies to greenness.

References

Bethea D, Parsons K (2002) Development of a practical heat stress assessment methodology for use in UK industry. Loughborough University. http://www.hse.gov.uk/research/rrpdf/rr008.pdf

Bruse M, Fleer H (1998) Simulating surface-plant-air interactions inside urban environments with a three-dimensional numerical model. Environ Softw Model 13:373–384

Bugliarello G (2003) Large urban concentration: a new phenomenon. In: Earth science in the city: a Read-er. AGU, Washington, DC, pp 7–19

Carmin J, Nadkarni N, RhieC (2012) Progress and challenges in urban climate adaptation planning: result of a global survey. MIT, Cambridge, MA, p 30

Carter JG, Cavan G, Connelly A, Guy S, Handley J, Kazmierczak A (2015) Climate change and the city: building capacity for urban adaptation. Prog Plan 95:1–66

Conti S, Meli P, Minelli G, Solimini R, Toccaceli V, Vichi M, Beltrano C, Perini L (2005) Epidemiologic study of mortality during the summer 2003 heat wave in Italy. Environ Res 98:390–399

Cuvelier C, Segal A, van Steenhoven AA (1986) Finite element methods and Navier-Stokes equations. Kluwer, Dordrecht, p 504

Dickson E, Baker JL, Hoornweg D, Tiwari A (2009) Urban risk assessments understanding disaster and climate risk in cities, Urban Development Series. World Bank, Washington DC, p 256

Eurostat Census Hub (2011) The 2011 Census database. http://ec.europa.eu/eurostat/web/population-and-housing-census/census-data/2011-census

Fanger PO (1972) Thermal comfort: analysis and applications in environmental engineering. McGraw-Hill, New York, p 244

Fortezza F, Strocchi V, Giovanelli G, Bonasoni P, Georgiadis T (1993) Transport of photochemical oxidants along the Northwestern Adriatic coast. Atmos Environ Part A 27(15):2393–2402

Foster J, Lowe A, Winkelman S (2011) The value of green infrastructure for urban climate adaptation. Center for clean air policy. html://www.ccap.org. Accessed 16 Dec 2015

Georgiadis T (2015) Urban climate and risk. In: Oxford Handbooks Online, ed. XX. New York: Oxford University Press, (forthcoming)

Georgiadis T, Lokoshchenko MA, Screti C, Vagnoli C (eds) (2013) Two hundred years of urban meteorology in the heart of Florence. In: Proceedings of the international conference on urban climate and history of meteorology. Firenze, Italy, 25th/26th February http://www.bo.ibimet.cnr.it/repository/proceedings. Accessed 15 Dec 2015

Grimm NB, Faeth SH, Golubiewsky NE, Redman CL, Wu J, Bai X, Briggs JM (2008) Global change and the ecology of cities. Science 319:756–760

Helliwell JF, Layard R, Sachs J (2013) World happiness report 2013. New York: UN sustainable development solutions Network Editor, pp 154. http://faculty.arts.ubc.ca/jhelliwell/papers/WorldHappinessReport2013_online.pdf. Accessed 15 Dec 2015

IPCC Fifth Assessment Report (2013) Climate Change 2013 (AR5). Intergovernmental panel on climate change. https://www.ipcc.ch/report/ar5/. Accessed 15 Dec 2015

Kalkstein LS, Greene S, Mills MD, Samenow J (2011) An evaluation of the progress in reducing heat-related human mortality in major U.S. cities. Nat Hazards 56:113–129

Koppe C, Kovats S, Jendritzky G, Menne B (2004) Heat-waves: risks and responses. World Health Or-ganisation. http://www.euro.who.int/document/E82629.pdf. Accessed 15 Dec 2015

Levizzani V, Georgiadis T, Isard S (1998) Meteorological aspect of the aerobiological pathway. In: Mandrioli P, Comtois P, Levizzani V (eds) Methods in aerobiology. Pitagora Editrice, Bologna, pp 113–185

Matzarakis A, Mayer H, Iziomon MG (1999) Applications of a universal thermal index:physiological equivalent temperature. Int J Biometeorol 43:76–84

Matzarakis A, Georgiadis T, Rossi F (2007) Thermal bioclimate analysis for Europe and Italy. Il Nuovo Cimento 30C(6):623–632

Mills G (2014) Urban climatology: history, status and prospects. Urban Clim 10:479–489

Morabito M, Crisci A, Gioli B, Gualtieri G, Toscano P, Di Stefano V, Orlandini S, Gensini GF (2015) Urban-hazard risk analysis: mapping of heat-related risks in the elderly in major Italian cities. PLoS One 18(May):18

Niitsu N, Tokura M (2015) Greenery to citizens! Urban agriculture in Yokohama City, Kanagawa Prefecture. Japan for sustainability. http://www.japanfs.org/en/news/archives/news_id035384.html. Accessed 15 Dec 2015

Oke TR (1987) The surface energy budgets of urban area. In: Modelling the urban boundary layer. AMS, Boston, pp 1–52

Oke TR (1988) The urban energy balance. Prog Phys Geogr 12(4):471–508

Oke TR (2006) Towards better scientific communication in urban climate. Theor Appl Climatol 84:179–190

Oke TR, Johnson GT, Steyn DG, Watson ID (1991) Simulation of surface urban heat islands under "ideal" conditions at night. Part 2: diagnosis of causation. Bound-Layer Meteorol 56:339–358

Peng S, Piao S, Ciais P, Friedlingstein P, Ottle C, Bréon F-M, Nan H, Zhou L, Myneni RB (2012) Surface Urban Heat Island across 419 global big cities. Environ Sci Technol 46:696–703

Poumadère M, Mays C, Le Mer S, Blong R (2005) The 2003 heat wave in France: dangerous climate change here and now. Risk Anal 25(6):1483–1494

Rey G, Fouillet A, Bassemoulin P, Frayssinet P, Dufour A, Jougla E, Hemon D (2009) Heat exposure and socio-economic vulnerability as synergistic factors in heat-wave-related mortality. Eur J Epidemiol 24:495–502

Rockefeller Foundation ARUP (2014) City resilience index: research report volume 1. Desk Study. p 53

Roclöv J, Forsberg B, Ebi K, Bellander T (2014) Susceptibility to mortality related to temperature and heat and cold wave duration in the population of Stockholm County, Sweden. Glob Health Action 7:1–11

Sailor DJ, Lu L (2004) A top-down methodology for developing diurnal and seasonal anthropogenic heating profiles for urban areas. Atmos Environ 38:2737–2748

Semenza JC, McCullogh JE, Flanders WD, McGeehin MA, Lumpkin JR (1999) Excess hospital admissions during the July 1995 heat wave in Chicago. Am J Prev Med 16(4):269–399

Souch C, Grimmond S (2006) Applied climatology: urban climate. Prog Phys Geogr 30(2):270–279

Taha H (1997) Urban climates and heat islands: Albedo, evapotranspiration and anthropogenic heat. Energy Build 25:99–103

The urban Technologist (2014) 12 simple technologies for cities that are smart, open and fair. http://theurbantechnologist.com. Accessed 15 Dec 2015

Wesseling P (2001) Principles of computational fluid dynamics. Springer, Berlin, p 644

Resource Efficiency and Waste Avoidance

Esther Sanyé-Mengual, Joan Rieradevall, and Juan Ignacio Montero

Abstract This section focuses on the environmental dimension of implementing rooftop agriculture (RA) regarding energy efficiency, water footprint, use of **residual CO_2** from buildings and global environmental preservation. RA contributes to improving the energy performance of cities, buildings and food production systems. The selection of water-efficient techniques and the promotion of rainwater harvesting and greywater recovery systems is a critical issue for minimizing the water footprint of RA. Finally, RA can positively contribute to face global-scale environmental problems such as climate change.

Introduction

The expansion of urban agriculture and building-based forms are linked not only to a growing urban population but also to an increased environmental awareness of citizens, particularly regarding the food industry and the global environmental issues. Beyond improving urban food security, the implementation of Rooftop Agriculture (RA) initiatives is also linked to the environmental advantages of producing food in cities, namely reducing resources consumption and environmental impacts. Such positive aspects are commonly used for advertisment by current RA companies. Gotham Greens details their sustainability approach of RA production as follows: "Our specially designed re-circulating hydroponic methods save land, save water, eliminate agricultural runoff and chemical pesticides, and offer the benefits of efficient, high-yield, local, year-round food production" (http://gothamgreens.com). Lufa Farms defines sustainable agriculture as "recycling water, optimizing

E. Sanyé-Mengual (✉)
Research Centre in Urban Environment for Agriculture and Biodiversity, Department of Agricultural Sciences, Alma Mater Studiorum, University of Bologna, Bologna, Italy
e-mail: esther.sanye@unibo.it

J. Rieradevall
Department of Chemical, Biological and Environmental Engineering, School of Engineering (ETSE) - Institute of Environmental Sciences and Technology (ICTA), Universitat Autònoma of Barcelona (UAB), Cerdanyola del Vallès, Spain
e-mail: joan.rieradevall@uab.cat

J.I. Montero
Institute of Agri-food Research and Technology (IRTA), Cabrils, Spain
e-mail: juanignacio.montero@irta.cat

F. Orsini et al. (eds.), *Rooftop Urban Agriculture*, Urban Agriculture,
https://doi.org/10.1007/978-3-319-57720-3_16

energy use and growing without any synthetic pesticides, herbicides or fungicides" (http://lufa.com).

This section evaluates the environmental dimension of implementing RA in developed cities by discussing the potential benefits and showing quantitative studies regarding energy efficiency, water footprint and global environmental preservation. This chapter focuses on these three aspects since other environmental aspects are discussed in specific chapters of this book, such as biodiversity (see chapter "Biodiversity of Flora and Fauna"). According to the literature (Astee and Kishnani 2010; Cerón-Palma et al. 2012; Ackerman et al. 2014; Specht et al. 2014; Thomaier et al. 2015), beyond the common benefits of Urban Agriculture (UA) and local food production, RA is associated with specific advantages. Regarding energy, the creation of new green infrastructure elements on roofs (i.e., current vacant spaces within cities) can indirectly improve the urban heat island effect as well as the energy efficiency of buildings. In particular, integrated Rooftop Greenhouses (i-RTGs) can exchange the metabolic flows (including energy, water, and CO_2) with the building below to boost the efficiency of both systems. Concerning water footprint, multiple techniques and crops can be employed in RA, which seeks to maximize the water efficiency of their systems. Furthermore, research-oriented and innovative projects of RA are testing the use of alternative sources of water for crop production: rainwater harvesting and greywater recovery. Finally, RA contributes to the global environmental preservation by minimizing the food-miles and the related environmental impacts, such as climate change. Furthermore, RA shortens the distance between producers and consumers thereby boosting the freshness of the produce and reducing the generation of food waste along the supply-chain.

Environmental Benefits and Rooftop Agriculture Typologies

The environmental benefits depend on the typology of RA. About the environmental performance, we can differentiate three main typologies of RA according to the various technologies that can be employed: integrated rooftop greenhouses (i-RTGs), isolated rooftop greenhouses (RTGs) and open-air rooftop agriculture (open-air RA).

First, RA can be differentiated among protected and open-air systems. Protected RA employs the greenhouse technology to isolate the food production from the weather conditions to have a controlled environment, regarding temperature, humidity and pathogens. This technique is associated with higher crop efficiency (i.e., higher food production) and, thus, this type of RA contributes larger to the environmental benefits linked to local food (i.e., mitigate climate change, boost food freshness).

Among protected RA, rooftop greenhouses (RTGs) can integrate their metabolic flows with the building or can develop the food production activity apart from the building metabolism:

- Integrated Rooftop Greenhouses (i-RTGs) are rooftop greenhouses that exchange the metabolic flows (energy, water,CO2) with the building to improve the efficiency of both the food production and the building metabolism. Particularly, the use of the residual heat from buildings minimizes the energy consumption for acclimatizing the greenhouse. The integration of buildings' water flows (rainwater, greywater) into the greenhouse reduces the water footprint by closing cycles. Moreover, it uses **residual CO_2** from the building as a source to enrich agricultural production, thereby optimizing the crop yield. Examples of i-RTG designs are the Fertilecity project in Barcelona (Spain) (http://www.fertilecity.com), the Roof-Water Farm project in Berlin (Germany) (http://www.roofwaterfarm.com) and the ECF Farm systems technology (http://www.ecf-farmsystems.com).
- Isolated Rooftop Greenhouses (RTGs) are greenhouses that have an independent metabolism from the building. Depending on the geographic context of the initiative, this typology of RA requires energy to acclimatize the greenhouse to ensure the optimal thermal conditions for the crop development. Examples of existing RTG initiatives are Lufa Farms in Montreal (Canada) (http://lufa.com) or Gotham Greens in New York (United States) (http://gothamgreens.com).

Finally, the implementation of food production systems on roofs without greenhouses as protective element is classified as open-air RA:

- Open-air rooftop agriculture (Open-air RA) includes gardens on roofs that have no isolation to weather conditions and produce seasonally according to the climatic context and the crop specifications. The morphology of open-air gardens tends to occupy the entire roof in a similar way as green roofs, thereby maximizing the contribution to reducing the urban heat island effect and improving the energy efficiency of the building. Both in New York (USA), the Brooklyn Grange Farm (http://brooklyngrangefarm.com), Barcelona Open Air Roof top Farming/Social Orchard/Institut Municipal de Persones amd Discapacitat mental(IMPD) or the Eagle Street rooftop farm (http://rooftopfarms.org) are open-air rooftop farm examples.

Table 1 summarizes the contribution to the environmental benefits of the three types of RA that were distinguished above.

Energy Efficiency

The implementation of gardens on the roofs of buildings can have a positive global impact effect from an energetic perspective. At the city scale, the introduction of new spaces for vegetation can reduce the Urban Heat Island (UHI) effect. Furthermore, gardens can improve the energy efficiency of buildings and innovative rooftop greenhouses can boost these effects.

Table 1 Environmental benefits of rooftop agriculture types: integrated rooftop greenhouses, isolated rooftop greenhouses and open-air rooftop agriculture

	Protected RA		Open-air rooftop agriculture
	Integrated rooftop greenhouses	Isolated rooftop greenhouses	
Reduce urban heat island			
Improve energy efficiency			
Low-energy consumption (greenhouse heating)			
Close water cycles			
Mitigate climate change			
Boost food freshness			

Current practices can combine characteristics from different typologies of rooftop agriculture. The scale of the benefit is classified high (black), medium (dark gray) and low (light gray)

Minimizing the Urban Heat Island Effect

Rooftop Agriculture (RA) has benefits as a new typology of urban green infrastructure on the Urban Heat Island (UHI) effect, which increases the mean temperature of urban areas compared to the adjacent rural zones. Increasing the areas within cities devoted to vegetation can alter the global urban heat balance by two major phenomena: alteration of the solar radiation (reflection, diffusion and shadow) and decreasing the air temperature through plant evapotranspiration (Akbari 2002; Ackerman et al. 2014). However, this effect has only been evaluated in the literature for other typologies of green infrastructure: green roofs and green facades. In the case of RA, the effect on the urban heat island depend on the specific design (e.g., crops, cultivation techniques, garden form) which determines essential aspects such as the real evapotranspiration.

Susca et al. (2011) evaluated the benefits of vegetation in the city of New York and found out that vegetated areas were fresher and showed a decrease in temperature of 2 ° C as compared with non-vegetated ones, on average. Such study demonstrated the contribution of vegetated surfaces compared to man-built ones. Then, RA can reduce the UHI effect and have positive consequences regarding urban livability. At the micro-scale, the presence of rooftop gardens can provide higher thermal comfort for inhabitants, particularly in warm climate areas and during hot seasons. At the citywide level, a large-scale implementation of RA can, therefore, have a significant effect on the UHI effect (Saiz et al. 2006; Astee and Kishnani 2010; Ackerman et al. 2014).

Improving the Energy Efficiency of Buildings

The implementation of RA provides the building with extra insulation elements, which particularly affects the energy metabolism and efficiency of the building. Green roofs, green facades and rooftop gardens are new types of green infrastructure and have positive effects on the thermal insulation of buildings, as demonstrated in the literature (Ekaterini and Dimitris 1998; Wong et al. 2003; Saiz et al. 2006; Astee and Kishnani 2010; Castleton et al. 2010; Ottelé et al. 2011).

On the one hand, open-air rooftop initiatives have similar effects to green roofs. During summer, the vegetation layer reduces the rooftop surface temperature and the energy requirements for cooling the building. In contrast to temperate climates, vegetation can largely contribute to reducing the energy consumption for cooling buildings in the Mediterranean area (Ottelé et al. 2011). Furthermore, green roofs minimize the heat losses in cold seasons. As an example, Saiz et al. (2006) assessed the effects of implementing a green roof on a building in Madrid (Spain). The vegetated area saved 1% of annual energy use and up to 25% of cooling load in summer, thereby reducing the environmental impacts up to 5%. Notwithstanding that other studies found this effect similar to a layer of insulating material (Cerón-Palma 2012), one may consider that rooftop gardens have further functions and benefits than a constructive solution.

On the other hand, Rooftop greenhouses (RTGs) also act as an insulating layer to the building thereby reducing the heat losses and improving the energy efficiency for acclimatizing the building spaces. For the Mediterranean region, Cerón-Palma (2012) simulated the insulation effect of implementing a rooftop greenhouse on an office building, which was lower than 5%. Nevertheless, the actual potential of RTGs regarding energy efficiency is related to the metabolic integration with the building, as explained in the following section.

Rooftop Greenhouses: Low Energy-Consuming Acclimatization

The implementation of greenhouses on roofs has specific characteristics concerning the energy metabolism. Some studies highlighted the potential of rooftop greenhouses (RTGs) to save energy by acting as cooling, heating and energy recycling (Specht et al. 2014). The two most usual trends in this issue are using renewable energies and integrating the building and the greenhouse metabolisms.

Renewable Energy The demonstrative RTG of the "Science Barge" project grows various crops in a 120m^2 greenhouse that is completely independent of the power network of New York City (Nelkin and Caplow 2007). The company Gotham Greens employs climate-controlled RTGs to produce year-round vegetables and uses photovoltaic panels to supply part of the energy demand (http://gothamgreens.com). Therefore, renewable energy production is a way to minimize the energy footprint of isolated rooftop greenhouses (Fig. 1).

Integrated RTGs Closing the energy flow in RA is a common goal in multiple research projects and initiatives. Figure 1 shows the energy metabolism of both isolated RTGs and integrated RTGs. On the one hand, isolated RTGs, which currently dominate RA practices, requires external energy (i.e., both non-renewable and renewable) while energy from the greenhouse (e.g., heat) is diffused to the atmosphere. On the other hand, integrated i-RTGs aim to create a common energy

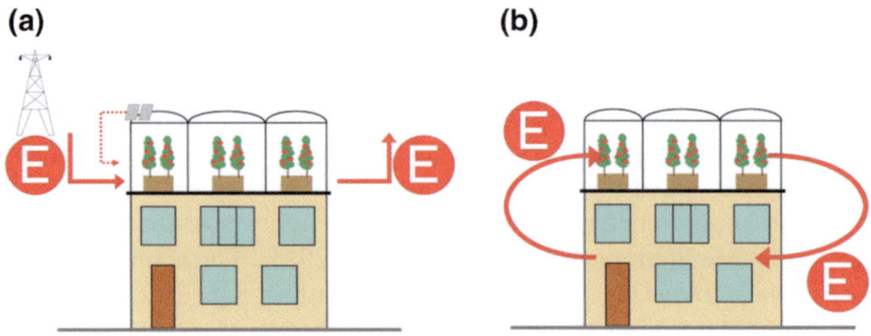

Fig. 1 Energy flows in isolated rooftop greenhouses (open-cycle) and integrated rooftop greenhouses (closed-cycle) (**a**) Isolated RTG, (**b**) Inegrated RTG (Image: E. Sanyé-Mengual)

metabolism thereby reducing both consumption and losses. First, the greenhouse can use the residual energy for improving the environmental conditions for food production (e.g., employing the waste heat from building spaces for heating the greenhouse in cold seasons). Second, the building can also employ the residual heat or cool from the greenhouse to improve the thermal comfort of the building without consuming energy (Sanyé-Mengual et al. 2014). In the Mediterranean area, Cerón-Palma (2012) simulated the effect of using the heat from a rooftop greenhouse in the energy requirements of an office building, which could be decreased up to 79%. Furthermore, the greenhouse can use the **residual CO_2** from the building as a source to enrich agricultural production (e.g., photosynthesis). However, metabolic synergies are especially limited for existing buildings and their architectural structures (Germer et al. 2011; Thomaier et al. 2015). The goal of the FertileCity project is to demonstrate these practices for the Mediterranean context (http://www.fertilecity.com) and the Fraunhofer Institute is working on integrated RTGs in continental climates (Fraunhofer UMSICHT 2011).

Water Footprint

The development of human activities has taken advantage from natural water resources leading to a water scarcity risk due to high consumption rates and contamination of water bodies. In particular, agriculture depends on the availability of water to ensure global food security. Thus, an improved water management to guarantee the water efficiency and to minimize the water footprint of agriculture is essential to satisfy future scenarios (de Fraiture and Wichelns 2010). The conventional way to obtain water in RA projects is tap water (Fig. 2a), leading to the environmental impacts related to the urban water supply. Initiatives of RA pursue water efficiency by employing the least water-consuming techniques or by promoting innovative water technologies, such as rainwater harvesting (Fig. 2b) or greywater recovery (Fig. 2c), as better explained in chapter “Water Management and Irrigation Systems”.

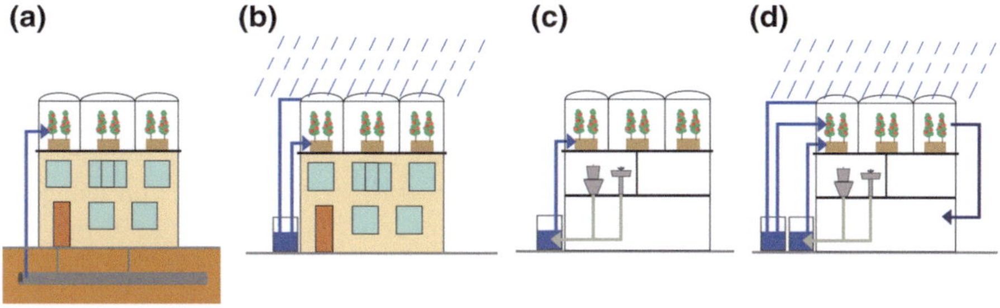

Fig. 2 Water flows in conventional rooftop agriculture and water-efficient solutions: rainwater harvesting, greywater recovery and integrated rooftop greenhouses (RTGs) (**a**) Conventional, (**b**) Rainwater harvesting, (**c**) Greywater recovery, (**d**) Integrated RTG (Image: E. Sanyé-Mengual)

Furthermore, the residual water from RA could be used for non-drinking purposes in the building, as proposed by the i-RTG concept (Cerón-Palma et al. 2012), which seeks to integrate all the water-efficient solutions into the design (Fig. 2d).

Water Efficiency

The selection of techniques, crops and garden design is determinant for the water efficiency of RA. According to Thomaier et al. (2015), hydroponics and recirculating systems are the most commonly used techniques in current RA initiatives to minimize water consumption. Some authors highlight the potential water savings of hydroponics. A recirculating hydroponic system can produce vegetables consuming between 5 and 10 times less water than conventional agriculture (Caplow 2009). Hydroponic rooftop gardens can produce similar yields to conventional farms with 75% less water (Astee and Kishnani 2010). Sustainable choices for water and irrigation management are thus essential (See chapter Water Management and Irrigation Systems for more details on water management) (Box 1).

Rainwater Harvesting and Greywater Recovery

Current practices reduce the water consumption by using rainwater harvesting systems and the building's greywater for irrigation purposes (Thomaier et al. 2015). These two sustainable strategies are of particular interest in urban agriculture, since using drinking water for food production can be prohibited by law and can result in an expensive input (Cerón-Palma et al. 2012).

Rainwater Harvesting Rainwater harvesting consists of collecting rainwater, as an endogenous and renewable resource, on the roof of the building for satisfying the irrigation requirements of the crop. In New York City, the "Science Barge" initiative is water self-sufficient through rainwater harvesting (Caplow 2009).

Box 1: Water Efficiency in the Mediterranean Climate

The water performance of RA strongly depends on the typology of rooftop agriculture and the techniques and crops that are included in the design. Table 2 compiles available data of water efficiency from quantitative studies. The water consumption has been quantified for open-air production using organic soil production, nutrient film technique (NFT) and floating hydroponic in Bologna (Italy) (Sanyé-Mengual et al. 2015b) and soil-less production in a rooftop greenhouse in Barcelona (Spain) (Pou 2015; Sanyé-Mengual et al. 2015a). Water efficiency depends on the technique, the design of the garden and the crop yield. First, soil-less techniques showed a better water performance than organic soil production, apart from the nutrient film technique (NFT) in which efficiency in water use was depleted by low crop yield. Second, leafy vegetables resulted in a higher consumption per product ($L{\cdot}kg^{-1}$) when the garden has a homogeneous design, where irrigation did not consider the specific requirements of each crop (e.g., lettuce and tomato have the same irrigation rate). Finally, a lower crop yield ($kg{\cdot}m^{-2}$) leads to a worse environmental performance and a lower water use efficiency ($m^3{\cdot}kg^{-1}$). The evaluation of the water flow in the Mediterranean climate is of great importance due to the water scarcity in these areas and the climate change risks (e.g., drought periods).

Table 2 Water consumption for various types of RA, crop techniques and crops

Type of RA	Crop technique	Crop	Water consumption [m^3/kg]
Open-air production	Organic production	Chili pepper	0.16
		Eggplant	0.05
		Lettuce	0.39
		Melon	0.08
		Tomato	0.09
		Watermelon	0.07
	Nutrient film technique (NFT)	Lettuce	0.124
	Floating hydroponic	Lettuce	0.056
Rooftop greenhouse	Soil-less production	Tomato	0.030
		Lettuce	0.034

Sanyé-Mengual et al. (2015a, b)

The FertileCity project in Barcelona (Spain) evaluates the potential water self-sufficiency of food production in an integrated RTG in the Mediterranean context (http://www.fertilecity.com). Preliminary results suggest that rainwater could cover more than 60% of the water demand for lettuce and tomato production (Sanyé-Mengual et al. 2014). However, the self-sufficiency level of the production system strongly depends on the climate conditions and the rainwater availability.

Greywater Recovery A source of renewable water that seems more stable than rainwater regarding supply is greywater from human activities. When the RA is placed on a residential building, greywater could be treated and used as a source for irrigating the plants (Cerón-Palma et al. 2012). Such strategy would reduce the impacts related to urban wastewater management and freshwater consumption. Cerón-Palma et al. (2012) accounted for the potential contribution of household greywater to irrigation demand and average figures suggested that greywater recovery can significantly contribute to the design of self-sufficient and sustainable RA systems. The "roof-water farm" project in Berlin (Germany) (http://www.roofwaterfarm.com) works on developing the integration of the greywater in rooftop agriculture with the aim of creating urban agriculture system independent of the water grid thereby facing current water crises. Notwithstanding the potential of greywater recovery, further research is needed for developing treatment technologies and testing safety limitations.

Global Environmental Preservation

As a new type of local food system, rooftop agriculture (RA) has some implications at the global scale. The implementation of RA within cities minimizes the distribution needs for food products by shortening the distance between producers and consumers. Even more, some types of RA are for self-production where the producer becomes the consumer. Main global environmental implications of promoting local food systems are the mitigation of climate change by minimizing the food logistics and the reduction of food waste generation.

Climate Change Mitigation: Local Production and Food-Miles

The use of non-renewable energy sources in transportation is one of the leading causes of greenhouse gas emissions, which contribute to climate change. The current environmental situation urges society to minimize the causes of climate change. For example, in the last COP21 in Paris (France) some "climate change mitigation goals" were established. Within the food industry, globalization and the development of transportation technologies have led to the enlargement of distances between production and consumption.

In this context, local food movements have created alternative food networks to shorten these long distances while reducing the contribution to climate change (Edwards-Jones et al. 2008). The concept of "avoided food-miles" has been used in the literature to evaluate the different environmental impacts of imported and local food supply chains, mainly regarding energy consumption and climate change (Edwards-Jones et al. 2008). Even more, local food systems are also known as "Zero-km agriculture." The use of vacant spaces within cities like roofs can lead to the actual development of Zero-km systems. The reduction of the carbon footprint

Box 2: Seasonality and Food Miles

The growing interest in local food production is also linked to a return to the traditional way of producing and consuming food. The recovery of traditional varieties transports sensibility and the consumption of seasonal products are highly valued by consumers of environmentally-friendly local food. Seasonality can significantly affect the "avoided food-miles" and the environmental benefits of local food products. Sanyé-Mengual (2015) observed the variance of the implication of increasing the local food production in Barcelona (Spain). When consuming city-produced food, citizens substitute their shopping in supermarkets and thus the imports of food from the global food industry. The "avoided food miles" and "avoided global warming" were evaluated according to the statistics of the food distribution center of Barcelona (MercaBarna) (considering food market data from 2010 to 2014) (Table 3). In the case of tomato, seasonal production (during summer) has the lowest avoided foodmiles and CO_2 emissions of the year, apart from August where the market requires importing from The Netherlands to supply a rise in the demand. In the case of lettuce, this can be produced year-round in the study area, but the low crop yield in winter boosts the imports and, thus, local production can avoid larger foodmiles and CO_2 emissions. For both products, avoided food-miles and global warming savings vary up to 50%.

Table 3 Avoided food-miles [km] and CO_2 emissions [g] per month and annual average for tomato and lettuce sold vegetables in the food distribution center of Barcelona (MercaBarna)

	Jan	Feb	Mar	Apr	May	Jun	Jul	Aug	Sep	Oct	Nov	Dec	Year avg
Tomato													
[km]	698	690	709	734	771	630	525	756	543	594	625	627	659
[g CO_2]	118	116	120	123	125	97	83	102	83	96	104	105	106
Lettuce													
[km]	395	415	405	406	356	302	272	304	291	314,8	339	407	351
[g CO_2]	67	71	69	69	59	50	45	51	49	53	57	69	59

Sanyé-Mengual (2015)

of RA products by shortening the supply chain is one of the most valued environmental opportunities of this type of urban agriculture (Cerón-Palma et al. 2012). Furthermore, changing the scale from global to local is seen as a chance towards the decoupling of the resources consumption from the economic activity (Cerón-Palma et al. 2012). Sanyé-Mengual et al. (2013) compared the local supply-chain of tomatoes produced in Barcelona (Spain) with the conventional supply-chain from the South of Spain, identifying environmental benefits up to 440 g of CO_2 per kg of product. Finally, the use of local resources can positively contribute to the reduction of the environmental burdens of food production while promoting the development of a circular economy, such as the employment of urban wastes as substrate instead of imported Rockwool, perlite or coco fiber (Grard et al. 2015) (Box 2).

Boosting Freshness: Food Waste Reduction and Environmental Implications

According to the Food and Agriculture Organization (FAO), around 30% of the food that is produced is wasted (FAO, 2011), accounting for 1.3 billion tons of food waste per year. According to the EU project FUSIONS, 88 million tons of food waste are annually generated in Europe, which cost 143 billion euros (Stenmarck et al. 2016). Thus, food waste has become a hotspot for global food security. Plans and programs are being designed and implemented to promote the reduction of food waste generation at the production and at the consumption stages. Local food production can positively contribute to this purpose by minimizing the supply chain of food products.

Gotham Greens company describes its produce as "Extraordinarily fresh produce, grown in extraordinarily fresh places" (http://gothamgreens.com). The freshness of the produce is valued by the consumers regarding quality, as the product can be harvested just some hours before the consumer purchase. From an environmental perspective, a fresh produce and a reduced supply-chain mean a minimization of food waste. Reducing food waste positively affects the environmental performance of the entire life cycle of food products. When reducing the food waste generation, we are avoiding the production, distribution and retail of an extra amount of food that is finally not consumed. Furthermore, reducing food waste implies avoiding the GHG emissions resulting from food waste management (Box 3).

Box 3: A Case Study of Food Waste Generation

Minimizing food waste generation is an added-value of local food and RA. Sanyé-Mengual et al. (2013) accounted for the environmental impact of the conventional supply-chain of tomato from Almeria to Barcelona and compared it to a hypothetical local supply-chain of tomatoes from a Rooftop Greenhouse (RTG) in Barcelona. Figure 3 details the food waste generation in the conventional supply-chain of tomato, which can be up to 21%. Compared to these figures, a local supply-chain of tomatoes from an RTG is linked to the generation of 0% food waste as no transportation is performed, local food market has no aesthetical restrictions (size, color) and current practices use overripe products for producing added-value products.

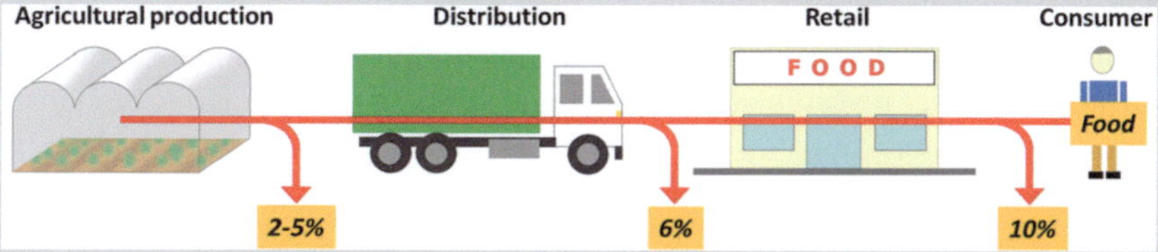

Fig. 3 Food waste generation in the conventional supply-chain of tomato: Barcelona-Almería (Image: E. Sanyé-Mengual) (Sanyé-Mengual et al. 2013)

Conclusions

The environmental benefits identified in this chapter are policy hotspots and can be the object of funding opportunities for RA promoters. Currently, policy-makers are interested in creating new economies, managing storm water, improving buildings' energy efficiency, reduction local and global emissions or boosting urban biodiversity (Thomaier et al. 2015). From a sustainability lens, we have to consider that the environmental benefits detailed in this chapter are linked to economic benefits (e.g., minimizing distribution costs) (Cerón-Palma et al. 2012) that can make the local food activity more economically sustainable. In social terms, RA can further develop local economies and, in particular, support the creation of green economy and circular economy networks. Furthermore, RA is commonly related to social and educational activities, which at the same time can promote environmentally-friendly habits with further effects at the global scale.

Bullet Points

- Rooftop agriculture can positively contribute to minimize and use local resources (residual heat,**rainwater** harvesting) and the environmental impacts of local food products.
- Since RA increases the vegetated spaces of cities, the urban livability can be improved by decreasing the urban heat island effect.
- Buildings that host rooftop gardens can boost their energy efficiency by increasing the thermal insulation of the building.
- Rooftop greenhouses are an innovative way to implement agriculture in urban environments to **reduce water and energy consumption as well as CO_2 emissions**
- Rooftop greenhouses to create a common metabolism between the building and the greenhouse maximize the energy savings resulting from the **exchange of flows between** the greenhouse and the building (for heat and/or cold)
- RA practitioners aim to design roof gardens as water efficient as possible by choosing water-friendly techniques and implementing technological solutions, such as rainwater harvesting and greywater recovery.
- Finally, the promotion of local food systems reduces the contribution to climate change and minimizes the food waste generation of the food supply-chain. However, further research and demonstrative projects are needed to demonstrate the feasibility and quantify the benefits of innovative solutions, such as integrated RTGs or greywater recovery.

References

Ackerman K, Conard M, Culligan P et al (2014) Sustainable food systems for future cities: the potential of urban agriculture. Econ Soc Rev (Irel) 45:189–206

Akbari H (2002) Shade trees reduce building energy use and CO2 emissions from power plants. Environ Pollut 116:119–126

Astee L, Kishnani N (2010) Building integrated agriculture: utilising rooftops for sustainable food crop cultivation in Singapore. J Green Build 5:105–113

Caplow T (2009) Building integrated agriculture: philosophy and practice. Urban futures 2030 Urban Dev. Urban lifestyles future. Heinrich-Böll-Stiftung, pp 48–51

Castleton HF, Stovin V, Beck SBM, Davison JB (2010) Green roofs; building energy savings and the potential for retrofit. Energ Buildings 42:1582–1591. doi:10.1016/j.enbuild.2010.05.004

Cerón-Palma I (2012) Strategies for sustainable urban systems: introducing eco-innovation in buildings in Mexico and Spain. Universitat Autònoma de Barcelona, Barcelona

Cerón-Palma I, Sanyé-Mengual E, Oliver-Solà J et al (2012) Barriers and opportunities regarding the implementation of Rooftop Eco.Greenhouses (RTEG) in Mediterranean cities of Europe. J Urban Technol 19:87–103. doi:10.1080/10630732.2012.717685

De Fraiture C, Wichelns D (2010) Satisfying future water demands for agriculture. Agric Water Manag 97:502–511. doi:10.1016/j.agwat.2009.08.008

Edwards-Jones G, Milà i Canals L, Hounsome N et al (2008) Testing the assertion that "local food is best": the challenges of an evidence-based approach. Trends Food Sci Technol 19:265–274. doi:10.1016/j.tifs.2008.01.008

Ekaterini E, Dimitris A (1998) The contribution of a planted roof to the thermal protection of buildings in Greece. Energ Buildings 27:29–36

FAO (2011) Global food losses and food waste – extent, causes and prevention. FAO, Rome

Fraunhofer UMSICHT (2011) Harvesting on urban rooftops – demo center at Fraunhofer inHaus. Fraunhofer UMSICHT, Oberhausen

Germer J, Sauerborn J, Asch F et al (2011) Skyfarming an ecological innovation to enhance global food security. J für Verbraucherschutz und Leb 6:237–251. doi:10.1007/s00003-011-0691-6

Grard BJ-P, Bel N, Marchal N et al (2015) Recycling urban waste as possible use for rooftop vegetable garden. Futur Food J Food Agric Soc 3:21–34

Nelkin J, Caplow T (2007) Floating hydroponics in the Big Apple. Grow Edge 18:38–40

Ottelé M, Perini K, Fraaij AL et al (2011) Comparative life cycle analysis for green façades and living wall systems. Energ Buildings 43:3419–3429. doi:10.1016/j.enbuild.2011.09.010

Pou M (2015) Rooftop greenhouses (RTGs) as local production systems in Barcelona: LCA and LCC of lettuce production. Universitat Autònoma de Barcelona (UAB)

Saiz S, Kennedy C, Bass B, Pressnail K (2006) Comparative life cycle assessment of standard and green roofs. Environ Sci Technol 40:4312–4316

Sanyé-Mengual E (2015) Sustainability assessment of urban rooftop farming using an interdisciplinary approach. Universitat Autònoma de Barcelona, Barcelona

Sanyé-Mengual E, Cerón-Palma I, Oliver-Solà J et al (2013) Environmental analysis of the logistics of agricultural products from roof top greenhouses in Mediterranean urban areas. J Sci Food Agric 93:100–109. doi:10.1002/jsfa.5736

Sanyé-Mengual E, Llorach-Massana P, Sanjuan-Delmás D, et al (2014) The ICTA-ICP rooftop greenhouse lab (RTG-Lab): closing metabolic flows (energy, water, CO2) through integrated rooftop greenhouses. In: Roggema R, Keefer G (eds) Finding spaces product. Spaces 6th AESOP Sustain. Food Plan Conf VHL University of Applied Sciences, Velp, pp 692–701

Sanyé-Mengual E, Oliver-Solà J, Montero JI, Rieradevall J (2015a) An environmental and economic life cycle assessment of Rooftop Greenhouse (RTG) implementation in Barcelona, Spain Assessing new forms of urban agriculture from the greenhouse structure to the final product level. Int J Life Cycle Assess. doi:10.1007/s11367-014-0836-9

Sanyé-Mengual E, Orsini F, Oliver-Solà J et al (2015b) Techniques and crops for efficient rooftop gardens in Bologna, Italy. Agron Sustain Dev 35:1477–1488. doi:10.1007/s13593-015-0331-0

Specht K, Siebert R, Hartmann I et al (2014) Urban agriculture of the future: an overview of sustainability aspects of food production in and on buildings. Agric Hum Values 31:33–51. doi:10.1007/s10460-013-9448-4

Stenmarck A, Jensen C, Quested T, Moates G, Buksti M, Cseh B, Scherhaufer S (2016) Estimates of European food waste levels. FUSIONS project report

Susca T, Gaffin SR, Dell'osso GR (2011) Positive effects of vegetation: urban heat island and green roofs. Environ Pollut 159:2119–2126. doi:10.1016/j.envpol.2011.03.007

Thomaier S, Specht K, Henckel D et al (2015) Farming in and on urban buildings: present practice and specific novelties of Zero-Acreage Farming (ZFarming). Renew Agric Food Syst 30:43–54. doi:10.1017/S1742170514000143

Wong NH, Tay SF, Wong R et al (2003) Life cycle cost analysis of rooftop gardens in Singapore. Build Environ 38:499–509. doi:10.1016/S0360-1323(02)00131-2

Community and Social Justice Aspects of Rooftop Agriculture

Kathrin Specht, Kristin Reynolds, and Esther Sanyé-Mengual

Abstract This chapter examines the community and social justice aspects of rooftop agriculture (RA), focusing on cities in the Global North. The goal is to provide an overview of the social aspects of diverse RA typologies and the potential community and social justice effects, from the individual level to the city scale. We show that, like urban agriculture overall, RA may have multiple benefits in the urban setting (such as improving community food security, providing educational opportunities, or fostering neighborhood participation). However, we argue that, like urban agriculture overall, RA is not in and of itself a sustainable or socially just practice. The chapter discusses these dynamics with examples from several Global North cities. We conclude with generalizable policy, funding, and design recommendations for RA that advances community well-being and social equity goals. While focused on the Global North context, the principles behind these recommendations are also applicable in Global South regions.

Introduction

Rooftop agriculture (RA) is a growing phenomenon in cities in the Global North (e.g., Thomaier et al. 2015; Gorgolewski et al. 2011). In North America, commercial rooftop farms exist alongside not-for-profit educational projects and gardens at low-income and mixed income residential facilities, some of which incorporate RA into programs for residents (Cohen et al. 2012). In Europe, RA is also expanding to

K. Specht (✉)
Leibniz Centre for Agricultural Landscape Research, Institute of Socio-Economics, Müncheberg, Germany
e-mail: specht@zalf.de

K. Reynolds
Environmental Studies and Food Studies Programs, The New School, New York, NY, USA
e-mail: kristin.reynolds@newschool.edu

E. Sanyé-Mengual
Research Centre in Urban Environment for Agriculture and Biodiversity, Department of Agricultural Sciences, Alma Mater Studiorum, University of Bologna, Bologna, Italy
e-mail: esther.sanye@unibo.it

F. Orsini et al. (eds.), *Rooftop Urban Agriculture*, Urban Agriculture,
https://doi.org/10.1007/978-3-319-57720-3_17

include health or community oriented projects (e.g., rooftop community gardens; gardens atop hospitals), high-tech start-ups (e.g., for-profit aquaponics) and agronomic research (e.g., integrated rooftop greenhouses). Indeed, types of RA are diverse, ranging from "micro-gardens" (in which food is grown in crates, buckets, bags, or other reclaimed materials) to full-scale greenhouses and green roofs with specialized soil media cultivated on an half a hectare or more. RA can also include beehives and chicken coops. RA complements more traditional in-ground urban agriculture (UA), particularly in dense cities with high real estate costs and competitive land markets. As with UA overall, RA is understood by supporters as a beneficial use of urban space, with multiple social and environmental functions including increased food security, mitigation of urban heat island effects, and storm water absorption. However, like UA overall, RA is not in and of itself a sustainable or socially just practice. Without attention to social equity, it can exacerbate economic and environmental disparities, and the tendency for RA to take place in privately-held spaces raises questions about community access to the sites. In order for RA to produce the social benefits that are necessary in building resilient cities, practices, funding programs, and policies must consider, and actively address, the socio-political complexities of the urban system in question. This chapter begins with an overview of benefits and challenges of UA more generally, and then examines these community and social justice aspects of RA, drawing from examples in several cities in the Global North.

Potential Community and Social Justice Benefits and Drawbacks of Urban Food Production

Social Impacts on the Individual and Community Level

In general, urban food production can provide social benefits at the individual and community level, in terms of food security, health, and empowerment (Armstrong 2000; Mok et al. 2013; Lovell and Taylor 2013)(see Box 1). These social impacts are typically more pronounced in community gardens than in commercial farms (the latter of which have financial solvency as a primary objective) and can be described as follows:

– *Food security*. Food security can be enhanced as UA provides new areas for food production. Users of private and community gardens highlight increased access to healthy and local food, lowering their food expenses (Carney 2011). Moreover, the production is often sufficiently high that food is shared with neighbors or further processed into value-added products (e.g., fruit preserves). However, the actual production quantity can be limited due to lack of agricultural knowledge among participants or limited space in the gardens. In for-profit projects, high food prices can limit accessibility for those with low incomes.

- *Health improvement.* Health of community members, including mental and physical health, can be enhanced via UA. This, since engagement in gardens can improve participants' diets by increasing vegetable and fruit consumption and variety (Alaimo et al. 2008; Centrone et al. 2014). Furthermore, gardening is linked to mental improvement by decreasing stress, improving self-confidence, providing therapeutic spaces for those facing trauma, and feeding imagination and inspiration (Adevi and Mårtensson 2013; Söderback et al. 2009). Finally, gardens can serve as spaces for leisure and physical activities for certain groups (e.g., elderly). Conversely, some consumers are concerned that urban-grown food products can have health risks related to soil, water and air contamination (Specht and Sanyé-Mengual 2017).
- *Community empowerment.* Gardens are potential tools for community building and community empowerment. They serve as a meeting point for community- and building residents, and can facilitate social cohesion in terms of networks for community problem solving (Armstrong 2000; Teig et al. 2009). Furthermore, community engagement in urban gardens can increase educational and economic empowerment through improved knowledge and skills (e.g., agricultural and management techniques) (Block et al. 2011; Mees and Stone 2012).

Box 1: Via Gandusio, Community Rooftop garden, Bologna (Italy)
Via Gandusio is a social housing project in Bologna built in the 1960s to host socially disadvantaged families and migrant workers that mainly came from Southern Italy. Today, the city council uses the available apartments as temporary housing for international migrants. Community cohesion has been a challenge due to the different cultures and age groups of residents, as well as the transitory nature of the resident population. A community rooftop garden was set up by the city council, Bologna University and the association BiodiverCity with the objective of creating a space for social interaction towards community building, social inclusion, and empowerment. Today, a group of residents self-manage the crop production of diverse vegetables, herbs and fruits, including some African and Asian species and varieties. The users identify not only the individual benefits that the garden provides (food self-production, physical exercise, hobby and leisure time) but also community building and empowerment. The creation of a group of gardeners has led to cultural and knowledge exchange. Group members have also taken ownership over certain aspects of building renovation such as the building's elevators. News media have highlighted the novelty of the rooftop garden in Via Gandusio, which also hosts cultural events involving neighborhood associations.

For more information:

See chapter "Soil Based and Simplified Hydroponics Rooftop Gardens": Rooftop agriculture experiences across the world, http://www.comune.bologna.it/casa/servizi/8:6436/20704/

Social Impacts on the Broader Neighborhood/City Level

Beyond the individual and community-level effects of food production in cities, UA has potential social justice benefits at citywide and societal scales (see Box 2). These include:

- *Public health.* Public health disparities within cities have been well documented (e.g., Story et al. 2008). The wider causes of such disparities are beyond the scope of this chapter. Still, to the extent that UA helps urban residents to access healthful foods, nature, and safe open spaces, the existence of farms and gardens in parts of a city where these may be lacking can help advance health justice at a city-wide scale.
- *Resilience and environmental justice.* Urban farms and gardens also form a part of a city's green infrastructure and, as such, they can be part of a city's strategy for environmental and social resilience (McPhearson et al. 2014; Kremer et al. 2013; Taylor and Lovell 2013; Barthel and Isendahl 2012). UA can reduce urban heat island effects and absorb stormwater, thereby reducing combined sewer overflow, and, through filtering effects of vegetation, can improve local air qual-

Box 2: Seeds to Feed Rooftop Farm, Brooklyn, New York (USA)
Seeds to Feed Rooftop Farm is a community farm situated atop Georgia's Place, a seven-story housing facility for formerly homeless, mentally ill, and low-income adults in Crown Heights, a historically low-income neighborhood in Brooklyn, New York. The facility and farm are operated by the not-for-profit social service organization Community Counseling and Mediation (CCM), which has provided mental health, supportive housing, and youth services for New York City residents since 1982. The farm project began with one small container garden in 2009 and has since expanded to encompass 284 gardening boxes, with a membrane layer protecting the roof.

Seeds to Feed Rooftop farm fulfills several social justice and community needs. It provides fresh produce to the facility's 48 residents, many of whom are from the West Indies, including crops selected according to residents' preferences. The farm also provides mental health benefits to residents and members of the surrounding community through opportunities to spend time outdoors in this green and safe space. Additionally, the farm produces flowers, which residents can pick and bring into the housing facility to brighten the indoors – an especially important benefit in the dense urban environment. In a city that has been a leader in commercial RA, Seeds to Feed is an example of community-oriented RA in one of New York's lowest income communities.

For more information:
http://seedstofeedrooftopfarm.tumblr.com/; http://ccmnyc.org/

ity. The distributional dimensions of these effects have not been studied to-date, but when many farm and garden sites are situated in neighborhoods with high concentrations of industry and/or low concentrations of green space (often lower income neighborhoods), they may help address environmental disparities, which is a core environmental justice concern.

– *Political engagement and control over public space.* Urban farms and gardens can also provide opportunities for community access to and control over public space (Eizenberg 2012). This has important social justice benefits when accessible to a broad socioeconomic diversity of city residents. Moreover, farm and garden spaces themselves can act as venues for community organizing on political issues, providing informal opportunities for policy advocacy that can be particularly important for those who have been historically under-represented in formal policy making (Cohen and Reynolds 2014; Reynolds and Cohen 2016). UA thus has potential socio-political benefits that are important social justice concerns at a citywide scale. While UA does not necessarily bring these benefits to all urban contexts (Reynolds 2014; McClintock 2013), they are in the realm of possibility.

Interacting with the Public and Potential Consumers

Additionally, UA can have both benefits and drawbacks vis à vis consumers' familiarity with food production (see Box 3). Due to its location in cities, UA projects can increase consumer awareness of agriculture and its fundamental role in the food

Box 3: ECF ROSTLAUBE Containerfarm, Berlin (Germany).

The ECF Containerfam is situated on a former industrial area of an abandoned inner-city malt factory in Berlin. It serves as a prototype of a hydroponic farm, producing fish and vegetables in a discarded shipping container. The container is designed to be placed on roofs in cities with climatic conditions that do not allow year-round open-air production. The project's founders call it a "modern allotment garden for the city of the twenty-first century." They are engaged in marketing, but also in the project's educational functions improving consumer awareness. They offer tours, lectures, workshops and various events for school classes, families, and the general public. The container thereby serves as a meeting point for many activities related to local food production, nutrition, harvesting and cooking events.

For more information:

ECF- Efficient | City | Farming: http://www.ecf-farmsystems.com/ and Rostlaube Containerfarm: http://www.malzfabrik.de/en

system (Gorgolewski et al. 2011; Steel 2009). In some cases, consumers are directly involved in the production process as participants, such as in the case of community gardens. Yet, agriculture in cities may also produce tensions or conflicts. Previous studies have revealed that stakeholders associate a number of risks with UA (e.g. lower quality of the products or growing techniques that are considered as an "unnatural" way of producing food) (Kaufman and Bailkey 2000; Sanyé-Mengual et al. 2015a; Specht et al. 2015a). Furthermore, potential consumers tend to be more critical and set higher quality standards for UA compared to rural agricultural products (Specht et al. 2015a).

The benefits and drawbacks of urban agriculture more generally have been reviewed here. The next section discusses those specific to rooftop agriculture.

Characteristics and Potential Social Benefits of Different Types of RA

Like UA more generally, RA projects have several distinct, yet often overlapping characteristics that are linked to their aim, productivity, accessibility, technology, and management. These characteristics vary among six main types of RA. Based on Thomaier et al. (2015), these types are as follows.

- *Commercial RA* refers to for-profit projects. By producing and selling food products, commercial RA contributes to local community development in terms of job creation and economic growth. Due to its high productivity, it can contribute to food security and access to healthy food, although product prices determine the affordability and thereby the target consumers (i.e., low-income; middle/upper income). Some projects combine food production with educational and training programs. The provision of environmental services depends on crop management and agrobiodiversity.
- *Life quality RA* includes private recreational spaces in residences or workplaces. Social benefits are mainly at the individual level (e.g., personal empowerment). Users can improve their access to healthy food, education, and physical activity.
- *Social RA* has social and educational purposes and encompasses *community-based and institutional* activities, such RA at schools or hospitals. This type of RA offers social benefits at the community level, although institutional projects are commonly limited to certain societal groups (e.g., children; elderly people). Community-based initiatives offer opportunities for personal and community empowerment and food security. Social benefits in institutional gardens can be less pronounced since their focus is often on physical activity and education.
- *Innovation RA* tests and demonstrates technological and environmental innovations (e.g. gardens attached to research centers). Immediate social benefits are limited to education, local development, and ecological improvement, though they may lead to broader social benefits in the long-run.
- *Image-oriented RA* is mostly used to add value to existing food businesses (e.g., restaurants, cafés). It has benefits similar to commercial projects.

Table 1 Characteristics and social benefits of different RA types: commercial, quality of life, social-(community and institutional), innovation and image

		RA types					
		Commercial	**Life quality**	**Social-community**	**Social-institutional**	**Innovation**	**Image**
Characteristics	**Aim**	For-profit	Social	Social	Social	Research	For-profit
	Productivity	High	Low-Medium	Low-Medium	Low	High	Medium
	Accessibility	Limited	Limited	Open	Limited	Limited	Limited
	Technology	Low to High	Low	Low	Low-Medium	Medium-High	Low to High
	Management	Private	Private	Communal	Public/Private	Public/Private	Private
Social benefits	**Food security**	major	medium	major	medium	minor	medium
	Empowerment	minor	medium	major	medium	minor	minor
	Access to healthy food	medium	major	major	medium	minor	major
	Physical activity	minor	major	major	major	minor	minor
	Food justice	minor	medium	major	medium	minor	minor
	Social inclusion	minor	minor	major	medium	minor	minor
	Environmental justice	medium	medium	major	medium	minor	minor
	Education	medium	major	medium	major	major	medium
	Local development	major	minor	medium	minor	medium	medium
	Environmental improvement	medium	medium	major	medium	medium	medium

Key: The table presents the 6 standard types of RA, although real-world RA projects often combine characteristics from multiple typologies (see Thomaier et al. 2015). Social benefits are identified as major (horizontal pattern), medium (vertical pattern) and minor (solid), based on current practices and individual project goals

The potential social benefits of RA also vary among these six types, ranging from food security to environmental improvement. A matrix of the six types of RA, their characteristics, and potential social benefits is shown in Table 1.

Social Issues Specific to RA

Social and Environmental Justice Benefits Specific to RA

Among the benefits specifically associated with RA, the main ones are the following:

– *Creation of green spaces.* Interest and investment in green roofs has expanded in recent decades, and RA is one form of this type of green infrastructure (Orsini et al. 2014). At the city scale, green spaces in urban environments are associated with improved air quality and reduced heat island effect, thereby increasing the liveability of cities. These benefits are notable in urban forests, which provide with large canopies and have a long lifespan (Baró et al. 2014). RA also contributes to these benefits, although the effects depend on the typology and design of the RA project. In particular, open-air and polyculture gardens may contribute the most to green spaces and biodiversity (Lin et al. 2015).

- *Access to green spaces.* Access to green spaces can be linked to improved quality of life due to several health and environmental benefits (Reklaitiene et al. 2014; Tamosiunas et al. 2014). Increased access to green spaces through RA is of particular interest in highly dense cities, with a lower number of green areas per capita, and in low-income neighborhoods, which often have higher concentrations of contaminated sites.
- *Access to fresh food.* RA can also increase access to fresh and healthy food in low-income communities and integration of such spaces into social housing can provide residents with spaces for food production. Furthermore, similarly to UA (Block et al. 2011; Calvet-Mir et al. 2012), socially-oriented rooftop gardens can become a tool to increase community food sovereignty by providing access to produce one's own food, if desired. Such effects have been observed in the community rooftop garden at Via Gandusio (see Box 1). Still, the extent to which RA increases food access in low income communities depends on the goals of the operators - whether social or for-profit (see next section).
- *Increased urban resilience.* At the city scale, roofs can become key spaces for expanding UA. There is often competition for land use in urban contexts, with real estate development limiting spaces for food production. RA may avoid conflicts with other land uses since these projects take advantage of unused spaces in the built environment (Sanyé-Mengual et al. 2015b). Additionally, increasing food production can make cities more resilient to crisis. RA can help cities maintain food security when faced with extreme weather events, economic crisis, and social conflict (Dubbeling and de Zeeuw 2011). At a broader scale, RA can, in a small, but important ways, help to mitigate impacts of global warming by developing local food systems (see chapter "City Resilience to Climate Change").

Social and Environmental Justice Drawbacks Specific to RA

Despite the many potential benefits of RA, there are specific aspects of the practice that need to be considered when evaluating its social and environmental justice effects:

- *Costs and access to funding.* The costs of building and maintaining intensive rooftop farms (i.e., those with soil depth of 15 cm or more) can be prohibitive. Brooklyn Grange, one of the leading rooftop farm businesses in New York City, estimated actual costs to build its nearly one-acre site atop a commercial roof in the borough of Queens at around 54 USD m^{-2}, totalling 200,000 USD for the whole installation, not including volunteer labor. The group estimated costs for its second site (at a shipyard and industrial park in Brooklyn) at 180 USD m^{-2}, which accounted for a warranty and maintenance on the roof itself as well as hired labor (at fair market prices) (Ben Flanner, personal communication,

12/13/2011). BrightFarms, a for-profit company that builds and manages several rooftop greenhouses in New York and Chicago, estimates the cost of a one-hectare rooftop greenhouse at 5 million USD (Rifkin 2011). These figures do not include building rent.

- In New York, groups have afforded these high start-up costs through public grants. In partnership with the Brooklyn Navy Yard, Brooklyn Grange received a New York City Department of Environmental Protection (DEP) grant for 592,730 USD in 2011 to help fund construction of the farm at that site (New York City Department of Environmental Protection 2011). One school in the Greenwich Village neighborhood of Manhattan funded construction of its educational rooftop garden through a combination of public and private donations totalling more than 1 million USD (Decker 2012). Lenox Hill Settlement House near Manhattan's Upper East Side received a 40,000 USD grant, also from DEP, to help build its green roof, which covers 483 m^{-2} of impervious area and includes vegetable production for children's education and meals for residents (New York City Department of Environmental Protection n.d.).

- Yet, despite the success of these groups in financing their projects through grants, small, not-for-profit groups often have difficulty financing such capital-intensive farms, since private funding is often awarded to groups with previous track records of receiving large grants (Cohen and Reynolds 2015). The risk of RA being dominated by large enterprises due to the investment costs is also one major concern of local stakeholders in Berlin and Barcelona (Sanyé-Mengual et al. 2015a; Specht and Sanyé-Mengual 2015; Specht et al. 2015b). At a city-wide scale, this can have the effect of concentrating ownership and operation of rooftop farms and gardens among already well-connected and well-resourced groups. Moreover, without oversight of the ways that funding patterns affect spatial distribution of rooftop farms throughout a city, these patterns may result in a concentration of RA in wealthy areas, which often already have higher concentrations of parks and green spaces.
- *Access to RA*. RA may also be less physically accessible to community members than are in-ground farms and gardens. In many cities in the Global North, there is a precedent for land owned by the city or by non-profit land trusts being available to residents for gardening and farming through permits or lease agreements. Moreover, community gardens on public land are often required to hold open hours for non-gardeners to enjoy the space (Cohen et al. 2012). Rooftops, being located above the ground level where most people may encounter them, are less readily accessible to non-participants and this raises important questions about the social and community benefits of RA. While these farms and gardens may provide food and environmental benefits to some urban residents, the extent to which they provide opportunities for those not directly connected to the farms or gardens - whether through ownership or residency in the building where rooftop farms and gardens are located - is constrained. These possible drawbacks give pause to the idea that simply increasing the overall presence of RA in a given city will produce social and environmental justice benefits. Finally, the extent to which RA provides the environmental benefits of other forms of green infra-

structure depends on its size, form, and management, so individual projects need to be evaluated for their potential contribution to urban resilience.

Social Acceptability of RA

Regarding the general perception of RA, the following elements need to be considered:

- *Perception and acceptance.* For many city inhabitants and local stakeholders, the idea of RA is new and, for some, even unknown. RA has only recently been introduced or re-introduced to cities in the Global North. At this stage of development, social acceptability of RA is a key component of future implementation. Experiences from different countries have shown that, on the one hand, RA creates a large "hype" and projects are receiving an increasing amount of attention from the public media, politicians and funding agencies (Reynolds 2014; Reynolds and Cohen 2016). Recent studies have revealed that local stakeholders (like urban planners, activists or policy makers) associate many potential benefits to RA and continue to promote its implementation as an element of multifunctional urban development (Specht et al. 2014, 2015b). On the other hand, studies have also revealed a number of risks that stakeholders attach to RA and that might negatively affect their social acceptability (Sanyé-Mengual et al. 2015a; Specht and Sanyé-Mengual 2017; Specht et al. 2015a). These risks are related to the production systems and the RA-related technologies or the RA products themselves.
- Moreover, there are general conflicts with respect to the perception of RA since agriculture is traditionally associated with rural and in-ground production (see Kaufman and Bailkey 2000). While supporters of urban agriculture in general have had to face this challenge of public perception vis à vis the legitimacy of growing food in cities, RA faces the additional challenge of being on top of buildings – an even more unconventional place for agricultural production. Thus, in addition to the potential benefits of RA, future practitioners, planners, and policy makers interested in RA need to recognize these potential drawbacks in order to create sustainable and socially just projects.
- *Risks related to RA technologies and products.* As for production systems, studies from Berlin and Barcelona (Sanyé-Mengual et al. 2015a; Specht et al. 2015b, 2016) have shown that potential conflicts arise through the fact that there is often limited acceptance of soil-less growing techniques, which are mostly proposed for large-scale and commercial RA. Particularly for the case of rooftop greenhouses, community members may perceive technologies as being too complex and expensive to achieve feasible for-profit businesses. Regarding potential products, some people perceive soil-less growing techniques, which are often employed in RA, as an "unnatural" way of producing food and they therefore expect the products to be of lower quality compared to those which come from "real soil". Additionally, stakeholders expect an increase in noise and light pollution as well as odors due to production activities. Another major barrier to

social acceptance is the common assumption that products from urban areas are more likely to be contaminated by air pollution compared to those from rural areas (Specht and Sanyé-Mengual 2017), although evidence of the opposite are actually found in literature (e.g. Antisari et al. 2015). In fact, in RA, health risks due to soil contamination are potentially reduced, because RA uses commercial soil or soil-less growing techniques.

Conclusions

This chapter has discussed how many of the well-described social benefits of UA in general - such as community building, health awareness, or health improvements - can also apply to RA. The chapter has also argued that, compared to on-ground UA, RA has some specific limitations in terms of community and social justice concerns. These are particularly linked to its limited accessibility, exclusivity (due to high investment costs), inequalities in access to funding, and perceived risks. These potential limitations need to be understood and addressed by those who want to build, fund, or politically support RA. Some of the social drawbacks of RA stem from the relative newness of the practice – community gardens, for example, have a longer history in cities, while policies and programs supporting RA are only nascent. As noted in the beginning of this chapter, RA does not create social benefits *per se* and the integration of social goals into RA are not a matter of course. RA has significant potential to benefit community members, but whether or not RA contributes to social justice in cities depends then on the specific project types. Designing socially just RA requires planning to ensure that equity is addressed, along with economic, sustainability, and resilience goals.

Bullet Points

In terms of community empowerment and social justice, RA shows a range of potential benefits but also drawbacks. Guidelines for the future development of RA that supports social justice goals need to be based on the principles of accessibility, social equity, and integration. We here offer several recommendations to improve the current conditions towards changing the potential theoretical benefits to real implementation:

- *Community involvement in decision making.* Planning for RA needs to involve residents of the community in question so that it addresses community needs, and so that if there are alternative needs that supersede RA, these are prioritized. Public funding programs for RA aimed at producing social and environmental benefits should prioritize projects in low income neighborhoods, *if* RA is desired by residents. Moreover, perceived risks of RA should be examined and addressed in project planning stages.

- *Consideration of, and research on health risks.* Communication about health risks of urban food production may highlight the potential of RA, which is expected to have a lower exposure to certain contamination sources (i.e., contaminated soil). Nevertheless, the question of contamination in RA has not been sufficiently addressed in the research community and needs further scientific investigation. Quality labelling schemes could help to improve consumers' acceptance of RA products. Yet, for food production on roofs, decisions about where to locate should also take into account the air quality at potential sites.
- *Integration of RA into urban policies.* City policies should include a framework for developing RA, particularly to solve food system inequalities such as access to healthy food. For example, RA might be required in new public housing designs to provide low-income families with spaces for food production as a complement to other provisioning sources. Examples of such projects exist in New York City, and could provide a model for other cities. Further, RA may be integrated into local-level policy commitments as a tool to increase local food production and address environmental concerns.
- *Integration of RA and environmental justice into green infrastructure and environmental policies.* Policies to support RA as green infrastructure with multiple social and environmental justice benefits need to be developed around a framework of urban resilience that includes social equity. Green infrastructure strategies have existed for decades in the Global North, and RA needs to be considered as an element of designing for resilience at city scale. One can imagine a future reality in which more public buildings have urban rooftop farms that are accessible to the public, in the way that public museums or libraries currently are, or in which public funding targets more RA projects in low income neighborhoods as a part of planning for urban resilience. Policy measures should facilitate such initiatives.

References

Adevi AA, Mårtensson F (2013) Stress rehabilitation through garden therapy: the garden as a place in the recovery from stress. Urban For Urban Green 12(2):230–237. http://doi.org/10.1016/j.ufug.2013.01.007

Alaimo K, Packnett E, Miles RA, Kruger DJ (2008) Fruit and vegetable intake among urban community gardeners. J Nutr Educ Behav 40(2):94–101. http://doi.org/10.1016/j.jneb.2006.12.003

Antisari LV, Orsini F, Marchetti L, Vianello G, Gianquinto G (2015) Heavy metal accumulation in vegetables grown in urban gardens. Agron Sustain Dev 35(3):1139–1147. http://doi.org/10.1007/s13593-015-0308-z

Armstrong D (2000) A survey of community gardens in upstate New York: implications for health promotion and community development. Health Place 6(4):319–327

Baró F, Chaparro L, Gómez-Baggethun E, Langemeyer J, Nowak DJ, Terradas J (2014) Contribution of ecosystem services to air quality and climate change mitigation policies: the case of urban forests in Barcelona, Spain. Ambio 43(4):466–479. http://doi.org/10.1007/s13280-014-0507-x

Barthel S, Isendahl C (2012) Urban gardens, agriculture, and water management: sources of resilience for long-term food security in cities. Ecol Econ 86:224–234

Block DR, Chávez N, Allen E, Ramirez D (2011) Food sovereignty, urban food access, and food activism: contemplating the connections through examples from Chicago. Agric Hum Values 29(2):203–215. http://doi.org/10.1007/s10460-011-9336-8

Calvet-Mir L, Gómez-Baggethun E, Reyes-García V (2012) Beyond food production: ecosystem services provided by home gardens. A case study in Vall Fosca, Catalan Pyrenees, Northeastern Spain. Ecol Econ 74:153–160. http://doi.org/10.1016/j.ecolecon.2011.12.011

Carney M (2011) Compounding crises of economic recession and food insecurity: a comparative study of three low-income communities in Santa Barbara County. Agric Hum Values 29(2):185–201. http://doi.org/10.1007/s10460-011-9333-y

Centrone SM, Humphries D, Kline R (2014) Investigation of an urban farm intervention for a low income Hispanic population with multiple risk factors for diabetes (624.3). FASEB J, 28(1_Suppl)

Cohen N, Reynolds K (2014) Urban agriculture policy making in New York's "new political spaces": strategizing for a participatory and representative system. J Plann Educ Res 34(2):221–234

Cohen N, Reynolds K (2015) Resource needs for a socially just and sustainable urban agriculture system: lessons from New York City. Renewable Agric Food Syst 30(1):103–114

Cohen N, Reynolds K, Sanghvi R (2012) Five borough farm: seeding the future of urban agriculture in New York City. Design Trust for Public Space, New York

Decker G (2012) At rooftop garden party, candidates tout budding principal. Gotham Schools. Accessed 10 Apr 2013. http://gothamschools.org/2012/09/21/at-rooftop-garden-party-2013-candidates-tout-budding-principal/

Dubbeling M, de Zeeuw H (2011) Urban agriculture and climate change adaptation: ensuring food security through adaptation. In: Otto-Zimmermann K (ed) Resilient cities. Springer, Dordrecht, pp 441–449. http://doi.org/10.1007/978-94-007-0785-6

Eizenberg E (2012) The changing meaning of community space: two models of NGO management of community gardens in New York City. Int J Urban and Reg Res 36(1):106–120

Gorgolewski M, Komisar J, Nasr J (2011) Carrot City: creating places for urban agriculture. The Monacelli Press, New York

Kaufman J, Bailkey M (2000) Farming inside cities: entrepreneurial urban agriculture in the United States. Retrieved from http://www.urbantilth.org/wp content/uploads/2008/10/farminginsidecities.pdf

Kremer P, Hamstead Z, McPhearson T (2013) A socio-ecological assessment of vacant lots in New York City. Landsc Urban Plan 120:218–233

Lin BB, Philpott SM, Jha S (2015) The future of urban agriculture and biodiversity-ecosystem services: challenges and next steps. Basic Appl Ecol 16:189–201. doi:10.1016/j.baae.2015.01.005

Lovell ST, Taylor J (2013) Supplying ecosystems services through multifunctional green infrastructure in the United States. Landsc Ecol 28:1447–1463

McClintock N (2013) Radical, reformist, and garden-variety neoliberal: coming to terms with urban agriculture's contradictions. Local Environ 19(2):147–171

McPhearson T, Hamstead Z, Kremer P (2014) Urban ecosystem services for resilience planning and management in New York City. Ambio 43:502–515

Mees C, Stone E (2012) Food, homes and gardens: public community gardens potential for contributing to a more Sustainable City. In: Viljoen A, JSC W (eds) Sustainable food planning. Academic Publishers, Wageningen, pp 431–452

Mok HF, Williamson VG, Grove JR, Burry K, Barker SF, Hamilton AJ (2013) Strawberry fields forever? Urban agriculture in developed countries: a review. Agron Sustain Dev 24(1):21–43. http://doi.org/10.1007/s13593-013-0156-7

New York City Department of Environmental Protection (2011) DEP awards $3.8 million in grants for community-based green infrastructure program projects. Press Release. June 9, 2011

New York City Department of Environmental Protection (n.d.) Project profile: Lenox Hill Neighborhood House. www.nyc.gov/html/dep/pdf/green.../lenox_hill_profile.pdf. Accessed 17 Aug 2016

Orsini F, Gasperi D, Marchetti L, Piovene C, Draghetti S, Ramazzotti S et al (2014) Exploring the production capacity of rooftop gardens (RTGs) in urban agriculture: the potential impact on food and nutrition security, biodiversity and other ecosystem services in the city of Bologna. Food Sec 6(6):781–792. http://doi.org/10.1007/s12571-014-0389-6

Reklaitiene R, Grazuleviciene R, Dedele A, Virviciute D, Vensloviene J, Tamosiunas A et al (2014) The relationship of green space, depressive symptoms and perceived general health in urban population. Scand J Public Health 42(7):669–676. http://doi.org/10.1177/1403494814544494

Reynolds K (2014) Disparity despite diversity: social injustice in New York City's urban agriculture system. Antipode 47(1):240–259

Reynolds K, Cohen N (2016) Beyond the kale: urban agriculture and social justice activism in New York City. University of Georgia Press, Athens

Rifkin G (2011) Cash crops under glass and up on the roof. New York Times (International edition; Business section)

Sanyé-Mengual E, Anguelovski I, Oliver-Solà J, Montero JI, Rieradevall J (2015a) Resolving differing stakeholder perceptions of urban rooftop farming in Mediterranean cities : promoting food production as a driver for innovative forms of urban agriculture. Agric Hum Values. http://doi.org/10.1007/s10460-015-9594-y

Sanyé-Mengual E, Cerón-Palma I, Oliver-Solà J, Montero J, Rieradevall J (2015b) Integrating horticulture into cities: a guide for assessing the implementation potential of Rooftop Greenhouses (RTGs) in industrial and logistics parks. J Urban Technol 22(1):87–111

Söderback I, Söderström M, Schälander E (2009) Horticultural therapy: the "healing garden"and gardening in rehabilitation measures at Danderyd hospital rehabilitation clinic. Sweden, Pediatric Rehabilitation

Specht K, Sanyé-Mengual E (2017) Risks in urban rooftop agriculture: Assessing stakeholders' perceptions to ensure efficient policy-making. Environ Sci Policy 69, 13-21. https://doi.org/10.1016/j.envsci.2016.12.001

Specht K, Siebert R, Hartmann I, Freisinger UB, Sawicka M, Werner A et al (2014) Urban agriculture of the future: an overview of sustainability aspects of food production in and on buildings. Agric Hum Values 31:33–51. http://doi.org/10.1007/s10460-013-9448-4

Specht K, Siebert R, Thomaier S (2015a) Perception and acceptance of agricultural production in and on urban buildings (ZFarming): a qualitative study from Berlin, Germany. Agric Hum Values. http://doi.org/10.1007/s10460-015-9658-z

Specht K, Siebert R, Thomaier S, Freisinger U, Sawicka M, Dierich A, Henckel D, Busse M (2015b) Zero-acreage farming in the City of Berlin: an aggregated stakeholder perspective on potential benefits and challenges. Sustainability 7:4511–4523. doi:10.3390/su7044511

Specht K, Weith T, Swoboda K, Siebert R (2016) Socially acceptable urban agriculture businesses. Agron Sustain Dev 36(17):1–14. http://doi.org/10.1007/s13593-016-0355-0

Steel C (2009) Hungry city: how food shapes our lives. Vintage Books, London

Story M, Kaphingst KM, Robinson-O'Brien R, Glanz K (2008) Creating healthy food and eating environments: policy and environmental approaches. Annu Rev Public Health 29:253–272

Tamosiunas A, Grazuleviciene R, Luksiene D, Dedele A, Reklaitiene R, Baceviciene M et al (2014) Accessibility and use of urban green spaces, and cardiovascular health: findings from a Kaunas cohort study. Environ Health Glob Access Sci Source 13(1):20. http://doi.org/10.1186/1476-069X-13-20

Taylor JR, Lovell ST (2013) Urban home food gardens in the Global North: research traditions and future directions. Agric Hum Values 31(2):285–305. http://doi.org/10.1007/s10460-013-9475-1

Teig E, Amulya J, Bardwell L, Buchenau M, Marshall JA, Litt JS (2009) Collective efficacy in Denver, Colorado: strengthening neighborhoods and health through community gardens. Health Place 15(4):1115–1122. http://doi.org/10.1016/j.healthplace.2009.06.003

Thomaier S, Specht K, Henckel D, Dierich A, Siebert R, Freisinger UB, Sawicka M (2015) Farming in and on urban buildings: present practice and specific novelties of Zero-Acreage farming (ZFarming). Renewable Agric Food Syst 30(1):43–54. http://doi.org/10.1017/S1742170514000143

Designing Green Corridors Network Within Cities: A Case Study in Vienna

Maeva Dang

Abstract The notion of green cities is not green enough. There is a necessity to readapt consumption modes and rethink urban spaces in order to ensure the long-term viability of the built environment. Urban agriculture on roofs could play a great role in this transformation process. The question is: where and in what conditions would rooftop gardens have the most significant and positive influence on our existing urban system? This chapter presents a method, using Geographic Information Systems (GIS) and parametric modelizations, which provides an effective planning strategy of the green corridors network within a city. This interdisciplinary research is based on different parameters: the lead angle of the Viennese rooftops, the existing urban green spaces and the pollinator's flight foraging distance. After importing the suitable rooftop surfaces into Grasshopper (Algorithmic modeling for Rhinoceros 3D, a CAD software), the model identifies the key surfaces capable to create large green corridors network and to connect existing green spaces in Vienna.

Introduction

Increasing urbanization and the growth of the world population in the last 50 years generate important challenges for tomorrow. According to Despommier by the year 2050, the earth's human population will have increased by around three billion and 80% will live in urban centers (Despommier 2010). In fact, global sustainability depends on how urban systems will be managed in the twenty-first century (Ferrão et al. 2013). Developing holistic approaches for urban planning is one of the necessities to improve the built environment.

Rooftop gardens provide a large range of benefits from enhancing biodiversity in the city to contributing to more sustainable processes, including the ones necessary for food production and the improvement of quality of life (Khandaker 2004). Focusing on the benefits of intensive greening on roofs of Vienna, the present work examines the existing surface opportunities within the city. The idea is to consider

M. Dang (✉)
Department of Industrial Building and Interdisciplinary Planning, Vienna University of Technology – Institute for Interdisciplinary Building Process Management, Vienna, Austria
e-mail: maeva.dang@tuwien.ac.at

F. Orsini et al. (eds.), *Rooftop Urban Agriculture*, Urban Agriculture,
https://doi.org/10.1007/978-3-319-57720-3_18

rooftop gardens as a key solution to convert existing roof areas into a contribution for a more sustainable urban structure. This chapter presents an overture of how flat roof landscapes could be planned, by identifying their potential connections with the existing green spaces and creating optimal networks to link them. The resulted model is fully parametric; therefore it can be applied to any city shape and can spot which flat roofs are the best located to enlarge the existing green corridors network of a given urban space.

"The Map Is Not the Territory"

Wrote Korzybski to express the unfeasibility of an accurate representation of the space (Korzybski 1933). The difficulty of modelization stands in the transcription of reality through this unavoidable prism of perception. Creating a city model is a complicated task: it is impossible to include all relevant parameters that are making a city what it really is; a highly complex and dynamic system. Despite these facts, modelization is a medium for planners and scientists to analyze and investigate urban morphologies. Valery poetically observed "Everything simple is false. Everything which is complex is unusable" (Valery 1942). In other words, the model needs to give a fair perspective that is accurate enough to be useful. "Remember that all models are wrong; the practical question is how wrong do they have to be to not be useful" (Box et al. 1987).

Keeping in mind these constraints and limits, the present study tries to give a representation of what cities could offer with the use of intensive greening on roofs. The strategy consists in taking in account those isolated spaces and trying to integrate them into a single model matrix. By considering all horizontal surfaces of the city landscape (from ground floor to rooftops) on a same physical elevation, the model gives an overview of what potential benefits those "extra" urban green areas could provide.

The physical organization of the city connections such as roads, streets and underground subways enables people to circulate within the urban space. Planning and activating those linkages must be coordinated with social behaviors. As a matter of fact the architect and design theorist Alexander wrote "for the human mind, the tree is the easiest vehicle for complex thoughts. But the city is not, cannot and must not be a tree. The city is a receptacle for life" (Alexander 1965). The idea is to promote city's physical connections in accordance to what "life" would need and not what the built environment organization would dictate. The way that species other than human perceive and respond to urban landscapes may be very different from the way people perceive the same landscape. So, what if we shift our vision on how we use our city topography to the way pollinator species discern it? How could green roofs facilitate pollinator's dispersion so, thus promote biodiversity in the city? Where are those "other" urban linkages and how could we plan their physical organization for ensuring the long-term security and resilience of urban biodiversity?

Green Corridors Network

In 1960, ecologists McArthur and Wilson developed the theory of insular biogeography. "Islands" are considered as any area of habitat suitable for a particular ecosystem which is surrounded by unlike ecosystems, such as (in this case study) human land development (MacArthur et al. 1967).

The location of green spaces within the urban matrix is an important factor for city biodiversity. Indeed the two main threats to biodiversity are fragmentation and habitat loss (Wilcove et al. 1998). Biodiversity movements are highly influenced by the connectivity of the landscape (Schippers et al. 1996). This suggests that strategic green locations may ensure a good connectivity of the "islands" and create an effective network for organisms.

In 1984, Merriam presented the concept of landscape connectivity as "the degree to which absolute isolation is prevented by landscape elements which allow organisms to move among patches" (Merriam 1984). Numerous researches have been conducted in the past decades in order to analyze (with the use of graph theory) the connectivity of green spaces in different cities. In 1995, Forman proposed the patch-corridor-matrix model that gives a representation of the landscape as a mosaic of three entities: patches, corridors and matrix (Fig. 1). Green corridors are identified as a series of connected green areas within an urban region, usually consisting of patches linked by corridors (Forman 1995; Forman et al. 1986). The urban landscape is considered as a model where the space is transformed into a network of habitat patches connected by green links. In graph theory, the patches are depicted

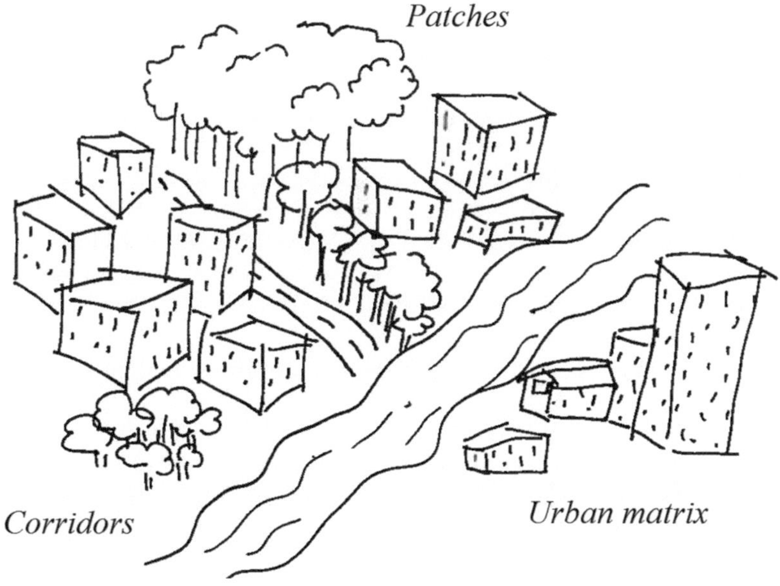

Fig. 1 Patch, corridor and matrix (Source: Dang 2016)

by nodes and represent the natural habitats located in the city (for instance parks, cemeteries, etc.). The corridors are the linkages that enable the dispersion of urban biodiversity between the green nodes. In reality, those linkages are not continuous but a series of "stepping stones" or other patches that connect larger green areas.

To optimize the current patch-corridor-matrix of Vienna there would be three options:

1. Increasing the weight and/or attraction potential of the nodes (enlarging the green areas such as parks…);
2. Enlarging the network by creating new bridges (or stepping stones) between the isolated green nodes;
3. Strengthening the nodes connectivity (multiplying the possibilities to access to a green node even if a linkage is already existing).

This chapter focuses on studying the second option. In particular, the parametric model investigates how green patches can be all connected to allow a higher dispersion potential of pollinators with the use of extra "stepping stones", which are the rooftop gardens. This approach gives the basics of a reflection on how green roofs can be linked to the green corridors network and can enlarge the existing urban biological system.

This proposal is adapting the idea suggested in the research of 2014 in the city of Bologna, where scientists analyzed the potential impact of a green corridors network connecting all flat rooftops within a certain flight foraging distance (Orsini et al. 2014).

Study Method

Assumptions

The present model has been adjusted for Anthophilous pollinators providing a specific range of flight foraging distance as a parameter. It has been considered that when the patches are located within a distance of 500 m then the two islands are close enough for the pollinator's dispersion. This represents a suitable distance for most common pollinators since their flight foraging distance is measured between 750 and 1500 m (Gathmann et al. 2002).

The hypothetical rooftop gardens are designed to grow food and provide substantial pollen and nectars. Natural barriers such as hills or buildings could be an obstacle for the insects to fly throughout the city, but there is no proof that such barriers would stop them from reaching high points when hard to reach green areas provide them substantial food. Another point is that numerous bee hives are located on rooftops. This shows the importance of reconsidering higher surfaces into the green corridors network.

Thanks to the use of a grasshopper modelization it is possible to adapt quite a few parameters such as, for instance, the flight foraging distance, the size and type of patches, the network topology and the corridor width.

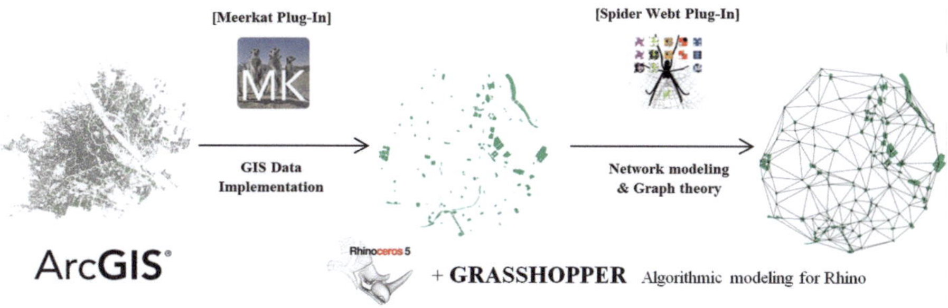

Fig. 2 Tools and workflow (Source: Dang 2016)

Tools

Figure 2 shows the digital tools combination and workflow used in this study. The workflow method consists of three stages as follows:

1. Identification of all flat roofs and quantification of the potential surfaces that could be converted into Rooftop gardens with Geographic Information Systems techniques (the roof lead angle must be inferior to 5 degrees inclination) (ARCGIS 2014),
2. Implementation of the geoformation data into Grasshopper, a graphical algorithm editor tightly integrated with Rhino's 3-D modeling tools. The GIS data is converted into a shapefile to be read in Grasshopper with the Meerkat Plugin (Lowe 2015). Thanks to this plugin, it is possible to keep all the attributes of the polygons (shape, address, location and other information provided within the initial data). In this case, the polygons represent the green areas and the flat rooftops elements.
3. Last but not least, the last step of the study is the identification of the green rooftops "Hotspots" with the Spider Web Plug-In (Schaffranek 2016). The investigation is made through their location potential within the existing green corridors network. The idea is to obtain a visualization of the intensive greening potential of city roofs landscapes.

Preparation of the Data with Geographic GIS Techniques (ARCGIS): Identifying the Green Spaces in Vienna

With 41,487 ha Vienna has 45.5% of its whole surface dedicated to green areas. The city is surrounded by a green belt formed by the Viennese Woods, the Lainz Game Preserve, the Donau-Auen National Park and other green surfaces. Across the city flows the Danube that occupies 1913 ha, thus 4.6% of the total area of the city of Vienna.

Green Space Categories	Elements
Land	498
Cemetery	58
City garden	1814
Water surface	349
Market garden, Orchard	257
Park & Green area	551
Forest	760
Wine garden	178
Meadow	1120
Total	**5585**

Fig. 3 Distribution of the selected *green* spaces for the case study in Vienna (Source: Dang 2016)

The urban space is attractive for pollinator species for two main reasons: the plant diversity and the absence of insecticides, fungicides or fertilizers. Biodiversity is richer inside Vienna than on the borders of the city. It is difficult for pollinators to get substantial food in cultivated areas such as corn and wheat fields, where no proper type of pollen and nectar is available in large quantities ("green deserts" as called by specialists and beekeepers). The Fig. 3 shows the distribution of the green spaces selected for the case study. Water surfaces are also included in the modelization ("blue space elements"). By using the data of the city of Vienna (Magistrate Wien 22 2008) 5585 green and blue space elements are identified and imported into Grasshopper.

Intensive Greening Potential

1078.7 ha that is 21% of the entire rooftop surface in Vienna which is adapted for intensive green roofs. This potential is equivalent of three times and a half the surface of the historical city center (called the Innere Stadt). The data preparation is similar than the previous step on ARCGIS: imported are all the suitable roof surfaces for intensive greening (surfaces with a roof lead angle inferior to 5 degrees) (Fig. 4).

Parametric Modelization with Grasshopper – Case Study

Green Corridors Network Analysis

Grasshopper is a visual programming tool for designers and architects to generate new shapes using generative algorithms and it does not require any knowledge of scripting. It has been chosen because it is a platform which integrates

Intensive greening potential selection in Vienna	
Roof Lead Angle	< 5°
Minimum Roof Surface	5 m²
Total Flat Roof Surface	**1078,7 ha**

Fig. 4 Flat roofs identification and mapping with ARCGIS (Source: Dang 2016)

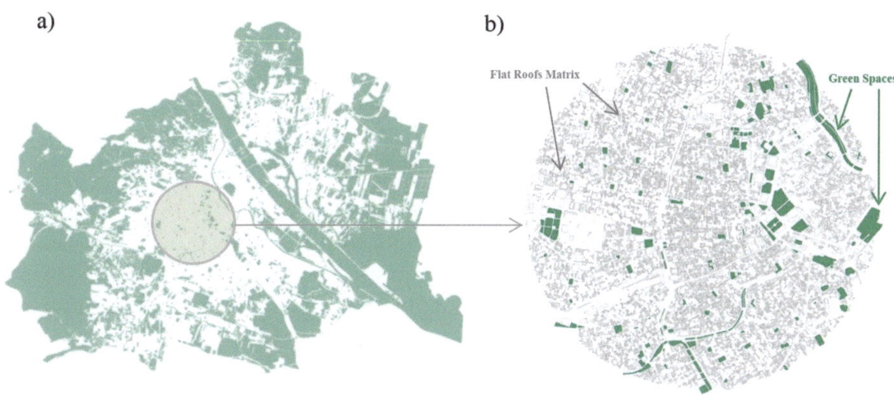

Fig. 5 (**a**) Case study focus in Vienna (**b**) *Green* spaces & rooftops locations in the selected area (Source: Dang 2016)

multidisciplinary modules for interdisciplinary projects. Since this research is based on different domains such as city planning, applied mathematics (graph theory), geo-information analysis and landscape ecology, the need of this parametric design platform seemed like a self-evident requirement. After the implementation of the pre-filtered geoformation data into Grasshopper, the data is processed and analyzed.

The city of Vienna contains 23 districts with different sizes and degrees of urbanization. For this case study the focus is made on a highly urbanized zone that regroups fully or partly the 1st, 2nd, 3rd, 4th, 5th, 6th, 7th, 8th, 9th, 15th, 16th, 17th and 18th Districts of Vienna (see Fig. 5). The characteristics of the selected area are presented in Table 1. Inside the focus area, 144 green spaces are identified and 10,210 flat rooftops elements are considered as "opportunity" to close the network. The aim of this case study is to find out how many surfaces would be needed to build up a closed green corridors network.

Table 1 Characteristics of the case study area

Area statistics		Elements	Surface (ha)
Total surface			**5027**
Green surfaces		**144**	308
Largest green space	*Stadtpark*		26.2
Smallest green space	*Neusserplatz*		0.1
Flat rooftops		**10,210**	

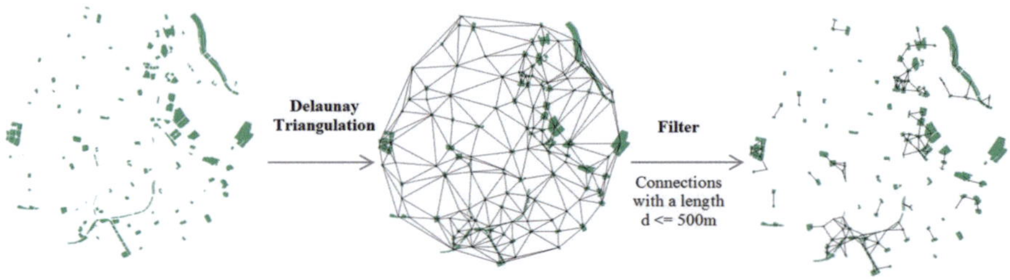

Fig. 6 Network analysis & connectivity of the *green* areas (Filtering the linkages with a length inferior or equal to 500 m) (Source: Dang 2016)

1ST STEP: Green Corridors Network Typology

The first step in Grasshopper is to give the input of the geometry inside the model with the Meerkat Plug-In. The network is then modeled by connecting the centers of each green space with the Delaunay triangulation method. Since the flight foraging distance parameter is set on 500 m, the network is filtered: only the adapted linkages remain. This gives a visualization of the existing connectivity of the green spaces for the pollinator's species (Fig. 6).

The network observed is a fragmentation of green patches inside the studied area. Several single green elements are not connected with the rest of the network. There is a need to produce extra connections in order to promote the dispersal ability of the pollinators (Fig. 7).

A few potential network typologies could be generated and evaluated in order to close the green corridors network. The purpose here is to build up a set of edges connecting all nodes (green spaces) such that the overall sum of the edge length is minimized. To achieve this objective, the minimum spanning tree with the greedy Kruskal's algorithm is chosen.

Kruskal first described it in 1956: "Perform the following step as many times as possible: Among the edges […] not yet chosen, choose the shortest edge, which does not form any loops with those edges already chosen" (Kruskal 1956). This means that the algorithm finds an edge of the least possible weight that connects any two trees in the forest (Cormen et al. 2009). The Spider Web Plugin can generate this network typology with Grasshopper and gives a direct modelization of the possible corridors.

The following figure shows the fragmented network and the minimum spanning tree obtained after the application of the Kruskal's algorithm (Fig. 8).

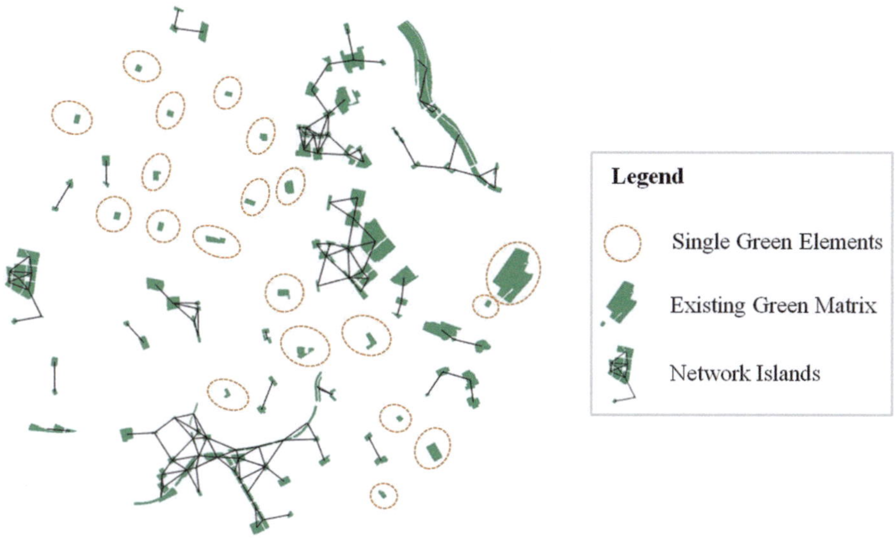

Fig. 7 Fragmented *green* corridors network (Delaunay triangulation method) (Source: Dang 2016)

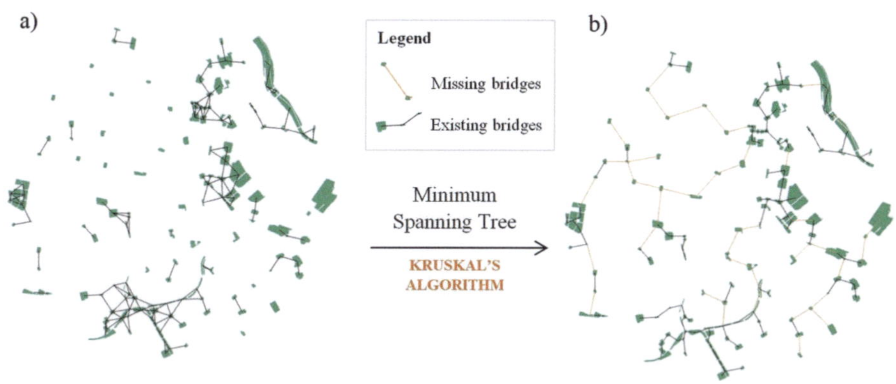

Fig. 8 (**a**) Fragmented green network (Delaunay triangulation method) (**b**) Generated *green* corridors network with the Kruskal's Algorithm (Source: Dang 2016)

After the first study step, there are 39 missing bridges (with a total length of 25.04 km) that are needed to close the minimum spanning tree network (Table 2).

2nd STEP: Finding the Suitable Bridges

Figure 9 shows the process of filtering and identifying the suitable flat roofs that would be needed to connect the green nodes. This filter is adjusted on the proximity of the so-called "missing bridges" (see the orange linkages in the Fig. 9). It is

Table 2 Statistics of the generated green corridors network with the Kruskal's Algorithm

	Bridges elements	Length (km)
Minimum spanning tree network (Kruskal's algorithm)	143	55.07
Existing bridges (length between two nearby green units = < 500 m)	104	
Existing network length		30.03
Missing bridges (length between two nearby green units >500 m)	39	
Missing bridges length		25.04

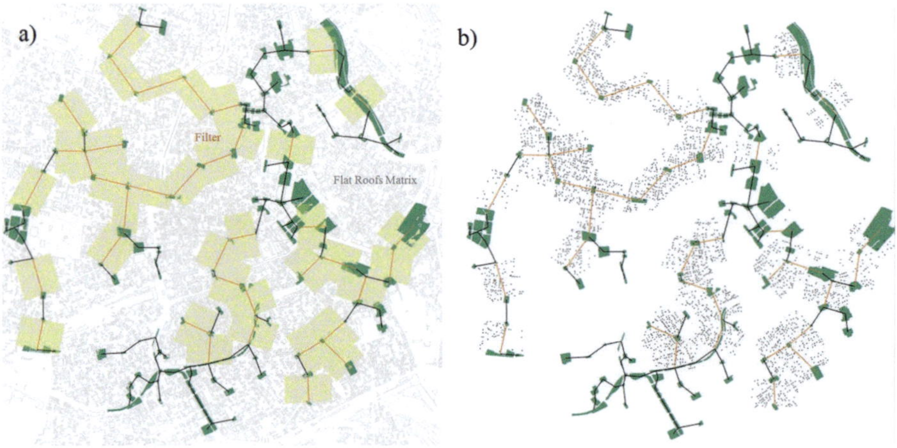

Fig. 9 (**a**) Implementation of the flat roofs filter (**b**) Visualization of the selected roofs that could be used as "bridges" (Source: Dang 2016)

important to filter this data volume since the model would require too much time to find the closest roof path for the 3rd study step.

3378 flat roofs out of the 10,210 roofs of the studied area are identified as "suitable" (which means located within 500 m from the green areas centers). It is therefore interesting to find out which ones would create the smallest bridge between the solitary green nodes.

3rd STEP: Selecting the Closest Flat Roofs to Create the Shortest Bridges

By generating a graph from the green spaces centers with the Spider Web plugin, it is possible to calculate the shortest path between the critical points. It has been previously calculated that 39 bridges (with an overall distance of 25.04 km) are missing to close up the network. Since one stop is calculated per rooftop between each isolated green area, each constructed bridge is composed of two small bridges. In this "shortest path" modelization 81 bridges with an overall distance of 25.10 km have been identified to construct the minimum spanning tree network. The resulted green corridor network is 55.13 km long.

Results can be seen in Figs. 10 and 11 and Table 3.

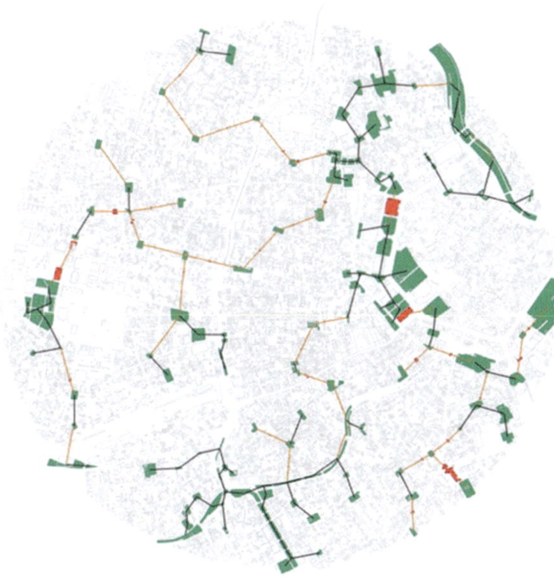

Fig. 10 Final *Green* Corridors Network (Source: Dang 2016)

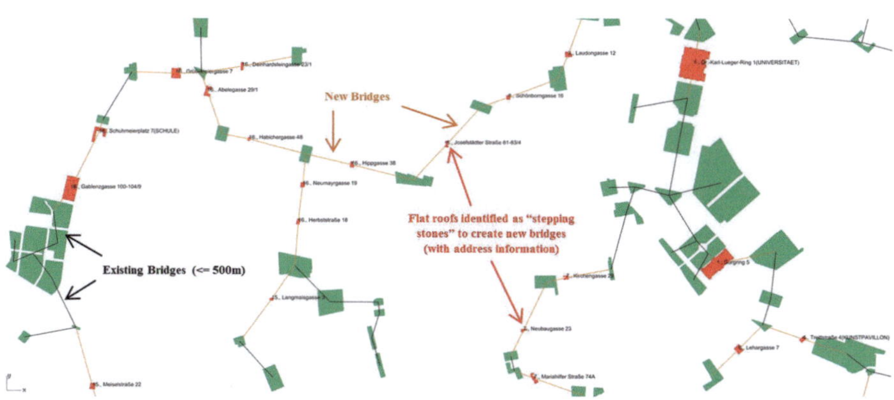

Fig. 11 Zoom on the flat roofs identification (Source: Dang 2016)

Table 3 Results of the case study

Results		
Selected flat rooftops	**42**	**Elements**
New bridges	**81**	**Elements**
Bridges created with 1 roof	79	Elements
Bridges created with 2 roofs	2	Elements
Overall length of the new bridges	25.1	km
Existing bridges	**104**	**Elements**
Overall length of the existing bridges	30.03	km
Final green corridors network	**55.13**	**km**

Discussions

Thinking Heavily with Light Models

One difficulty of the modelization process lies in the big amount of information Grasshopper has to compute. The exact GIS locations (street number, district, etc.) of each rooftop as well as their shape have been used as initial data. ARCGIS processes quite fast the data but it is long to generate graphs in Grasshopper. In cause the parametrization of the network that requires a calculation of every possible path between the polygons (Fig. 12).

Grasshopper, a Platform Which Integrates Multidisciplinary Modules

For this modelization it was relevant to test the plugin Meerkat GIS in order to keep all the attributes of the green spaces and the flat roofs. Grasshopper gives an alternative to the usual GIS Software because it proposes several free-access Plugins such as Meerkat GIS, Heron (generates Grasshopper geometry from GIS shape files) (Washburn 2015) and Elk (generates map and topographical surfaces using open source data from OpenStreetMap) (Logan 2016). This enables flexibility in the modelization process and the possibility to work on an interdisciplinary level.

Parametrization of the Model (City Shape & Species)

In this research the model created is parametric and flexible. It is important to highlight the fact that the same landscape may have different degrees of connectivity for different species (Kindlmann et al. 2008). It is a highly complex exercise to create connected green space networks. There are still many issues and questions about the

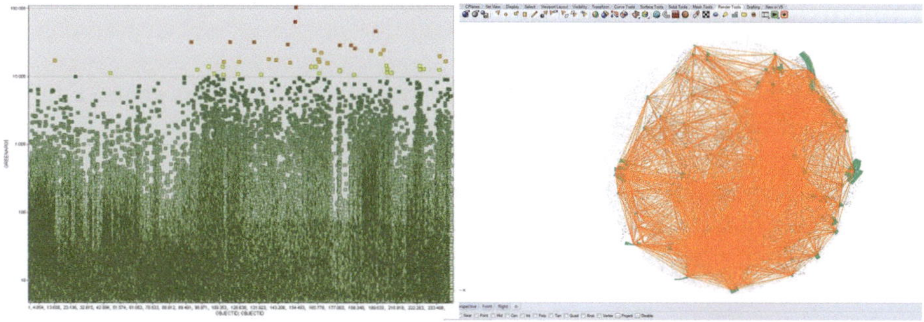

Fig. 12 Visualization of the flat roofs locations data on ARCGIS & Paths test between the *green* areas on Grasshopper (Source: Dang 2016)

planning strategies and the way it could be optimally designed. The important questions are "which species is the corridor for?" and "how would it possible to optimize green planning according to the city shape?" The city of Seattle started a project called "The pollinator Pathway" (Bergmann 2012) that aimed to create paths for pollinators by encouraging people to green their private spaces on a selected track. This illustrates the great potential of collaborative initiatives between private and the public entities in order to apply green corridors planning strategies.

Invasive Alien's Species

Apart from the results, it is important to underline a non-negligible aspect of the green corridors: the problematic of invasive species. When the dispersion of beneficial species such as pollinators can be facilitated by the improvement of green nodes connectivity, it also allows pest organisms to move around the green patches. When modifying any existing ecosystem structure and organization, other consequences have to be considered and managed.

Network Typologies

The minimum spanning tree has been chosen for this study. Different network typologies could have been generated according to other modelization's objectives such as the Hierarchical network or the Least Cost to User (Hellmund 1989). Another point that has not been taken into account is the weight of the green nodes: this weight can be calculated according to the surface and the potential attractiveness.

Conclusions

The aim of the present study case was to build up a set of edges connecting all nodes (or green spaces) such that the overall sum of the edge length is minimized. The model gives a possible solution of selecting the hotspots rooftops. Forty-two roofs have been identified on 5027 ha of a highly urbanized area in Vienna. These results highlight the existing opportunities in the focused surface and give a clear quantification of the green areas that are lacking to construct the green corridors network. This amount of roofs to convert into gardens seems realistic and feasible on the scale of 5000 ha built area. However, it would be interesting to investigate the application of this model in reality in order to monitor the possible improvements on site.

Defining the hotspot rooftops to construct viable green corridors network would contribute to a more efficient green planning strategy. Depending on the urban space and the flight foraging distance considered it is possible to design precisely the green corridors network. This study shows the potential opportunities of a city landscape for urban biodiversity with the use of intensive greening. It gives a possible mapping tool of pollinators' interactions towards an urban space.

The modelization works here as a platform encouraging city planning strategies to rebind fragmented landscapes. By promoting a way of designing across "unusual" networks, the idea is to look at our city landscapes as a driver for environmental sustainability.

Bullet Points

- The identification of the existing green corridors network in a highly urbanized zone of Vienna revealed a significant degree of fragmentation;
- The present parametric model could be a support for decision-making in designing urban roof landscapes. Using it as a tool is made possible thanks to its flexibility to any urban geometry;
- The proposed green corridors network is based on a minimum spanning tree and allows few degrees of freedom such as: the flight foraging distance (thus the pollinator's species considered), the size of the patches and the width of the corridors. Investigating the green matrix of a city with the use of graph theory would be the basis of an efficient strategy to design new green environments on roofs. Generating other typologies and weighing the green nodes according to their attraction potential could also give a stronger dimension to the model;
- However, this design should include many other aspects (linked to rooftop gardens construction): for instance the structural and exploitation costs remain crucial parameters. In collaboration with experts from other disciplines, this model should be developed (network typologies, rooftop gardens density and dimensions) in order to become more practical.

Acknowledgements I would like to thank my Professor, C. Achammer for his supervision and his precious advices. Appreciation is also given to the Magistrate 22 of the city of Vienna for providing me the necessary Data to begin the study. A lot of the inspiration was drawn from my colleague Rüdiger Suppin with whom I could learn and develop this Grasshopper modelization.

References

Alexander C (1965) A city is not a tree. Architectural Forum 122(1):58–62

ARCGIS (2014) Version 10.3. Accessed 11 Oct 2016. www.arcgis.com

Bergmann S (2012) The pollinator pathway. Accessed 11 Oct 2016. http://www.pollinatorpathway.com/

Box GEP, Draper NR (1987) Empirical model building and response surfaces. Wiley, New York, p 74

Cormen T, Leiserson CE, Rivest RL, Stein C (2009) Introduction to algorithms, 3rd edn. MIT Press, Cambridge

Despommier D (2010) The vertical farm: feeding the world in the 21st century. St. Martin's Press, New York

Ferrão P, Fernández JE (2013) Sustainable urban metabolism. MIT Press, Cambridge

Forman RTT (1995) Land mosaics: the ecology of landscapes and regions. Cambridge University Press, New York

Forman RTT, Godron M (1986) Landscape ecology. Wiley, New York

Gathmann A, Tscharnkte T (2002) Foraging ranges of solitary bees. J Anim Ecol 71:757–764

Hellmund PAC (1989) Quabbin to Wachusett wildlife corridor study. Harvard Graduate School of Design, Cambridge

Khandaker MSI (2004) Rooftop gardening as a strategy of urban agriculture for food security: the case of Dhaka City Bangladesh. Acta Hortic 643:241–247

Kindlmann P, Burel F (2008) Connectivity measures: a review. Landsc Ecol 23:879–890

Korzybski A (1933) Chapter IV: on structure. In: Schuchardt C (ed) Science and sanity: an Introduction to non-Aristotelian systems and general semantics. Institute of general semantics, New York, p 58

Kruskal JB (1956) On the shortest spanning subtree of a graph and the traveling salesman problem. Proc Am Math Soc 7:48–50

Logan T (2016) Elk version 2.2.2., Accessed 11 Oct 2016. http://www.food4rhino.com/project/elk?etx

Lowe N (2015) Meerkat GIS version 1.6, Accessed 11 Oct 2016. www.food4rhino.com/project/meerkatgis?etx

MacArthur RH, Wilson EO (1967) The theory of island biogeography. Princeton University Press, Princeton

Magistrate Wien 22 Umweltgut (2008) ViennaGIS database

Merriam G (1984) Connectivity: a fundamental ecological characteristic of landscape pattern. Paper presented at the 1st international seminar on methodology in landscape ecological research and planning. Roskilde, Denmark

Orsini F, Gasperi D, Marchetti L, Piovene C, Draghetti S, Ramazzotti S, Bazzocchi S, Gianquinto G (2014) Exploring the production capacity of rooftop gardens in urban agriculture: the potential impact on food and nutrition security, biodiversity and other ecosystem services in the city of Bologna. Food Sec 6(6):781–792

Richard S (2016) Spider web version 4.1. Accessed 11 Oct 2016. http://www.food4rhino.com/project/spiderweb?etx

Schippers P, Verboom J, Knaapen JP, van Apeldoorn RC (1996) Dispersal and habitat connectivity in complex heterogeneous landscapes: an analysis with a GIS-based random walk model. Ecography 19:97–106

Valery P (1942) Oeuvres II. 863. Gallimard, Paris
Washburn B (2015) Heron version GHA 0.2.0.0. Accessed 11 Oct. http://www.food4rhino.com/project/heron?etx
Wilcove DS, Rothstein D, Dubow J, Phillips A, Losos E (1998) Quantifying threats to imperiled species in the United States: assessing the relative importance of habitat destruction, alien species, pollution, overexploitation, and disease. Bioscience 48(8):607–615

Part V
A Geography of Rooftop Agriculture in 20 Projects

Henk de Zeeuw, Francesco Orsini, Marielle Dubbeling, Giorgio Gianquinto

A Geography of Rooftop Agriculture in 20 Projects

Henk de Zeeuw, June Komisar, Esther Sanyé-Mengual, Rémi Kahane, Giorgio Gianquinto, Emmanuel Geoffriau, Ching Sian Sia, Alfredo Rodríguez-Delfín, Salwa Tohmé Tawk, Heshem el Omari, Saumil Shah, Juan Ignacio Montero, B.N. Vishwanath, Rajendra Hegde, Luana Iori, Jessie Banhazl, Christopher Horne, Saber Osman, Carl Philipp Schuck, Viraj Puri, Bryna Bass, Edwin "Pope" Coleman, Chris Somerville, Pol Fabrega, Mat Pember, Amelie Asselin, Ricardo Omar, Sergio Eiji Nagai, Lyvenne Chong-Phoon, Allan Lim, Maria Lloyd, Shuang Liu, Gloria Samperio Ruiz, and Arlene Throness

Abstract In this chapter, selected cases of rooftop agriculture across the world will be presented, explaining their organisation, technical design and operation, their business model and main functions, lessons learned during establishment and operation, their productive and societal results and their policy relevance. The owner or manager of the rooftop garden or farm and an independent researcher were involved in documenting the cases, When selecting the cases, we tried to include examples of the various types of rooftop agriculture presented in previous chapters. We also sought to include cases from all continents. Accordingly, a comprehensive list of most relevant rooftop agriculture experience across the world is presented, following an alphabetical

---

The original version of this chapter was revised.
An erratum to this chapter can be found at https://doi.org/10.1007/978-3-319-57720-3_21

H. de Zeeuw (✉)
RUAF Foundation, Leusden, The Netherlands
e-mail: h.dezeeuw@ruaf.org

J. Komisar
Department of Architectural Science, Centre for Studies in Food Security, Ryerson University, Toronto, ON, Canada

E. Sanyé-Mengual • G. Gianquinto
Research Centre in Urban Environment for Agriculture and Biodiversity Biodiversity (ResCUE-AB), Department of Agricultural Sciences, Autonomous University of Bologna, Bologna, Italy

R. Kahane
Department Persyst, Centre of International Cooperation in Agricultural Research for Development, Montpellier, France

E. Geoffriau
Research Institute on Horticulture and Seeds, Angers, France

F. Orsini et al. (eds.), *Rooftop Urban Agriculture*, Urban Agriculture,
https://doi.org/10.1007/978-3-319-57720-3_19

order by city. For each case, the names of main informants (case study representatives) are listed altogether with the name of the author that coordinated data collection.

Introduction

The preparation of this chapter also generated some insights regarding regional differences in the development of rooftop agriculture, which are presented below.

North America

Rooftop agriculture has become a quite popular phenomenon in Canada and the USA. Numerous rooftop-farming initiatives have taken place in the last decade. Various cities are now supporting rooftop agriculture, often as an integrated part of their local or regional food policy, and are adapting building regulations and planning codes to enable rooftop gardening while safeguarding associated risks. Initially, this support was restricted mainly to non-productive green roofs and socially oriented productive rooftop gardens, but more recently also commercially oriented rooftop farms are widely accepted and various smaller and larger-scale commercial rooftop farms have become well known examples.

C.S. Sia
Department of Architecture, School of Design and Environment, National University of Singapore, Singapore, Singapore

A. Rodríguez-Delfín
Hydroponics and Mineral Nutrition Research Center, La Molina Agrarian National University, Lima, Peru

S.T. Tawk
Faculty of Agriculture, Lebanese University, Beirut, Lebanon

H. el Omari
Urban Agriculture Bureau, Greater Amman Municipality, Amman, Jordan

S. Shah
EnerGaia Co. Ltd., Bangkok, Thailand

J.I. Montero
Institute of Agri-food Research and Technology (IRTA), Cabrils, Spain
e-mail: juanignacio.montero@irta.cat

B.N. Vishwanath • R. Hegde
Garden City Farmers (GCF), Bengaluru, India

L. Iori
Community rooftop garden Via Gandusio, Bologna, Italy

J. Banhazl
Green City Growers, Boston, USA

C. Horne
Urban Development for Deutsche Gesellschaft für Internationale Zusammenarbeit (GIZ), Koeln, Germany

Increasingly, facilities for rooftop gardening are included in schools and university buildings, social housing projects, condominiums for elderly people and projects directed at specific underprivileged groups (recent migrants, jobless youth and physically challenged people). Also food preparation (restaurants, hotels) and selling (supermarkets) establishments have started to use their rooftop space to produce fresh healthy food for their customers.

Europe

The development of rooftop agriculture is not as advanced as in North America and some Asian countries. Most of the rooftop agriculture initiatives in Europe to date are taken by civil society groups that are promoting local food production for ecological or social reasons and by research institutes for experimentation and

S. Osman • C.P. Schuck
Climate Change component, Participatory Development Programme in Urban Areas (PDP), GIZ, Cairo, Egypt

V. Puri
Gotham Greens, Chicago, USA

B. Bass
Rooftop Garden School, Rothenberg Preparatory Academy, Cincinnati, USA

E. "Pope" Coleman
Rothenberg Rooftop Garden, Cincinnati, USA

C. Somerville
Food and Agriculture Organisation of the United Nations (FAO), Rome, Italy

P. Fabrega
Rooftop Republic, Hong Kong, China

M. Pember
Little Veggie Patch Co., Melbourne, Australia

A. Asselin
Communications and Public Affairs, Palais des congrès, Montreal, Canada

R. Omar • S.E. Nagai
Shopping Eldorado, Sao Paolo, Brazil

L. Chong-Phoon
Spectra Secondary School, Singapore, Singapore

A. Lim • M. Lloyd
Comcrop, Singapore, Singapore

S. Liu
UrbanFarmers de Schilde, The Hague, The Netherlands

G.S. Ruiz
Mexican Hydroponic Association, Toluca, Mexico

A. Throness
Ryerson Urban Farm, Toronto, Canada

technology development. Also some commercial enterprises have created some pilot farms/showcases, but fully commercial rooftop farms are still quite rare.

In Germany and the UK, rooftop gardens have been established mainly for environmental and landscaping effects (most of these developed by design companies), whereas multi-functional and sustainable food production rooftop gardens have been introduced in France, Switzerland and the Netherlands, and open-air rooftop gardens for socio-economic purposes are common in Italy and Spain.

Policy support for rooftop agriculture is still in early stages of development and each productive green roof initiative has to obtain local government support case by case, and only few cities have developed general guidelines and regulations.

Some cities are strong supporters of socially oriented rooftop agriculture, while others mainly look at its ecological functions (reduction of urban heat and water runoff). Recently, also high-tech rooftop farms capture the attention of city authorities mainly for their iconic function for the city's public relations and their role in retrofitting former offices or industrial buildings.

Asia

In some Asian cities such as Tokyo, Hong Kong and Singapore, rooftop agriculture is already spreading quite quickly because of the scarcity of land within such metropoles and the growing interest of their citizens in local food production, the need for green meeting places and/or the fear of food contamination (e.g. in Chinese cities and in Hong Kong, which is importing a major part of its food from China).

Several small-scale businesses have developed rooftop gardening initiatives, often applying innovative technologies or interesting new business models.

Also several large-scale enterprises have set up rooftop gardens as part of their environmental sustainability plan and/or as an amenity for their staff and/or customers, e.g. the commuters of East Japan Railways (EJR) who can rent a plot in allotment gardens on top of one of the EJR stations.

In other parts of Asia such as in several states of India and Nepal, local authorities in cooperation with local NGOs have adopted open-air rooftop gardening in containers or raised beds (there mainly known as "terrace gardening" including rooftops and larger balconies) as an important means to improve urban food security, enhance recycling of household wastes, promote rainwater collection and storage, and reduce urban heat.

Latin America

In various countries of Latin America such as Brazil, Peru, Colombia, Chile, Argentina and Mexico, small-scale rooftop gardens have been created especially in low-income areas of the cities to enhance food security, reduce food expenses and possibly generate a small additional income. These rooftop gardens are mainly of the container type, using recycled materials, sometimes with lightweight growing media and drip irrigation or another simplified hydroponic system (*"hidroponia popular"*) as has been promoted for many years by the Food and Agriculture Organisation (FAO) office in Latin America and several local NGOs and universities. More commercial and larger-scale rooftop gardens have not yet been encountered in this region.

Africa and the Middle East

In Africa, some projects have been undertaken by international development organisations such as the FAO and the German Agency for International Cooperation (GIZ) working with local NGOs to introduce on a pilot scale simplified hydroponics and aquaponics in cities such as Gaza and Cairo, mainly as a means to create a means of subsistence for urban families severely affected by a crisis situation. Although these projects were very important experiences from a learning perspective, they have not led to widespread proliferation of these models.

In these and other cities such as Dakar and Durban, mainly NGOs have been promoting open-air container rooftop farming. Examples of commercial rooftop agriculture of any scale have not been identified.

Local authorities of most sub-Saharan African cities are not yet aware of the potential of productive roofs for urban food security and resilience. In the Middle East and Northern Africa, some cities (e.g. Amman, Casablanca and Gaza) are currently developing more interest in rooftop agriculture.

Oceania

In Australia and New Zealand, hardly any example of productive rooftop gardens or farms could be found. Many local authorities do support the development of green roofs for environmental reasons but do not allow any productive types of rooftop agriculture. The few small multifunctional gardens that exist have not yet secured sustainability.

Cases

In the following pages, twenty most representative rooftop projects are presented in alphabetical order of the cities where these projects are to be found, as also illustrated in Fig. 1. For each case study, names of main informants and data collectors are included, and a description (generally including data on history and size of the project, number of participants, main aims and functions, images) is provided.

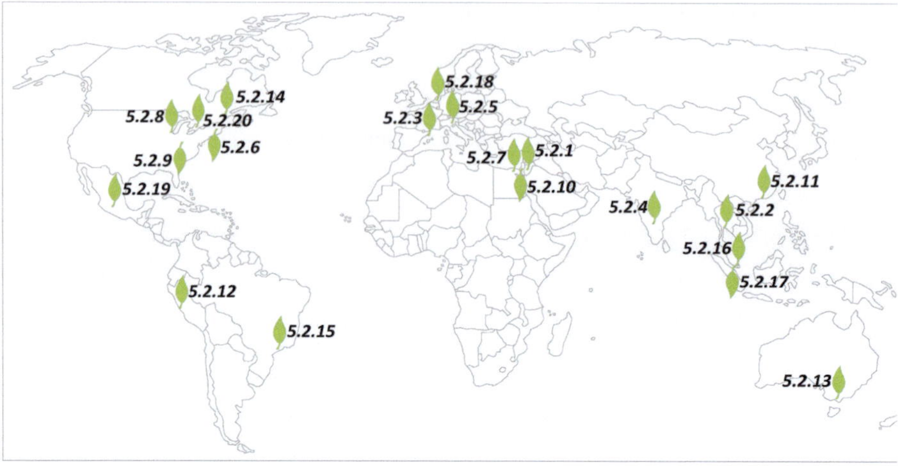

Fig. 1 Geographical distribution of the rooftop agriculture projects described in the present chapter

Amman, Jordan – Mixed Micro-Farming on Top of Residential Houses

Heshem el Omari, Salwa Tohmé Tawk

Introduction

Municipal support to the development of urban agriculture in Amman started in 2007, when the Environment and Sustainable Development Unit (ESDU) of the American University in Beirut, Lebanon, supported by the RUAF Foundation , a Global Partnership on sustainable Urban Agriculture and Food Systems, conducted an exploratory study on urban agriculture in Amman. This was followed by a project in cooperation with the Greater Amman Municipality that enabled the establishment of a multi-stakeholder forum on urban agriculture (which is now known as the "Committee for Green Amman"), the joint development of a City Strategic Agenda on the sustainable development of urban agriculture, and the setting up of a specialised Urban Agriculture Bureau with dedicated human and financial resources.

As part of the implementation of the City Strategic Agenda in 2010 (Tohmé Tawk et al. 2011) a municipal rooftop-gardening programme was initiated, providing inputs (seeds and fertilisers) and training/technical advice (e.g. on crop production and protection, greywater recycling and use) to households interested in creating a rooftop garden. One of the households in the rooftop-gardening programme was that of Mr and Mrs Kamal As'hab in the Swayleh area of Amman. Gradually, this rooftop garden became a demonstration unit in the programme, and its owner – an electrical engineer – became an active participant in workshops to share his experience and to provide practical advice on establishing and managing rooftop gardens. By 2014, 240 productive rooftops had been established in Amman (GIZ and ICLEI 2014).

The Rooftop Farm

The rooftop farm consists of three production units:

- A greenhouse (6 × 4 m) equipped with a drip-irrigation system and made of galvanised iron, covered with nylon or plastic film in the winter and with a green net for shading during the hot summer season (April to end of August) (Fig. 2). The production takes place in boxes on tables made of recycled materials and plastic filled with growing substrate made of commercial compost mixed with fermented local cow manure. The main crops planted are vegetables (tomatoes, squash), leafy vegetables and herbs (thyme, mint, rosemary) and some strawberries and eggplant among others.
- Six beehives on 6 m^2
- One chicken coop on 2 m^2 with four layers.

The main water source is recycled greywater obtained from a trial unit for recycling greywater (2 m^2). Initially, household organic wastes were composted along with organic wastes from the crops on the rooftop. However, this practice was discontinued after the first season.

Fig. 2 The greenhouse with visitors from local organisations (Credit: Heshem el Omari)

The total investment costs were about €1850, distributed as follows: 650 JOD (Jordanian Dinars; 1 JOD = €1.28) for the garden area structure, plastic and shade net; 120 JOD for the chicken coop; 400 JOD for the beehives; and 1200 JOD for the greywater recycling unit.

The investment was made by Mr and Mrs As'hab themselves. The running costs are 15–30 JOD per month.

The Results Obtained

The production is satisfactory for this household, made up of two persons only, and accounts for 10% of their annual consumption of fresh vegetables. Their savings in food expenditures are equivalent to almost 250 JOD per year; moreover, they get to share some of the surplus produce with friends and relatives. As for the honey, the production started with 30 kg per season in 2013 and reached 60 kg per season in 2016. They managed to sell most of their production, the market value of honey being 10–15 JOD /kg. The chicken coop did not function after a while, but the owners plan to restart it soon, despite the fact that the building regulations forbid animal rearing. However, as long as neighbours do not complain and oppose such activities, the authorities overlook them.

The rooftop farm is being upgraded at the moment. Inside the greenhouse, three planting levels were constructed along the sides (Fig. 3).

The initial trial recycling unit was replaced in May 2016 by an improved vertical closed system with lower emission of bad odours; the recycling unit consists of four sub-units or containers in which greywater is filtered through white sand; the filtered water is connected to a water reservoir, pump and filter, and then to an automated drip-irrigation system (Fig. 4).

Fig. 3 The newly installed greenhouse with 3 levels of planting areas on the sides (Credit: Heshem el Omari)

Fig. 4 The new greywater-recycling unit (Credit: Heshem el Omari)

In addition, a hydroponic unit for crop production will be installed together with a solar-energy panel to generate power for the water pump of the greywater recycling unit.

The owners also plan to further expand the honey production, as it is showing high production and benefit.

Lessons Learned and Policy Relevance

The combination of different micro crop- and animal-production units adds substantially to the relevance of the rooftop farm for the household and enhances its sustainability (combining food for self-consumption and additional income-earning opportunities).

The interest in gardening and the will to invest from the beneficiaries' side are important factors for the success of the rooftop farm. This ensures an involvement in both sustaining and gradually improving the farming activities. Support organisations can play an important role in training, technical advice and providing quality seeds, which are required at the start of the project, as well as spot interventions during the following years to remedy problems and further support the development of the micro-farm.

This mixed rooftop micro-farm with greywater recycling has become an important showcase and training ground that is visited by many organisations and institutions with a view to replication, which may lead to substantial impacts at city level over time (improved nutrition, higher household income, more organic wastes and wastewater recycling and use/lower municipal waste-management costs).

Regulations forbidding animal production in residential areas should be made more lenient and allow for various micro-production units/sizes that pose no harm to the surrounding households or the environment.

The multi-stakeholder forum has played an important role in establishing communication and cooperation beyond the sectoral divides and the integration of urban agriculture in the Greater Amman Municipal policies.

Bangkok, Thailand – Energaia: Rooftop Production of Spirulina

Saumil Shah, Henk de Zeeuw

Introduction

EnerGaia is a for-profit organisation, founded in 2009 by Saumil Shah and Ingo Puhl that applies commercial strategies to maximise improvements in human and environmental wellbeing. Since 2011, we operate a facility for the commercial production of the microalgae *Arthrospira platensis* better known as spirulina at the rooftop of the Novotel in the Siam area of Bangkok, whose manager also has strong concerns for sustainability and advocated for us the implementation of our system on the rooftop. *EnerGaia* leases the rooftop from Novotel with a two-year (renewable) contract.

Design of the Rooftop Production System

Our closed system, virtually able to operate on any unused space, includes two circles of each 50-food-grade 250-liter semi-transparent polypropylene tanks. Each production circle further includes an air hose, air blower, three-phase electric cables, PVC piping and fitting, aquarium pump and a harvester with bag located at the end of the circle (Figs. 5 and 6).

The tanks function as photo bioreactors: the spirulina biomass develops inside the tanks thanks to sun radiation that reaches the plant biomass through the

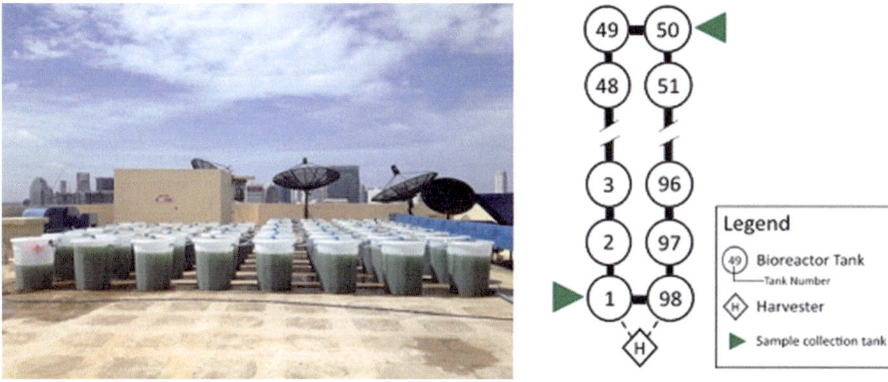

Fig. 5 The Energaia photo bioreactor system with connected harvester-bypass (Credit: Energaia)

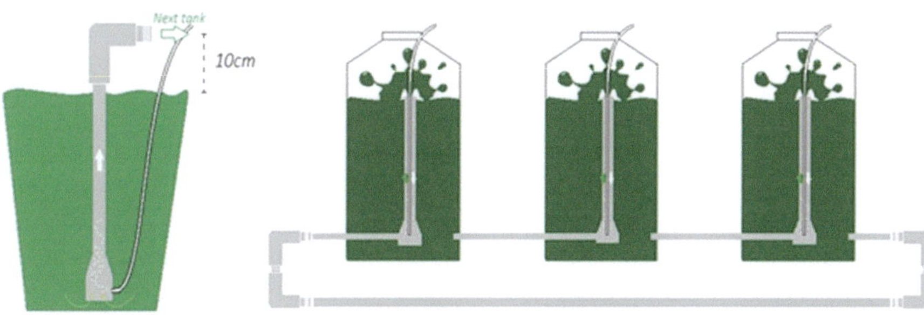

Fig. 6 Working schematics of the Mark III pumps (*left*) and the Mark IV pumps (*right*) (Credit: Energaia)

transparent walls and circulation of the nutritive media in the tanks. The functioning principle of this technical design is capillarity, enabling smooth stirring, solution circulation and nutritive media spreading thanks to outside air injection. In this way, the system allows the spirulina to double in biomass within 48 h.

The system's reliability stems from its simplicity and takes advantage of the physical conditions of the rooftop providing a flat area under ideal sunlight and CO_2 exposure in the Bangkok climate. Moreover, the closed system enables stable control of production parameters, prevents outside contamination and reduces dependence on weather.

The rooftop production system obtains water to fill the tanks and electricity for pump functioning from Novotel. The contract with Novotel includes, next to the rent, the payment for electricity, water and insurance of the rooftop against damage that may be caused by EnerGaia.

Since we are using only water as a production medium and no soil, the total weight of our system is 300 kg/m^2, a weight that most common flat roofs can withstand and thus no strengthening of the roof was required.

Production and Commercialisation Practices

Our facilities generate a global productivity of 4 t per year with an average productivity of 30 g dry weight/m^2/day, which is triple the productivity of the traditional open-pond production method, mimicking the African alkaline lakes where spirulina grows naturally.

Since our system is a closed system and effluents are recirculated, we generate very low nitrate and phosphorus wastewater, which is evacuated through evaporation. Evaporation is also possible since no pesticides are added to the solution thanks to the antibiotic pH of 10 suitable for spirulina cultivation. The required reactor cleaning is done every 3 months by emptying the reactor content on the roof floor, leading to its evaporation.

Our line of products includes dry powder spirulina, fresh paste, frozen paste, different varieties of pasta (fusilli, linguine, penne, and spaghetti), chocolate truffles, ice cream and gluten-free rice noodles, all fortified with spirulina. These products are distributed locally in Bangkok through a network of short channels fitting retailing habits in developing countries.

Main Lessons Learned and Challenges Encountered

The main lesson learned from our technology development is the way to strike a good balance for the classic low-cost trade-off, i.e. productivity versus reliability. The first-generation plastic tanks had a higher productivity but these tanks lasted only 9 months. The current tanks have a lower productivity but last up to five years, resulting in much lower replacement costs (materials, labour) and a higher net profit.

Our other main challenge was that local consumers are not familiar with spirulina and its nutritional properties. In order to overcome this challenge, we developed – together with local gastronomic chefs and food retailers – several products that meet consumers' palatability expectations.

A disappointment has been the sequestration of CO_2 from the city atmosphere. If all of the carbon used to grow the spirulina in our current production capacity came from CO_2, we would sequester 21 t. However, in our experience, spirulina cannot grow well on 100% CO_2 and the city atmosphere does not have a high enough concentration for optimal growth rates (we would need to be next to an industrial CO_2 emitter to use their waste CO_2, which we currently are not). So we currently use sodium bicarbonate powder (baking soda) as the primary inorganic carbon source that is fed to the spirulina, like all other commercial spirulina producers in the world.

Another future challenge is to multiply pilot farms in major cities in other developing countries metropoles in order to increase our system's adaptability. Our future partnerships with Kasetsart Bangkok Agricultural University, with the NGOs Antenna Technologies and Winrock International and with USAID will help us to reach those objectives.

Societal Impacts

The production and distribution of spirulina contributes to improve the nutrition of the urban population by providing an alternative nutritional resource with high protein (50–70%), lipid (7–16%), vitamin content (especially A and B complex) and high omega 6 (gamma-linolenic acid) and omega 3 and fatty acids content.

Economic Sustainability

Regarding economic sustainability, we recorded in past years an annual growth in sales of 200% and the breakeven point will be reached this year, thanks to new economies of scale leading to a USD 5 (€ 4.44)/kg dry weight price.

It should be noted that, from a legal point of view and in line with Thai agricultural tradition, our organisation has been recognised by the Board of Investment of Thailand as an innovative venture and, thanks to this, we benefit from taxation and ownership privileges.

Policy Relevance

Our spirulina rooftop production model can be considered a good model for replication in other cities in developing countries on account of its applicability on unused spaces such as rooftops, its flexibility (being a movable system), its low-cost technology, the production of highly nutritive food and the positive environmental impacts.

Barcelona, Spain – RTG-Lab, An Experimental Integrated Rooftop Greenhouse

Juan Ignacio Montero, Esther Sanyé-Mengual

Introduction

The Rooftop Greenhouse Lab (RTG-Lab) consists of two greenhouses, each with a surface of 125 m^2. The RTG-Lab was included in the design of the new building that hosts the Institute of Environmental Science and Technology (ICTA). Architects, engineers, agronomists and environmental scientists were involved in designing the greenhouse. Contrary to common practice, the metabolic flows (energy, water, CO_2) of the greenhouse are integrated with the metabolism of the building (Sanyé-Mengual et al. 2014).

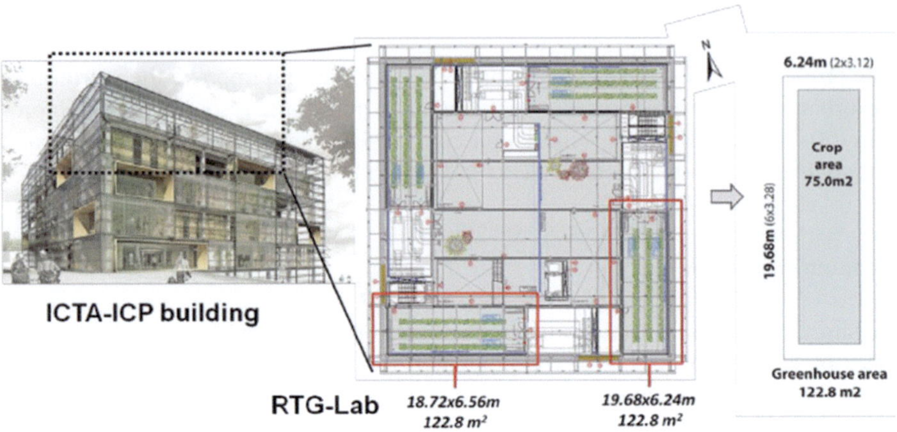

Fig. 7 Layout of the RTG-lab atop the ICTA building (Credit: ICTA-RTG Lab)

Fig. 8 Inside the two greenhouses of RTG-Lab (Credit: Pere Llorach-Masana)

The integrated greenhouse was used as a case study of the Fertilecity research project (http://www.fertilecity.com) that evaluated the feasibility of producing food in rooftop greenhouses in a Mediterranean climate (Fig. 7).

Design

The structure of the RTG-Lab is mainly made of steel, single-layer polycarbonate on the roof and polyethylene on the sidewalls. An air duct allows exchange of flows of energy and CO_2 within the building. Rainwater is harvested from the building's roof and stored for irrigating the crops. The greenhouse is facing southwest for more sunlight. This orientation compensates for the shade cast by construction elements on the crops. The greenhouse has roof and side ventilators; it relies on natural ventilation, not mechanical. The RTG-Lab climate is computer controlled (Fig. 8).

Production and Distribution/Commercialisation Practices

The RTG-Lab produces without soil by using perlite bags as substrate and fertigation (drip irrigation with NPK-Ca, as well as required microelements). Pests are managed ecologically to avoid the use of chemicals. Produce of the RTG-Lab is used for research and the surplus is distributed among the ICTA staff and students of the university. To date, the RTG-Lab has evaluated lettuce (multiple varieties) and tomato (a local high-valued variety).

Societal Impacts

From a life cycle perspective and considering the entire supply chain, local tomatoes from the RTG-Lab can decrease the environmental impact per kg up to 42% and become 21% cheaper (Sanyé-Mengual et al. 2015a).

Regarding the metabolic integration with the building, preliminary results highlight the potential water self-sufficiency through using rainwater-harvesting techniques, although in summer the savings are limited to 60%. The diffuse energy from the building improves the thermal conditions of the greenhouse in the autumn-winter period.

The ICTA staff and students directly benefit from the produce distributed among them. Furthermore, the ICTA community participates as evaluators of the produce through the "Quality and perception survey".

Beyond the direct advantages, the "Fertilecity" project aims to identify best practices and a rooftop greenhouse design that contributes to development of sustainable cities and mitigation of global concerns, such as urban food security or climate change.

Economic Sustainability

The investment in the greenhouse came from ICTA and public funding. Experiments have been funded by public research funding. Although the investment cost for building the RTG-Lab was three times higher than for a conventional greenhouse, tomatoes grown in the RTG-LAB are 21% cheaper than conventional ones, thereby making rooftop-greenhouse tomatoes a competitive product in the current market (Sanyé-Mengual et al. 2015a) as a result of the higher crop yield and lower distribution costs.

Main Lessons Learned/Policy Relevance

The RTG-Lab highlighted the legal and planning barriers that rooftop greenhouses may face in Europe. The design was modified to comply with structural and safety laws (Sanyé-Mengual et al. 2015a), and the local land-use code was modified to accept the use of the building as an experimental space for agricultural production.

Furthermore, architectural and agronomic demands were discussed and balanced during the design process. The RTG-Lab thus highlighted weak points in current policies and regulations that need to be further discussed with the aim of making the current legal framework more facilitative for implementing rooftop agriculture in Spain.

Bengaluru, India – An Organic Terrace Gardening

B.N. Vishwanath, Rajendra Hegde, Henk de Zeeuw

Introduction

Since 1995, the practice of terrace gardening in Bengaluru (formerly named Bangalore) was promoted initially by Kadur Agro and then on a regular basis by AME (Agriculture Man Ecology) under the leadership of Dr B.N. Viswanath. In 2006–2007, RUAF Foundation enabled promotion of terrace farming with various Resident Welfare Associations in Bengaluru, and *A handbook of organic terrace gardening* was published to serve as a practical guide for practising terrace gardening. In 2011, the NGO Garden City Farmers (GCF) was established in Bengaluru (www.gardencityfarmers.org). It organises seminars on organic urban farming and terrace gardening at regional and state level. In Bengaluru, GCF organises the quarterly *Oota from your Thota (*Food from your Garden) event with demonstrations, sharing of experiences, seed sharing, exhibitions of organic inputs and products and film shows on urban farming and terrace gardening.

Preparing and Operating a Terrace Garden: Two Practical Examples

Mr Harish Mysore Ramaswamy of Bilekahalli in Bengaluru is a software professional but from a farming family. He is around 60 years old and a full-fledged organic urban farmer with about 280 m^2 terrace garden where he grows a large variety (over 50 species!) of vegetables and fruits in containers. The containers he uses are both purchased and recycled, including plastic pots, grow-bags, cement pots, bathtubs, washing machine drums, water cans, paint drums etc. to suit the growth requirements of the various crops.

The wide terrace has no shade net but the perennial creeper beans and Indian spinach on the support systems sometimes provide some shade to the bushy and leafy vegetables (Fig. 9). Watering is not mechanised but designed so well that it is easy to water all plants with outlets at many points. Liquid fertilisers of natural origin are included in the irrigation water. Harvested rainwater and recovered excess irrigation water are filtered through a self-designed filter and collected in the basement in a tank to be recycled for watering. The growing media is a mixture of animal compost, vermicompost and coco peat in equal proportions and enriched with biofertilisers.

Fig. 9 View on part of the terrace garden of Mr Harish Mysore (Credit: Rajendra Hegde)

Crop planning is based on the seasons. While the bushy and vine crops are more common in the rainy winter season, leafy greens are grown in summer. Plant protection is practised, if required, only in summer in local, natural ways, with crop diversity being the best tool to manage the pests. This urban farmer harvests 7–10 kg of completely organic, safe and functional food each week, which is more than enough to feed his family; he shares the surplus with friends and relatives.

Both Mr Harish Mysore and his wife are excellent hosts, who welcome many visitors every day and week and he is also active on social media to promote organic terrace gardening.

Mr S. Laxminarayan (40 years old), located in the ISRO Layout of Bengaluru, worked in a multinational IT company until recently. He became a terrace gardener five years ago and is growing a variety of vegetables, fruits and medicinal plants (55–60 species!) on his terrace of about 40 m^2 in the firm belief to "grow what you eat and eat what you grow". He constructed raised beds and boxes of different sizes with iron bars and used wooden pieces collected at junkyards. Smaller containers are for vegetables and medicinal plants; in the larger containers, he planted perennials such as curry leaf, pomegranate, guava and grapes (Fig. 10). The growing medium is a uniform mix of soil, animal compost, vermicompost and coco peat in

Fig. 10 View on (part of) the terrace garden of Mr S. Laxminarayan (Credit: Rajendra Hegde)

equal proportions. The plants are fed with natural origin biofertilisers on a regular basis to maintain plant health.

Next to growing his own food, he wanted to show that it is possible to realise substantial vegetable production on a small terrace. He is able to meet around 60% of the vegetable requirements of his family from this terrace. He became a member of Garden City Farmers Trust, taking an active role in sharing and spreading his experiences through the Organic Terrace Gardening Group on Facebook and other networks. Many interested urban farmers in Bengaluru and from elsewhere visit his terrace garden to discuss with him and to gain a first-hand idea about setting up their own terrace gardens.

Results at Farm Level

During the "Food from your Garden" event in February 2016, GCF interviewed 895 organic terrace gardeners from Bengaluru. About 63% of the gardeners reported that they were harvesting up to 30% of their weekly vegetable requirements, 26% were harvesting about 30–50% of their weekly vegetable needs and 11% reported that they harvested more than 50% of their vegetables from their own garden. Higher yields are always their concern, but the gardeners are happy with what has come with the fresh food they have produced themselves in their garden.

The survey results also underline the social role of terrace gardening: 94% of urban gardeners reported that they are practising organic urban gardening not only to obtain fresh healthy food and/or some economic benefits, but also to have a green space to relax from the pressures of urban living, to sit and communicate with family members, to have a place to enjoy the pleasure of growing crops and have some physical exercise, a place to let children experience gardening and develop understanding for ecology and food. About 36% of the terrace gardeners under 40 years of age have developed the terrace garden also to keep their elderly parents active and occupied in simple activities such as watering, weeding and harvesting. Their parents have village roots and love plants, but had not had a proper opportunity to practise gardening anymore in the city. Now, with the rooftop-gardening option, they are active and enjoying the garden. More laborious tasks are done by the youngsters during the weekends and holidays.

Impacts at City Level

The Facebook group on Organic Terrace Gardening now has over 27,000 members, of which about 20,000 are Bengalureans; of these, about 14,000 have a functional terrace or kitchen garden. Today, all parts of Bengaluru have a gardening group that meets at regular intervals, exchanges ideas and planting materials, organises events to showcase their gardens and bring more urbanites into gardening, organises a farmers market to sell surplus products or develops initiatives to convert the common apartment terrace into a garden.

To cater to their needs, many start-ups have emerged to provide various solutions including seeds, seedlings, nutrition supplements, plant protection materials, net houses, polyethylene houses and so on. New firms are developing products for organic gardening. The Department of Horticulture is providing essential materials to the needy urbanites under various programmes. The Solid Waste Management Round Table in Bengaluru has initiated a programme called *Swacha Graha* (Clean House), targeting waste segregation, kitchen-waste composting and urban gardening.

Bologna, Italy – The community Rooftop Garden of Via Gandusio

Luana Iori, Esther Sanyé-Mengual

Introduction

Via Gandusio is a social-housing complex in the north of Bologna (Italy) that hosts elderly Italians from the 1960s domestic migration flows and current international immigrants mainly from Africa and Asia. Inhabitants of this social housing have limited relations with each other. To solve this problem, the city council planned and funded a community rooftop garden as a new meeting point of the housing complex, in collaboration with the association BiodiverCity and the University of Bologna. In the implementation process, inhabitants of the social housing were involved. The 250-m^2 rooftop garden started in 2011 as the first of its kind in the city. The innovation of this not-for-profit project lies in the multi-actor design process and the social-inclusion function.

Design of the Garden

After improving the safety conditions of the terrace (e.g. protective fence), wooden containers were constructed from former pallets and distributed over the surface of the terrace; a pipe for hydroponic production was installed along the fence that protects the perimeter of the terrace. Tap water is used for irrigation, since the technical characteristics of the building were a limitation for setting up a rainwater-collection system. Three different cultivation techniques are employed in the garden: organic soil-based production in the containers, floating-root hydroponic production in containers, and nutrient film technique (NFT) in pipes (Figs. 11 and 12).

Fig. 11 View on a part of the garden (Credit: ResCUE-AB – University of Bologna)

Fig. 12 The containers constructed from recycled pallets (Credit: ResCUE-AB – University of Bologna)

Production and Distribution/Commercialisation Practices

The garden includes leafy species (lettuce, chicory), vegetables (tomato, pepper, melon, watermelon, eggplant) and herbs (basil, aromatics). Production is pesticide-free, and the participants collect the household wastes from the residents in the building and produce their own compost for organic practices. All the produce is for self-consumption and is distributed among the participants in the production process.

Societal Impacts

The community rooftop garden of Via Gandusio has around ten active participants, who are the direct beneficiaries of the garden as well as some other residents, mainly elderly people, who receive produce from the gardeners. The rooftop garden is also used as a place for events organised by the participants under the name Gandusio Green Actions or in collaboration with other associations in the neighbourhood. The university and the educational association L'Altra Babele also use the garden for educational purposes (Fig. 13).

At the city scale, the expansion of rooftop gardens on available roofs in Bologna could satisfy up to 77% of the urban vegetable demand (Orsini et al. 2014), according to the production efficiency evaluated in Via Gandusio. From a life-cycle perspective, fruit crops resulted in a lower environmental impact and economic cost

Fig. 13 An impression of one of the neighbourhood meetings on the rooftop (Credit: ResCUE-AB – University of Bologna)

than leafy vegetables, and organic soil production was the more eco-efficient solution (Sanyé-Mengual et al. 2015b).

Economic Sustainability

Public funds from the Bologna City hall supported the implementation of Via Gandusio. The initial project cost € 10,000, 60% for materials and 40% for staff. Operational costs were initially funded by the university as this was part of a research project. Costs are now carried by the actual participants in the gardening in a sharing-economy framework.

Main Lessons Learned

Several lessons were learned during the implementation of Via Gandusio. First, safety elements must be included in the design. Second, gardeners are open to use different cultivation techniques when practical information is provided. Third, the project unveiled the importance of creating a long-lasting gardeners group that successfully organises itself to run the garden. Setting up the garden required the intervention of a social worker from the Municipality to assist in creating the gardeners group and facilitating participation of diverse inhabitants in the garden activities. The garden is now managed by a small group of inhabitants; this highlights the limited participation of a large number of community members.

Policy Relevance

Via Gandusio demonstrated the potential use of roof gardens in the renovation of social housing and in the promotion of social inclusion. Bologna has a large number of flat roofs that could be used for rooftop agriculture, thereby increasing food self-sufficiency and urban biodiversity, as evaluated in Orsini et al. (2014).

Boston, United States – Fenway Farms: A Restaurant Garden on a Baseball Stadium Roof

Jessie Banhazl, June Komisar

Introduction

Fenway Farms is a spacious restaurant garden in Boston, Massachusetts, on the roof of the administrative offices of Fenway Park, the historic stadium of the Red Sox baseball team, one of the oldest ballparks in the US. The building is over 100 years old and located in the midst of Boston's dense urban core, squeezed into a large city block surrounded by both new and old mid-rise buildings. The garden project belongs to the Fenway Sports Group, a partnership that owns the Red Sox. This prominent garden is located at third base, visible from the EMC Club restaurant located within the ballpark structure, and was the initiative of Linda Henry, the wife of one of the partners, who began the project to promote healthy living and eating (Fig. 14).

Design of the Garden

Green City Growers and Recover Green Roofs installed the garden just before the 2015 baseball season. After the renovation of the roof over the office portion of the stadium, growing containers were placed in areas that could withstand a load of

Fig. 14 Fenway Farms: *Left*: View from garden; *Right*: View from EMC Club restaurant (Credit: Joe Nasr)

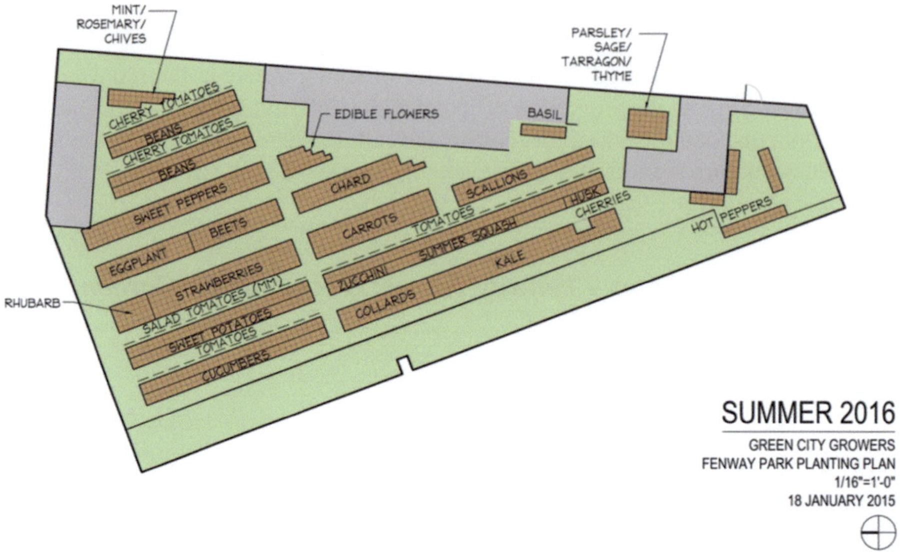

Fig. 15 The production plan of Fenway Park garden for 2016 (Credit: Fenway Farms)

about 500 kg/m^2. The almost 500 m^2 productive roof is covered with four long rows of containers, each four containers wide, about 1900 in total (Fig. 15). They are standardised five-gallon (19-l) plastic containers that resemble standard North American milk crates – about 45 × 33 × 28 cm – but are "Recover Aerated Media Modules" (RAMM) made with 50% recycled plastic. They are lined with food-grade grow bags and filled with 25 cm of locally-sourced compost-rich soil added to a lightweight mix including peat moss and vermiculite designed specifically for roofs. This combination helps to slow the compaction problem common to a lightweight mix when used alone.

Before installing the containers, a roof contractor installed a new waterproofing system as a base for them. Synthetic grass beneath the long stretches of plastic containers provides a uniform and tidy bright green look for the baseball fans to see as well as a comfortable surface for the farmers to stand on. To prevent problems with high winds, the rows of containers are secured to each other. They can be delinked and re-assembled easily, as needed, for flexibility. The containers can also accommodate season extension by the installation of small plastic hoop structures directly on the rows of containers. Hardy plants like kale and spinach are started very early with this system.

The final aspect of the project was a smart, electronically controlled drip-irrigation system installed to use potable city water through a connection with the stadium waterlines. The irrigation system is designed for flexibility and can be moved if the containers are relocated.

Further integration of food garden and restaurant, by placing additional raised-bed plots as dividers in the outdoor lounge seating area for the Fenway Park patrons, is being considered.

Production and Commercialisation Practices

Green City Growers (GCG), a for-profit company with high social and environmental standards, is contracted by the team owners to manage Fenway Farms. They plan the garden by working closely with the restaurant's chefs, matching plants that will thrive on a New England roof with seasonal menus. They plan crop successions, strategically inter-plant crops to maximise space in the containers, and use only organic, OMRI[1]-approved products for fertility and pest management. At this time, they do not have a composting operation but rather obtain compost from nearby farms.

The sunny location, rich soil and skilful gardening combined to supply over 1800 kg of greens, herbs, vegetables and fruit. Peas, rosemary, eggplant, kale, broccoli, sweet potatoes and strawberries were grown in abundance during the first season.

GCG supplies both the EMC Club restaurant and the "concessions" – the take-out food stands – in the baseball park, so all patrons have the opportunity to taste the fresh produce.

Economic Sustainability

This farm was not designed as a money-making operation in itself, but is an enhancement to the Fenway experience. Revenue comes from selling the tasty local food in the restaurant, and from the tours, which cost USD18 (€ 16) per ticket.

Societal Impacts

From its conception, the purpose of the farm was to provide "hyper-local" food on site for the restaurant as well as to serve as an interactive educational space for fans. This dovetails with the mandates of the farm managers, Green City Growers (GCG). Undoubtedly, this site is GCG's most visible showcase. Thousands of people see the garden every week because of the inclusion in the official Fenway Park tour, one of the most popular tourist attractions in Boston. Visitors, from school groups to baseball fans, are treated to an introduction to the possibilities of rooftop growing. This roof serves not only as a public good, introducing a variety of people to growing vegetables in containers, but also as a way to introduce tasty, healthy, local food to the patrons.

Fenway Farms not only greatly reduces storm-water runoff from the roof and keeps the offices below cooler in the summer months, but it provides zero-food-mile organic produce and, through the large number of visitors to the farm, contributes significantly to public awareness of the possibility to grow local organic produce only metres from farm to table.

---

[1] OMRI: Organic Materials Review Institute

Cairo, Egypt – Rooftop Farming in Informal Settlements

Christopher Horne, Saber Osman, Carl Philipp Schuck

Introduction

In 2013, a study (Laban and Osman 2013) commissioned by the Participatory Development Programme in Urban Areas (PDP)[2] in Cairo identified urban agriculture as a potential participatory-adaptation measure to improve the urban dwellers' socio-economic conditions (income, food security) while at the same time having microclimatic effects improving the living conditions in the informal settlements and enhancing urban resilience.

Given the residential ownership and building structures in Cairo, the large amount of unused flat rooftop space appeared promising for rooftop farming. To explore local challenges and opportunities, PDP partnered with Schaduf, a local social enterprise specialised in urban agriculture, to implement a pilot project in 2014 to test the feasibility of rooftop farming and to gain knowledge on urban agriculture as part of adaptation to climate change, in order to enable further projects to succeed on a larger scale.

The informal settlement of Ezbeth El-Nasr in the southeast of Cairo was chosen as the project site because of its high vulnerability to climate change, manifested in the low awareness level of its residents, the particularly poor economic conditions [most residents in the neighbourhoods earn between EGP 600 (€ 75) and EGP 1000 (€ 127) a month] and the pressing need to reduce the heat trapped in the informal settlements during the long and dry summer periods (see Fig. 16).

Local charity-oriented NGOs selected the participants and hosted trainings. Criteria for selecting participants included: strong interest in home-based food production, access to a rooftop with at least 12 m^2 available for farming, willingness and ability to share 10% of the equipment costs to ensure local ownership (approx. € 13 – 26 per household, depending on plot size), and commitment to attend three training sessions before the implementation phase. The project attracted both male and female local residents as participants.

Spatial clustering of rooftop farms was highly promoted in order to support shared learning and to boost the microclimatic effect.

Schaduf's role was to provide the initial training, install equipment and provide marketing support for the participants during the one-year pilot project and thus back its economic sustainability. After the pilot ended, Schaduf continued the project with social responsibility funding by the 7 Up company.

---

[2] PDP is implemented by the Deutsche Gesellschaft für Internationale Zusammenarbeit (GIZ) with funding from the German Federal Ministry for Economic Cooperation and Development (BMZ)

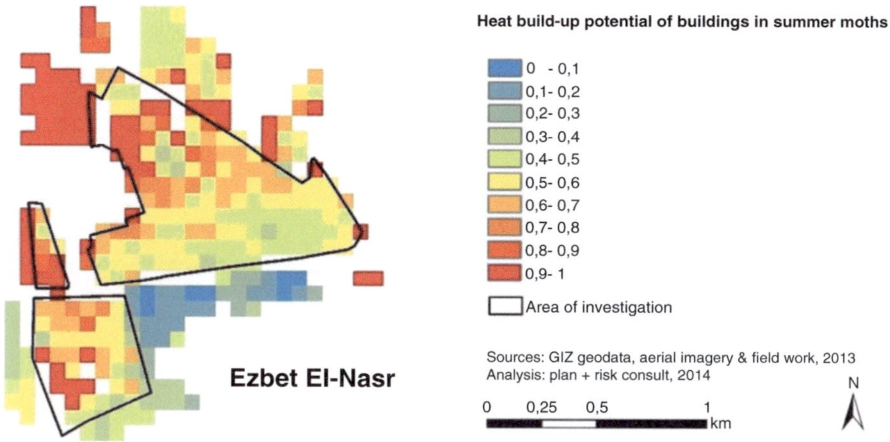

Fig. 16 Urban heat island effect in Ezbeth El-Nasr (Credit: PDP)

Fig. 17 *Left*: Ms Mariam and her children in front of her production unit; *Right*: Preparation of the farming tanks/containers (Credit: Saber Osman)

The Rooftop Farms

On each rooftop, a simple hydroponic system was installed that consisted of 3–4 water beds made of wooden frames, plastic sheets, foam panels and cups filled with peat moss and pyralite substrate (3.75 m^2). The beds were filled 15 cm deep with water supplied through a water pipe and were further provided an electricity connection from below and maintained by a water pump and a water filter (Fig. 17).

Fig. 18 Produce made ready for the market by Schaduf (Credit: Schaduf)

After the pilot project ended, Schaduf continued experimenting with geotextile panels to improve the provision of continuous water supply.

In total, this model cost about € 16 /m^2 (three beds: about € 178; four beds: about € 222) and included successive acquisitions such as fertilisers, seeds and pesticides as well as technical support for 6 months. The geotextile panels added by Schaduf to improve the provision of continuous water supply raised the initial costs by approx. € 4.4 /m^2.

The initial training offered by Schaduf consisted of three sessions covering subjects such as farming conditions on rooftops, management of the hydroponic system, growing techniques, management of pests and growth problems. During implementation, also hands-on training in "green-production" techniques was given to reduce fertiliser and pesticide use.

Schaduf also organised the input supply to the participants and the marketing of the produce. They bought the products at a pre-fixed price from the participants, packaged and labelled the products, and sold these in a "Mahali" specialty food shop (Fig. 18).

Some participants were trained to become coordinators among all farmers of the settlements and then acted as focal points vis-à-vis Schaduf. This was of benefit, as it helped tackle common problems shared by the participants and facilitated mutual learning.

The Results Obtained

Of the 24 households trained, only six households installed the growing unit during the pilot project, while another eight households installed such a unit in the second season.

Initially, EGP 300 EGP (about USD 30) was targeted as individual monthly income. This target was not met (around EGP 150/participant was realised) due to typical challenges associated with the ups and downs of pilot projects: the first crops often saw failure, mainly due to a still low level of understanding, skills and dedication among the participants, miscommunications between participants and Schaduf, as well as some technical problems (e.g. discontinuities in the water flow). The geotextile panels added by Schaduf later on increased the residents' income by 25%, because this shortened the plant cycle. The geotextile panels currently used could be technically further improved.

Since most of the production was sold, the effect of the rooftop units on household nutrition and food security was low (part of the income earned may have been used for buying food but this was not measured). According to research by the Cairo-based Climate Laboratory Centre, rooftop farms reduce the day temperatures in the upper floor of the building by 7 °C on average.

Out of the total number of pre-selected participants, one third did not take part in the implementation phase, either for economic reasons – these residents did not see immediate benefits of their 10% contribution to the equipment costs – or on account of unclear rights of access to the particular rooftop. During the implementation phase, there were no dropouts. The 18 participants continue cultivating their rooftop farms in coordination with Schaduf.

Main Lessons Learned

Despite the limited economic results, strong increase of awareness among residents about what urban agriculture can offer has been noted: The project initiated further interest among a broader number of residents due to its income-generating and beautification character. However, all participants without prior farming experience confirmed that the setup of a rooftop farm requires stronger endurance than they had anticipated.

Group formation and training on entrepreneurial capacities may lower the residents' dependency on Schaduf as their only technical support and marketing channel. Selectively addressing women and also youth during selection and training could widen their chances to enjoy the economic and environmental benefits of urban agriculture.

The monitoring of this kind of project should also include impacts on household nutrition/food security, resource use and micro-climatic effects (e.g. reduction in temperatures and water runoff).

Policy Relevance

Though small in scale, the pilot project generated valuable insights on opportunities and challenges for home-based rooftop farming under the perspective of socio-economic and microclimatic improvements. The knowledge gained allows a promising outlook on a project on rooftop farming in informal settlements of larger scale and impact.

Chicago, United States – Gotham Greens: The Largest Rooftop Greenhouse in the World

Viraj Puri, June Komisar

Introduction

Gotham Greens is a commercial enterprise that since 2011 completed three rooftop greenhouses in New York: one atop the two-story Greenpoint Manufacturing Design Center, a non-profit for start-up companies; another on a Whole Foods supermarket; and the third atop a four-story factory building. Their most innovative farm is their newest facility built in 2015 on top of Method Products manufacturing plant (producing eco-friendly cleaning products) on a site with parkland and wetlands in the historic Pullman area of Chicago's South Side (Fig. 19).

Fig. 19 Bird's eye view of Gotham Greens and Method Products manufacturing plant (Credit: Gotham Greens)

Fig. 20 Gotham Greens at Pullman from the inside (Credit: Gotham Greens)

Design of the Greenhouse

Gotham Greens and Method Products agreed to combine a new manufacturing facility and a new greenhouse farm into one integrated environmentally responsible building. The building, designed by the sustainable design architecture firm of William McDonough + Partners, is the world's first LEED-Platinum certified manufacturing plant in its industry combined with an almost 7000 m^2 greenhouse on top (the largest rooftop greenhouse farm in the world).

In the greenhouse, crops are grown in a highly controlled environment. Bathed primarily in the wide spectrum of natural sunlight, plant nutrients are introduced to the root systems in water that recirculates and the environment is monitored electronically for optimal heating, cooling and humidity conditions. Power comes, in part, by wind turbines and solar collectors, while waste heat from the manufacturing plant below also helps to warm the greenhouse space in cold periods. LED lighting, thermally efficient glazing and thermal curtains in the greenhouse also contribute to the building's energy efficiency (Fig. 20).

Production and Commercialisation Practices

The Pullman facility of Gotham Greens produces year-round, with a yield of up to 10 million heads of leafy greens and herbs (half a million kg) annually for retail grocers and restaurants across greater Chicago. This impressive yield is partly due to the ability to grow 25 crop turns a year (compared to the two or three turns per year on a traditional Chicago-area farm).

The Pullman farm grows a variety of lettuces including butterhead and romaine, tomatoes, and herbs such as basil, Crops are sold to several prominent supermarkets, including the local stores belonging to large North American companies such as Whole Foods Market, Peopod and Target, as well as smaller local supermarkets. They also provide food for local institutions such as Greater Chicago Food Depository, Greater Roseland West Pullman Food Network, Pilot Light and Chicago Botanical Garden's Windy City Harvest. Over half a dozen restaurants, including the very local Pullman café, serve produce from Gotham Greens.

Because it is a meticulously clean hydroponic facility, herbicides are unnecessary, and insect pests are kept at bay by introducing beneficial insects. In addition, workers are trained to spot any harmful insects. This clean way of growing also prevents pathogens such as E. coli and salmonella from invading the crops.

Economical Sustainability

The privately held company had the challenge of fundraising for constructing this ambitious new farm. The new Pullman greenhouse required an investment of over € 7 million.

Societal Impacts

The integration of a greenhouse into the design of the manufacturing facility has resulted in an important reduction in the ecological footprint of the manufacturing firm by recovering and reusing heat from the manufacturing plant.

The Pullman farm hired nearly 50 employees in its first year, generating jobs for the economically challenged in the neighbourhood.

Growing large quantities of fresh vegetables in the city reduces the need to import food from far away, reducing "food miles" and related greenhouse gas emissions, and contributes to the local economy. The average head of lettuce takes about 3200 km from farm to table in Chicago; Gotham Greens delivers produce within a 100-km radius (Pirog et al. 2012). Unlike open-air farms, greenhouse rooftop farms may not help to absorb rainwater runoff, so the architects took care to design the site with a neighbourhood park that incorporates bioswales: simple landscaping features used to slow, collect and absorb rainwater runoff.

The establishment of the integrated building also contributed to the economic revitalisation of the historic Pullman Park district, which is one of America's first model industrial towns built by the Pullman Palace Car Company and recently declared a National Historic Monument.

Although Gotham Greens is a privately held commercial operation, it supports various community programmes. In addition to providing jobs and supporting the Greater Chicago Food Depository, Chicago's largest food bank, it also gives seedlings to school and community gardens and hosts field trips. This is part of Gotham Greens' philosophy that urban farming is "about re-connecting with our food supply, educating our youth and nourishing our souls".

Lessons Learned/Policy Relevance

A challenge in the company's growth is finding existing buildings that can handle the load of a rooftop greenhouse, what might be difficult in some cities. In such a situation, the Chicago facility shows the benefits of the integration of a rooftop farm from the start of an industrial building project. However, finding an environmentally responsible industry to work with may be quite a challenge. In the case of constructing on existing buildings, older factory building roofs that were conservatively designed (for extreme snow loads or possible additional floors) can provide the best possibilities for rooftop greenhouse sites.

Cincinnati, United States – The Rothenberg Rooftop Garden School

Bryna Bass, Edwin "Pope" Coleman, June Komisar

Introduction

In Cincinnati, Rothenberg Preparatory Academy's Rooftop Garden School illustrates the role that a garden can play in education. This public school, serving students from pre-kindergarten through elementary school, is located in the inner-city neighbourhood of Over the Rhine, an area settled in the 1800s. After a severe decline, a revival of the neighbourhood, including its beloved historic buildings, began in the 1990s, marked with the formation of the not-for-profit Over-the-Rhine Foundation.

The Rothenberg Preparatory Academy, a public school that is over 100 years old, is part of this legacy. 2013 marked both a revival of the building by WA Architects and the creation of the building's rooftop garden. The garden was included in the planning from 2008 onwards, when the building was saved from demolition and it was revealed that the original playground was on the roof. The roof was strong, and a high parapet already existed to protect students. With virtually no yard to turn into a garden space, the building's roof became the only viable option for a badly needed hands-on green space and learning garden.

Urban activist and project manager, Pope Coleman, began the fundraising effort for the new garden. Both the rooftop garden and the Rooftop Garden School are independently managed, supported by a Garden Guild, a volunteer group created by the Over-the-Rhine Foundation. This not-for-profit community group provides financing, guidance and operations for the school-garden activities, and funded the conversion of the roof to a garden. Through the cooperation of Cincinnati public schools, many neighbourhood groups from parents to nearby gardeners are involved in this far-reaching project.

Design and Functioning of the School Garden

After testing was done to confirm the strength of the roof, a green-roof installation company, Green City Resources, installed a new planter-ready roof on a nearly 800 m^2 surface, 32 wooden raised beds were set between concrete and rubber roof-pavers, filled with about 25 cm of lightweight growing medium (a mixture of nutrient-rich soil mixed with vermiculite or other light admixtures), which is a sufficient depth for a variety of vegetables (Fig. 21).

As part of the Edible Schoolyard Project – a network of schools all over the USA that integrate gardens into the curriculum – the garden is used for lessons across the curriculum, providing hands-on experience for around 450 to 550 students annually. The rooftop garden offers learning that contributes to science, technology, engineering and math education as part of the curriculum.

The students and community volunteers are the gardeners, as they build community and gain knowledge by doing. The Garden School Manager, Bryna Bass, prepares and leads the classes in the garden and works with teachers to create lesson

Fig. 21 View of the Rothenberg Rooftop Garden School (Credit: Rothenberg Preparatory Academy)

Fig. 22 How do they grow? (Credit: Preparatory Academy)

plans that integrate the garden classes into the curriculum. Together, the Manager and the teachers seek to develop critical thinking skills, an understanding of the environment and a sense of community (Fig. 22). Vegetables grown in the garden are served in the school cafeteria; this helps to promote healthy eating, addressing obesity and early-onset diabetes issues in the community.

Garden programming, particularly in the summer, involves the larger community in a variety of ways: Healthy Cooking Classes for the food insecure, organised by the local food bank, provides knowledge of how to process the produce and increase food literacy. In 2015, a mental health organisation used the garden to provide horticultural therapy sessions for high-risk children. Also continuing education classes for teens and adults from the neighbourhood are organised in the Garden School.

Societal Benefits

The Garden School is an important means to provide hands-on experience for inner-city children, as well as teens and adults in the neighbourhood. In this way, they engage with the natural environment, to realise the importance of preservation and conservation, as well as issues of nutrition, healthy eating and sustainability. Incorporating this garden with many of their academic courses enhances their critical thinking, science, math and literacy skills through hands-on application. The school garden also provides much needed green space in a dense concrete and masonry neighbourhood, critical for a sense of wellbeing among inner-city residents, and functions as a community learning centre for gardening and healthy food.

Economic Sustainability

The volunteers in the Garden Guild act as fundraisers in the local community emphasising that they are "schooling in a garden, not gardening in a school". In-kind donation of space, water and maintenance comes from the school district.

Lessons Learned; Policy Relevancy

This project demonstrates how gardens atop schools can be an integral part of the school and its teaching while managed professionally and independently. Like most school gardens in North American winter cities, one challenge is that the usual North American academic year is over when the garden needs its most urgent attention, which makes the robust summer programming essential. In addition, fundraising was and will continue to be a challenge that is being met by the community.

Gaza, Palestine – Rooftop Aquaponics for Family Nutrition in the Gaza Strip

Chris Somerville

Introduction

Land available for horticulture is extremely limited in the Gaza Strip and, with the current restrictions on the movement of products and people, fresh vegetables are expensive and hard to find. Furthermore, 97% of the Gaza Strip population are urban or camp dwellers, and therefore do not have access to land. Considering that lack of access to good agricultural land and water will continue to be a chronic issue within Gaza, aquaponic units on rooftops could be an appropriate food-production option able to provide nutritious fresh vegetables and fish (protein) and to generate additional income for poor and food-insecure households. Moreover, aquaponics is a water-efficient way of producing food, and the aquaponic units are easy to install on any flat, urban platform using local low-tech materials.

In the period 2011–2013, the Food and Agriculture Organisation of the United Nations (FAO), with funds from the Government of Belgium and in collaboration with local partners, implemented a project introducing various small-scale food production packages (e.g. micro poultry production, vegetable gardens). Within this project, FAO piloted aquaponic food production units (fish and vegetable growing) on the rooftops of 15 mostly poor female-headed households, making use of experiences gained during previous projects installing locally designed soilless rooftop systems. The 15 most successful households from these projects where selected for the aquaponics pilot, as they had some familiarity with soilless culture methods.

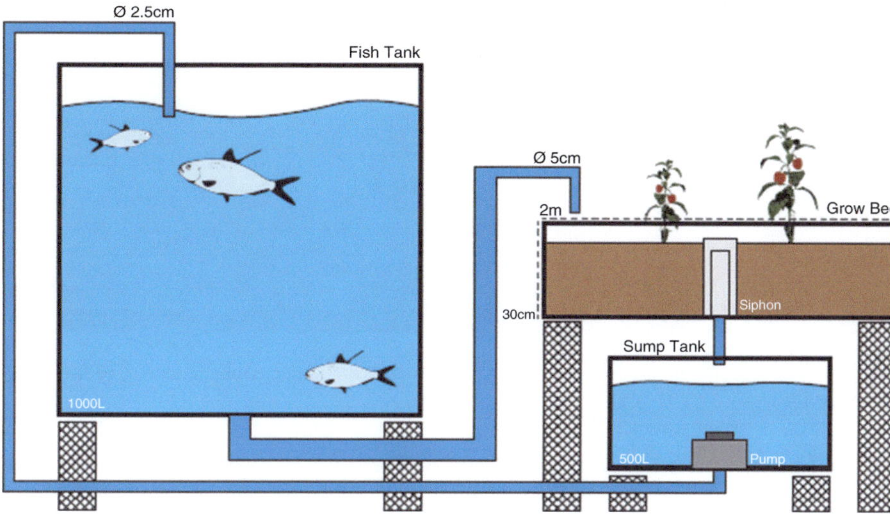

Fig. 23 Schematic overview of the aquaponic unit (Credit: Chris Somerville)

The Project

The project provided the participating households with an aquaponics rooftop starter kit in kind. This included a fish tank (locally made plastic container 1 m^3 in size), fibreglass grow beds with volcanic gravel, an electric pump (using electricity from the grid), simple water-quality monitoring kits, enough tilapia fingerlings and fish food for one seven-month growing cycle and assorted vegetable seedlings for one growing season. Households predominantly used groundwater from private wells to fill and replenish their units. Once assembled, each beneficiary had an aquaponic unit with a 4 m^2 growing space and a maximum fish-stocking density of 20 kg. Using a simple example, at maximum capacity, one unit could produce 20–25 lettuce heads per week and 30–35 kg fish per year, which is suitable for consumption by the household (usually at least six people in the Gaza Strip) (Fig. 23).

All beneficiaries received initial training (aquaponic ecosystem dynamics; fish care; crop management; planting times; seedling preparation; organic and natural pest control; and seedling protection) in order to maximise the use of inputs received. Most families grew lettuce, tomatoes, eggplants, cucumbers, hot peppers and rocket, but some families were also/instead growing culinary and medicinal herbs (i.e. basil and sage) within the first year.

Each aquaponic unit costs USD 1000–1500 (€ 888 – 1332) including the fish fingerlings and plants required for one season. Maintenance requirements are quite low, largely involving replacing the water pump every 2–3 years; the pump costs are 6–10% of the overall unit cost. All the input costs per year (water, electricity, seedlings, fingerlings, pump depreciation, water-testing kits) come to about half the

Fig. 24 The aquaponic unit in operation on the rooftop under a shade net (Credit: Chris Sommerville)

value of annual production. In practice, however, output varied widely since new users were encouraged to practise polyculture and grow what they need at home (Fig. 24).

The Results

The project was initially successful in providing a means for households with no land to grow fresh food. Beneficiaries enhanced their household's access to nutritious food and increased dietary diversity. It also allowed women to participate in productive activities at home. By the end of the first year, up to 13 households were still producing vegetables. Yet, soon after, an unusually cold winter storm killed most of the tilapia fish. As all were poor households, over half chose not to restock fish and plants, explaining that it was too expensive to continue. Poor groundwater quality leading to poor plant performance was also a key factor in their decision not to continue. Four households restocked and continued into Years 2 and 3.

When interviewed, households expressed great satisfaction with their new ability to produce green spaces within densely populated urban zones. Most found the daily practice of tending to their plants and fish a very peaceful and enjoyable experience, giving them a release from the stresses associated with daily life in Gaza. Households also highly valued the fact that chemical pesticides were not used. The interviews showed that families produced up to 15 kg of fish each within the first 6 months, all for household consumption. With regard to vegetables, during the first summer season, some households produced enough tomatoes and cucumbers to meet their families' needs for those months, making savings on their food bills (approx. 120 kg of fruit).

The initial package allowed mainly production for home consumption. It was expected that, if successful after the initial stage, participants would increase the capacity of their production unit by adding more vegetable-growing beds and fish tanks to start growing for the market. Yet only two of the 15 households expanded production using materials they had received from a previous FAO rooftop pilot project; although still not to a commercial scale. Monitoring indicated that the aquaponic units demand a higher educational capacity for operation, which is rather challenging for poor female-headed families in Gaza. The project sought to remedy this by making the training course and materials for each beneficiary as simple and as accessible as possible.

Unfortunately, power cuts led to some fish mortality, particularly during the hot summer months when the capacity for water to hold dissolved oxygen reduces as the water temperature increases above 30°C. Solar power or battery-powered air pumps can solve this issue, although solar power units were not provided for during the project.

Main Lessons Learned

Essentially, piloting and subsequent adoption of a new and complex production method such as aquaponics require time (years), extensive technical support, coaching and educated beneficiaries with good agricultural experience. However, given its nature as an emergency intervention, the project had to target poor, urban families who happened to have minimal farming experience. Thus, after the initial inputs provided by the project, many families did not have the resources to buy new inputs or a basic farming background to ensure good plant performance or access to technical support from local, experienced producers.

The Gaza experience demonstrates the need to work initially with a small number of families who have not only an entrepreneurial spirit but also a farming background to ensure that an aquaponic enterprise is profitable and sustainable in the local context. Once these families build capacity and prove working models, future projects can include other interested families, knowing that a local technical base can provide support in the future. For humanitarian projects, other micro farming practices (poultry production, vegetable gardens etc.) that are more familiar to the target population will have greater success, as they will demand less capacity building and technical support, allowing for successful implementation within the standard timeline of a humanitarian project (usually less than 9 months).

Finally, the environment and climate must be favourable for aquaponics for at least 9 months of the year. Other technical requirements include: adequate water quality for soilless culture, affordable access to 24-hr electricity (either via the grid or using solar power) and affordable access to key inputs including fish feed, fish fingerlings, simple water-testing kits, seeds, and electric water and air pumps.

Policy Relevancy

The lessons learned throughout this project are applicable in other cities with similar climate, water and soil constraints, assuming affordable access to key inputs. In terms of expansion within Gaza, there is potential considering the high rate of urban sprawl within the coastal enclave driven by the high fertility rate. Yet the main limiting factor is poor groundwater quality, particularly in urban zones, which can negatively affect plant growth to a greater degree in soilless culture systems when compared to soil production. As such, expansion should start in communities in North Gaza, where water quality is adequate.

Simplified hydroponic rooftop systems may be more relevant in Gaza, given the fact that grid electricity is currently available for only 6 hours a day, as they demand less energy compared to aquaponics. Further experimentation with salt-tolerant crops could also allow for expansion into zones with poorer water quality like Rafah and Khan Yunis. Finally, for scaling-up interventions, local non-governmental and community-based organisations and other stakeholders already working in the field of (urban) sustainable agriculture should be involved and trained at the start of the project to provide technical support and to constructively engage with the wider public on new agricultural technologies.

Hong Kong, China – Rooftop Republic at Fringe Club: An Educational Rooftop Garden

Pol Fabrega, Ching Sian Sia

Introduction

The Fringe Club in Central Hong Kong is an organisation that promotes Hong Kong artists through cultural exchange and provides rent-free facilities to young artists. Some years ago, the Kiehl Company created a rooftop garden on top of the heritage building of the Fringe Club to grow herbs and flowers for the production of Kiehl's facial and body products. When Kiehl stopped its gardening activities there, Fringe Club reached an agreement with Rooftop Republic (which also manages rooftop gardens atop of the Bank of America and the Confucius School) to use the rooftop space at no costs (plus free electricity and water supply) in exchange for delivering fresh vegetables and herbs for the Fringe Club restaurant from April 2015 onwards (Fig. 25).

The Farm

The rooftop garden has a gross area of approximately 90 m^2. The garden uses 30 plastic containers filled with organic soil that are lightweight, easy to assemble and customisable to fit any space. As crops are grown within the plastic containers, there was no need to waterproof the rooftop.

The rooftop garden does not use any chemical fertilisers and applies organic farming methods to grow the food crops. Wastes generated on site are all composted, also including coffee grounds and other organic wastes supplied by the Fringe Club's restaurant. A hired garden operator manages the crops. The Fringe Club

Fig. 25 Images of the Rooftop Republic Fringe Club rooftop farm (Credit: Rooftop Republic)

Fig. 26 Conducting a workshop at Rooftop Republic farm (Credit: Rooftop Republic)

supplies water and electricity to the garden as part of its agreement for maintaining the rooftop garden.

The garden grows mainly seasonal vegetables such as Italian basil, mint, lemon balm, Thai basil, cucumber, okra, eggplant, morning glory, Ceylon spinach, chilli peppers, bell peppers, and luffa squash. Most of the produce is supplied to the restaurant. Any surplus products are given to participants of workshops. But the most important products of the Rooftop Republic farm are not the crops grown, but the many workshops, cooking classes, educational tours and even yoga classes that are conducted at the rooftop garden (Fig. 26).

Societal Impacts

Since 2015, Rooftop Republic has conducted 32 workshops on the farm and provided education to approximately 640 people in the importance of local food production and consumption, how to grow and cook their own vegetables, thereby establishing a stronger connection between city dwellers and food.

Economic Sustainability of the Garden

The garden currently generates a monthly income of about HKD 10,000–15,000 from conducting workshops on the rooftop farm, and is currently making a profit.

Lessons Learned

Rooftop Republic experienced that this business model of growing food for a restaurant kitchen in exchange for rooftop gardening space to conduct educational activities has been their best business model as compared with the models applied for their farms atop Bank of America and the Confucius School. Bank of America currently pays the company to maintain the farm and the school had similarly engaged the urban farming start-up but is no longer doing so.

They also learned that, although education is their main objective and source of income, it will also be necessary to improve the production efficiency, e.g. by introducing sensors to monitor the moisture in the soils and to improve the precision of the irrigation.

Policy Relevance

Hong Kong is one of the most densely populated cities in the world and also home to the second highest number of high-rise buildings in the world, right after New York. With urbanisation taking place at a rapid pace, arable land within the city-state is being converted to other uses continually. This has resulted in Hong Kong's having to import 90% of its food supply to feed its population. With major food-exporting countries facing problems such as climate change, natural disasters and food shortage, this may result in Hong Kong's food supply being disrupted, since Hong Kong relies so heavily on food imports. Moreover, since most of Hong Kong's food is imported directly from China, there are also growing concerns on account of the increase in food-safety issues and heavy use of chemical fertilisers and pesticides in China. Rooftop Republic hopes that their workshops and other educational activities will contribute to raising the awareness of citizens and policymakers on the important benefits of supporting local food production rather than importing food of doubtful quality and at high environmental costs from China. However, a recent incident in which the rooftop of a building at City University of Hong Kong collapsed under the load of the rooftop garden, has led the local authorities to review guidelines pertaining to the greening of rooftops in Hong Kong.

Lima, Peru – Small-Scale Hydroponics in Villa El Salvador: The Case of Ms Esther Flores

Alfredo Rodríguez-Delfín

Introduction

In 2001, the Urban Work Programme (UWP) was created under the direction of the Ministry of Labour and Employment Promotion, with the aim of mitigating the effects of the severe economic recession facing the country. The programme's overall objective was to generate (temporary or self-) employment and income in urban areas with extremely high poverty levels. Most of the funded projects involved making roads, sidewalks, stairs, irrigation channels and similar works, but a small number of projects were of the productive type. One of them, called "Hydroponic Lettuce Production", was presented by the Women Popular Federation of Villa El Salvador (FEPOMUVES), a social organisation promoting organisational development and leadership capacities among women, as well as their rights and gender equity in this suburb of Lima.

The aim of the project was to jointly produce lettuce for sale and to use the proceeds to sustain part of the salaries of the women participating in the project, while the main part of their salaries was contributed by UWP. On the roof of four FEPOMUVES community centres, floating-root hydroponic systems were installed using wooden containers lined with black polyethylene sheet and Styrofoam plates (2.5 cm thick) to support the plants and river and quarry sands as the growing media[3].

When the UWP funding ended in 2005, FEPOMUVES decided not to continue the project on their rooftop and accepted the proposal of four of their members to sell to them the hydroponics materials and supplies to enable them to start their own small business on the roof of their own houses. These four members, led by Ms Esther Flores, formed the Crop Hydroponics Company Bio Nutri Verde S.A.C. and installed on the roofs of their homes a hydroponics module with a floating-root system because they saw a potential business in producing lettuce for the market. The Hydroponics Company existed till 2011, but then three of them had to cease their participation because of economic problems.

However, Ms Esther Flores continued her own business and even improved the infrastructure of her garden with the help of a loan from some family members. She changed the Styrofoam plates to PVC corrugated roof plates (see Fig. 27), since the Styrofoam plates frequently broke and their replacement led to higher production costs.

[3]This simplified hydroponics model was developed by CIHNM-UNALM in the context of the FAO-supported "Popular Hydroponic Garden Project" in the early 1990s.

Fig. 27 The rooftop hydroponics farm of Mrs Flores initially with floating-root system and PVC corrugated roof plates (Credit: Rodríguez-Delfín)

Fig. 28 The current farm with NFT system (Credit: Rodríguez-Delfín)

The Current Micro-farm of Ms Esther Flores

In 2014, Ms Esther Flores decided to change from the floating-root system to a modified nutrient film technique (NFT) system (see Fig. 28) developed by the Hydroponics and Mineral Nutrition Research Centre (CIHNM) of the National Agricultural University La Molina (UNALM), after several visits to the hydroponics modules installed at UNALM. To implement the NFT system, locally readily available materials such as PVC pipes 7.5 cm in diameter, PVC fittings and a

1000-litre polyethylene tank to store the nutrient solution were used and a pump with a power of 0.5 HP at a flow rate of 80 litres per minute.

The hydroponics garden now has an area of 130 m^2 divided into two parts: in the larger one (90 m^2) are three modules 12 m long, each with eight growing channels 7.5 cm in diameter; in the smaller one (40 m^2) are two modules 6 m long, each with nine growing channels.

The water to prepare the nutrient solution for the plant production is taken from the potable tap water of the local service. The pump operates ten minutes each hour and is switched on automatically, controlled with a timer; this is enough to oxygenate the nutrient solution stored in the polyethylene tank. Solid crop wastes such as leaves and roots are removed every second day and are picked up by the garbage truck that passes daily in front of the house.

Ms Esther Flores produces only hydroponic lettuce of the crisp and butterhead types and of heat-tolerant varieties, allowing her to produce all year round, even in summer conditions, when the demand increases. Fifteen days after sowing, she transplants the seedlings into a container 40 × 40 × 10 cm under a floating-root system. Another 30 days later, she transplants to the PVC pipes. From planting to harvest, it takes 60 days (50 days in summer conditions). Every 30 days, the plants that are growing in the channels are harvested.

After harvesting, the entire system is cleaned and disinfected to avoid root contamination by *Phytium* (a fungus that appears when hygiene is poor and oxygen is lacking in the nutrient solution) and in order not to lose the "good agricultural practice" certification required by the supermarket chain to which she sells the lettuce.

Results Obtained

The theoretical production capacity of Ms Esther's rooftop farm is 1980 lettuce plants per month but, in practice, she produces around 1680–1780 plants per month (about 10–15% loss due to various management problems).

The marketing and distribution is taken care of by her 22-year-old son, Manuel Flores, who delivers the lettuce every two days to two supermarkets nearby. The proceeds of the lettuce sales, amongst other things, finance Manuel's university studies.

As an initial investment when Ms Esther started her individual business, she obtained a loan of S/ (soles) 3000 (*ca* USD 1000) from some family members. In 2014, when she changed from the floating-root to the NFT system, she borrowed another S/ 3000 from the bank. To pay back the debt and the interest, she had to spend S/ 250 per month over 15 months.

According to Ms Esther, the production cost of a head of lettuce is S/ 0.40 (USD 0.15) and she sells the lettuce to the supermarket at S/ 1.20 (USD 0.40). When there is overproduction of lettuce, she sells at S/ 0.80 per head to her neighbours. On average, the gross income obtained from lettuce sales is S/ 1500 (USD 455) per month. Unfortunately, the supermarket pays only every 40 days which occasionally creates cash problems for Ms Esther.

Lessons Learned

Among the lessons learned by Ms Esther Flores is that the initial floating-root system had high running costs because the Styrofoam plates had to be replaced often and the system was less efficient (higher use of water and nutrients compared to the NFT system). Another disadvantage that she realised with the floating-root system is that it takes much time to wash and disinfect the containers and Styrofoam plates to start a new production cycle.

Ms Esther always tries to improve her lettuce production and often visits the Hydroponics Module at UNALM to see something new. Recently, she saw the new NFT cascade system that is being used there (see Fig. 29) and she is planning to modify her NFT system into a cascade type. In this way, she may increase the number of plants in the same roof area of 90 m^2 from 1980 to 2340 plants per month.

The case of Ms Esther Flores illustrates that hydroponic systems can be applied in poverty-stricken city suburbs in developing countries as a micro-enterprise that generates income and employment if simplified low-cost but productive hydroponic systems using natural and locally available growing media are used and if participants with an entrepreneurial attitude and strong motivation and discipline can be identified.

Fig. 29 The planned NFT cascade system (Credit: Rodríguez-Delfín)

Melbourne, Australia – Fed Square Pop-Up Patch: A Small-Scale Commercial Allotment Rooftop Garden

Mat Pember, Henk de Zeeuw

Introduction

Fed Square Pop-Up Patch is located on the rooftop of the Federation Square car park in the Central Business District of Melbourne, Australia. The garden takes up approximately 1000 m^2 and was officially opened on 12 October 2012 as a joint venture between the land owners, Federation Square, and Little Veggie Patch Co. (LVPC), a small business dedicated to helping people grow food in small vacant spaces in the city. The initial agreement with Fed Square was for 12 months and, since then, it has been renewed on a 6-month basis.

The Fed Square Pop-Up Patch project provides local residents and businesses the opportunity to grow food in a subscription-based model (Aus$ 108 per month, about €70) that entitles them to the rental of the 1.5 m^2 growing space prepared by LVPC; the rent also includes free seeds and seedlings, pest and disease control and hands-on advice/ education. There are two types of subscribers: domestic subscribers who grow their own food and restaurants/cafes in the precinct that grow food for use in their menus. If someone rents more than one container, the rent is lowered to Aus$ 80/month. The rooftop garden also organises educational workshops and hosts events (Fig. 30).

Fig. 30 Bird's eye view of the garden during an event (Credit: Mat Pember)

The Design

The 140 containers are built out of recycled apple bins: non-treated wooden crates that measure 1.2 × 1.2 × 0.73 m. As a result of the loading restrictions on the rooftop, we used Styrofoam pods to take up half of the crates' depth and then filled the top half with an organic growing mix: mostly compost, along with potting mix, pea straw, slow-release fertiliser and rock dust (Fig. 31). At the time of installation, we investigated the opportunity to collect rainwater from the rooftop to water the gardens; however, the cost was prohibitive for a 12-month project. The crates therefore need to be watered by hand – an onerous task in the warmer months. The nature of the rooftop – a hard concrete surface – makes drip irrigation difficult since it would need pipes running all over the space, and visitors walking through the garden might fall and eventual sue the garden for the resulting damage.

On the roof garden, there is a shipping container that doubles as a small garden shop cum information centre. During the event season, we recently installed a 15 × 15 m marquee on the roof to provide shade. The temporary nature of the license (6 months) makes investing in infrastructure very challenging.

Production and Distribution/Commercialisation Practices

Domestic subscribers tend to grow a large variety of vegetables and herbs, which they consume at home. However, the produce seems to be almost a by-product of the gardening. The number one reason why residents take up plots is for the social and health benefits of gardening and having a green open space to unwind in.

Fig. 31 The growing containers at the opening in 2012 (Credit: Mat Pember)

Restaurants, on the other hand, focus on growing mostly garnish produce; edible flowers, rarer greens and vegetables that are often expensive and difficult to source.

The Societal Impacts

The Pop-Up Patch has been recognised widely for its social impacts, and these have been the main reason for continuing beyond the original 12-month period. In 2014 and 2015, the project was a finalist in the City of Melbourne Awards for contribution to the community, and the space has become an important community conduit, connecting inner-city residents. It is ironic that, in the city, where so many people live, work and play, it is hard for residents to find meeting places and common threads. The garden space has become that for so many.

It has also become a refuge for inner-city workers, particularly for the kitchen staff of the restaurant members. The common practice is to send apprentice chefs/kitchen hands down to the patch to learn about the produce, but also to allow them to escape the hectic and often stressful kitchen environment.

The place also has become a testing ground for further innovations and educates the public on how to grow food in urban environments. Currently, we are testing wall gardens, wicking beds, aquaponic systems and a variation of soil mediums and growing strategies (Fig. 32).

Fig. 32 Kid's workshop in Pop-Up Patch (Credit: Mat Pember)

Economic Sustainability

The initial investment in the space was split by Fed Square and LVPC, both contributing AUS$ 40,000 for setting up the garden. Twelve months later, both contributed another AUS$ 20,000 and, since then, LVPC has invested another AUS$ 40,000 in physical costs. The main cost of the garden has been the labour of maintaining it. For the first three years, we had fulltime staff onsite, at a cost of AUS$ 60,000–70,000 per annum, but we have since scaled this back slightly. There are a few forms of income: subscription of the plots, small gardening shop/coffee, workshops and events. About 80 (of the 140 plots) are rented out permanently; in the top year, about 90% of the plots were rented out.

Despite being a for-profit enterprise, the Pop-Up Patch has run only at breakeven for most of its time. The major opportunity for its economic sustainability has always been the hosting of events/parties. However, since the original concept involved, our business supporting the tenants of Fed Square (one of them being an events business itself) meant that we needed to tread carefully when exploring commercial opportunities. It was only last year that we were finally allowed to explore a 6-month trial of events by collaborating with a catering partner Tommy Collins. The events proved successful and showed where the true commercial potential of the garden lies, although we can say that the story of the garden's members, along with the environmental and sustainability messaging, all have added to the appeal and the initial success of the events.

The 6-month license for the space at a time makes investment difficult to obtain or commit to. We are currently attempting to secure the space for a longer term. However, we are now in a position where we may fall a victim to our own success. The commercial opportunity that has been opened up through the events has highlighted to our landlords, Fed Square, that there is a commercial outcome possible from the space, too. As a result, the space is currently been open to public tender.

Main Lessons Learned in Establishing and Running This Rooftop Garden

The project has been a four-year lesson and continues to teach us. If we set up a comparable garden in future, we would definitely seek a space with a longer-term rent and be assessing all the important elements of such a space, from community engagement, to sustainability, commercial outcomes and basic garden infrastructure.

A long-term approach would allow us to find sponsorship and investment. It would make any future garden more environmentally sensitive and sustainable, in all senses of that word. It would help to future-proof it, by allowing us to pursue not only social and environmental outcomes but also commercial ones – these being essential if any project is to survive in the long term.

The concept of the garden has evolved and, if we continue with the space, we would give more emphasis to organising events in the rooftop garden. Also, we would not staff it fulltime, since the members we have there now are fairly autonomous and, other than a stock-up of seeds, seedlings and materials, they no longer need the assistance of a staff member.

Another learning point is the high turnover rate among the member-gardeners. Many people get excited by the concept but underestimate the work involved in caring for a garden and, since most of the members have busy lives, many end their membership after some time.

Policy Relevance

The Fed Square Pop-Up Patch proves the value of productive green spaces in the urban environment. It shows that cities are appealing to live in only if there are green meeting places built into it. Rooftop spaces are plentiful and the relevance of having green roofs are such that the City of Melbourne is preparing a new policy, and our space is an important project for shaping it.

Rooftop spaces are challenging for a number of reasons: accessibility, the harsh conditions and cost of construction, but they also present obvious opportunities. The tangible outcome of food is only half the story; it is the intangible outcome of growing food in the city that grounds us, helps to connect people through a common thread (our food culture is one of our strongest links) and, perhaps most importantly, provokes questions about how we could be doing things better in the city.

Montreal, Canada – Culti-Vert, the Productive Green Roof of Palais Des Congrès

Amelie Asselin, June Komisar

Introduction

Montreal's convention centre – Palais des congrès de Montréal – was built in 1983 and expanded in 2002, and is architecturally distinctive with multi-coloured glazing on the new façades. With a 32,000 m^2 roof, the Palais des congrès was retrofitted to reduce its environmental impact. To reduce the roof's heat-island effect, absorb rainwater and contribute to greening the city, the centre invested in converting part of the roof terraces to a productive green roof in 2010. The project was expanded in 2011 (Fig. 33).

Fig. 33 Views on and from the rooftop garden (Credits: *left*: Joe Nasr; *right*: Culti Vert)

Design of the Rooftop Garden

The Culti Vert project has three components:

1. *A productive garden* of 490 m^2 consisting of 450 plastic garden bins that produce vegetables and herbs and greens and 11 arches with grapevines, managed by Capital Traiteur, the food service provider of the Palais des congrès. Three different planter types were chosen:
   - Biotop (long, narrow sub-irrigated, semi-hydroponic plastic planters that hold 10 litres of water each): the planters can be linked together and fitted with watering hoses that enable timed irrigation (Fig. 34);
   - a more cubic-shaped sub-irrigated planter, designed by Alternatives, that enables the cultivation of larger crops (an instruction kit available on line enables people to construct similar do-it-yourself planters);
   - geotextile Smart Pots, soft-sided garden pots that let the soil and roots aerate, are quite lightweight and help prevent the overheating of plants that is sometimes an issue on roofs during hot summer days.
2. A *green-roof area* of 1390 m^2 planted with attractive vines that climb on arches, stonecrop plants (such as sedum) and grasses. In this area, five different green-roof technologies were applied (Fig. 34).
3. *Beehives.* Miel Montréal, a local beekeepers cooperative, is paid to manage the three beehives on the roof, and the honey produced is used by the Palais des congrès and its caterer Capital Traiteur, mostly as promotional gifts.

Production and Distribution Practices

The productive garden is managed by Capital Traiteur, assisted by volunteers who help to care for and harvest the vegetables, fruits, herbs and other diverse plants including vegetation chosen for use in a local workshop to dye fabric, medicinal herbs for an herbalist training centre as well as heritage vegetables that continue the

Fig. 34 The biotope-type planters (*left side*) and the green roof vegetation (Credit: Joe Nasr)

propagation of older species of fruits and vegetables, such as Montreal melons and lemon cucumbers. The first year, about 650 kg of fresh products was produced.

Capital Traiteur provides the harvested fresh food products to the restaurant of the convention centre. In addition, the restaurant Osco of the Intercontinental Hotel and Crudessence, a raw food restaurant, receive food grown in the Culti Vert rooftop garden.

Economic Sustainability

While the initial investment was from the Palais des Congrès itself, a grant from the Québec Department of Health and Social Services' Climate Change Action Plan and the Montréal Urban Ecology Centre helped the project expand in its second year. The main goal of the rooftop garden was not to be an economically self-sustaining productive roof, but to enhance the environmental sustainability of the building as part of the green-roof project, which also serves as a model for eco-education and local production of fresh food.

Societal Impacts

The green roof has reduced the temperatures on the roof and in the building and reduced rainwater runoff from the building. The green roof also contributed to making the Palais des congrès a widely recognised energy-efficient building.

Since its creation, the rooftop garden has been educating the public through tours provided by UQAM (Université du Québec à Montréal), which introduce visitors to container vegetable gardening and also expose them to a variety of extensive green-roof systems. Also the volunteers learn about gardening through their hands-on labour.

Further community outreach included growing some food for La rue des femmes, a non-profit organisation that helps women in distress and homeless women in Montreal.

The green roof also contributes to maintain culinary heritage, biodiversity and pollination.

Policy Relevancy

This Culti-Vert project shows what can be done when roofs are designed to withstand the weight of a growing medium and water and designed to accommodate gardeners, containers and a host of visitors. The ability to maintain a comfortable temperature in rooms below the roof seems to be an advantage. The field needs some longitudinal studies to be able to understand the extent of economic benefits of productive green roofs on the daily operations of the building below.

While not many urban roofs are as large as the Palais des congrès roof, when planning large buildings such as convention centres, office buildings, university buildings and the like, the roof can be engineered to accommodate container gardening. Best practices would include a water source via a hose bib, an area for tools and another area for shade and safe, high parapets for gardeners.

Sao Paulo, Brazil – The Eldorado Shopping Centre Rooftop Garden

Ricardo Omar, Sergio Eiji Nagai, Henk de Zeeuw

Introduction

Shopping Eldorado, a big shopping mall in the centre of Sao Paolo, initiated in 2011 a sustainability programme aiming to reduce the use of resources and enhance the recycling of waste materials. In 2012, as part of this programme, a rooftop garden was created that, in the following two years, was extended to 5000 m^2.

Fig. 35 Two steps in composting the wastes from the restaurant in Eldorado Shopping Centre (Credit: Shopping Eldorado)

Design of the Garden and Production Practices

Ecological production methods are applied to produce vegetables and aromatic and medicinal herbs in plastic boxes of 25 × 50 cm each.

The growing substrate in the boxes is compost produced in the cellar of the shopping mall, making use of a composting unit that processes about 1000 kg organic wastes per day collected in the restaurant in the shopping mall. The composter was designed by the technical firm Korin Meio Ambiente (KMA), making use of two enzymes and heating to accelerate the composting process to last only one day (Fig. 35).

In the garden, vegetables and fruits (lettuce, cabbage, tomato, aubergine, lady's finger, courgette, cucumber, pepper etc.) are produced in the boxes and raised beds, and aromatic and medicinal herbs (parsley, watercress, ginger, anise, mallow, sage, rosemary, lavender, basil etc.) in the pots. The garden is managed by an agronomist from Agro Garden Company Ltd, who also supervises the two assistants that operate the composting unit. The garden is irrigated with collected rainwater. The products are distributed among the employees of the shopping mall (Figs. 36 and 37).

Economic Sustainability of the Garden

The total investment in the composting and garden project during the first year was around USD 100,000 including the composting machine, preparation of the roof, laying out the garden, irrigation equipment, garden management and staff for the composting and gardening, and materials.

The annual maintenance and operation cost of the composting and gardening project is largely compensated by the now lower costs of waste disposal to the municipal landfill and lower energy costs for cooling the shopping mall.

Fig. 36 Bird's eye view of the "Telhado Verde" of Eldorado Shopping (Credit: Shopping Eldorado)

Fig. 37 The various growing systems applied (boxes, raised beds, pots) (Credit: Shopping Eldorado)

The Societal Impacts of the Garden and Policy Relevance

The rooftop garden and related organic-waste recycling have various important societal benefits**:**

- Reduction in the amount of food wastes deposited in the landfill, reduction in energy use related to transporting the wastes and reduction in methane emissions from the landfill
- Reduction in the temperature on the roof and in the building below, which reduces the amount of energy needed for cooling and refrigeration in the shopping mall
- Contribution to reducing living costs and improving diets of employees of the shopping mall by supplying fresh nutritious food and medicinal herbs to them (who are mainly in the lower-income categories)
- Creation of some additional jobs (in the composting and gardening work)
- Contribution to raising ecological awareness among schoolchildren (about 1200 visit the garden each year), students (doing their practical training or theses research in the garden), staff, consumers and similar enterprises.

The model applied by Shopping Eldorado can be easily applied by many other malls, restaurants, hotels, agro-industries and similar enterprises, which together could make a substantial impact on reducing urban-waste disposal, energy use and greenhouse gas emissions, while enhancing closed cycles, nutrient recovery and use (also reducing the use of energy and scarce resources involved in producing chemical fertilisers) and improving nutrition and health of their employees.

Singapore, Singapore – Spectra Edible Learning Rooftop Garden

Lyvenne Chong-Phoon, Ching Sian Sia

Introduction

As a highly urbanised city-state, Singapore has less than 1% of land area dedicated to farming and only half of that is used specifically for local food production. As a result, many young Singaporeans do not have any idea where the food they consume comes from, or how food is grown and harvested. This is a worrying trend evident in several developed countries where many children have seen food products only on supermarket shelves.

The Design of the Spectra Rooftop Garden

Spectra Secondary School is located in the northern part of Singapore. It is a specialised school that began accepting students in January 2014, admitting students 13–17 years of age who are eligible for the Normal (Technical) course, meaning that they are not academically inclined and are offered subjects that are more technical than theory based. During the design of the school, the architect of the building had proposed a rooftop garden. In 2013, the school building was completed and a rooftop garden with 11 concrete raised beds measuring $7 \times 2 \times 0.35$ m and $5 \times 2 \times 0.35$ m was installed. Geotextile is used to line the concrete raised beds and improve the drainage of the beds, which are filled with soil.

The 180 m^2 rooftop area is used for growing crops, with the main objective being to educate the students about where food comes from and how it is grown. The rooftop farm plays an important role in the school curriculum, which incorporates both academic and vocational learning by students (Fig. 38).

Fig. 38 *Left*: view of the Spectra school garden with raised beds; *right*: the rainwater-harvesting system (Credit: Lyvenne Chong-Phoon)

Production and Commercialisation Practices

The rooftop farm adopted a soil-based organic-farming approach to cultivate its food crops. Brinjals, winter melon, lady's finger, long bean, sugarcane, lettuce, water spinach, Chinese cabbage, sweet potato leaves, local lettuce, red spinach, moringa leaves, Thai basil, holy basil, chilli padi, lemon grass, turmeric, Roselle, white radish, bitter gourd, cucumber, French beans, bananas and papayas are grown on the rooftop farm. To water the plants, a rainwater-harvesting system is in place with two 250-litre water tanks to store collected rainwater that is distributed to the crops by a drip-irrigation system with a humidity sensor that will automatically stop irrigating crops for 24 hours if it has rained. To improve the soil quality, eight compost bins are used to compost organic wastes from the garden, and mulching is done to improve soil moisture.

In 2015, the school engaged Edible Gardens – an urban farm management consultancy firm – to manage the rooftop farm and run workshops to teach students different aspects of farming. After a year, the school took over from Edible Gardens and began running the workshops on its own, with volunteers – mainly relatives and residents living nearby – helping out on weekends.

A farmers market is held at the school every 3 months when students harvest their crops. Students help harvest, pack and sell the vegetables at a price slightly below the market rate, which makes it more attractive for the public to purchase them directly from the school (Fig. 39).

Societal Impacts

The key concept of the rooftop garden is "No one owes us a living – We work hard to put food on the table". Students are taught to work hard and be self-reliant and self-sufficient rather than depend on others.

Apart from learning the English language and some science through garden-based activities, students are taught societal values such as helping those in need, being responsible, working in a team, learning where food comes from and appre-

Fig. 39 Students harvesting the produce (Credit: Lyvenne Chong-Phoon)

ciating the amount of effort that goes into growing one's own food as well as to reduce food wastage. This is important, as most children have no idea where food comes from and are less likely to appreciate the food they consume on a daily basis.

Proceeds from the farmers market are used to assist students from underprivileged families. As 50% of the students from Spectra Secondary School receive government support for their study, students are instilled values to help others or to help themselves through the sales of the harvests at the farmers market. In 2015, more than SGD 2000 (€ 1330) was raised from the farmers markets held at the school to fund the school's needy students.

Economic Sustainability

Under the Skyrise Greenery Incentive Scheme by the Singapore National Parks Board (NParks), the government funds up to 50% of the costs to establish a green roof. The Spectra Secondary School also tapped into funds provided by Ministry of Education (MoE), which subsidised the installation of the rainwater-harvesting and drip-irrigation systems under the MoE Innovation Funds. The Spectra Secondary School's rooftop farm receives help from NParks and the Agri-Food & Veterinary Authority of Singapore (AVA) through providing seeds, training for gardeners and tips on how to improve the existing garden. To start the farm, SGD 3000 (€ 2000) was spent to purchase 70 potted plants. In the first year (2015), Edible Gardens was paid SGD 2000 (€1330) monthly. Since early 2016, the garden is managed by the school with help from the teachers and volunteers.

Lessons Learned

The architect and builders had little knowledge of what soil would be needed to grow crops and the contract did not specify this either. This was the reason why the raised beds on the rooftop were filled with clay with very low permeability, resulting in many problems such as root rot. Measures were then taken to improve the soil quality and porosity by removing a large amount of clay and mixing compost and sand into the soil.

Growing crops on exposed rooftops also implied that the rooftop farm was susceptible to pest infestations. To keep the farm organic, neem oil was used instead of pesticides to treat these infestations. Also the students were deployed to remove caterpillars whenever there was an infestation, which was effective and the students also enjoyed it.

Policy Relevance

The Spectra Secondary School rooftop farm shows that school gardens can play an important role in language and science training as part of the curriculum, as well as in enhancing students' (and parents') awareness of where food is coming from and how it is grown.

The school garden also plays a role in encouraging values like to work hard, be self-reliant and self-sufficient and in enhancing social responsibility (by supporting underprivileged students with the proceeds from product sales).

Singapore, Singapore – ComCrop, a Commercial Aquaponics Rooftop Farm

Allan Lim, Maria Lloyd, Ching San Sia

Introduction

Comcrop rooftop farm is a for-profit but social enterprise located on the rooftop of Scape Mall in the heart of Singapore's shopping district Orchard. The rooftop farm was set up in 2014 in partnership with the Scape Mall management and is Singapore's first commercial rooftop urban farm. Comcrop was able to build its farm on the rooftop of the Mall without having to pay any rent to them, as Scape Mall has the objective to support small businesses and provides working space for many start-ups in Singapore.

The Design of the Farm

The Comcrop rooftop farm occupies 372 m^2 which is about half of the total available space on the rooftop of the Mall. Aquaponics was adopted as the method of production: 12 vertical A-frames for growing vegetable crops are each connected to

Fig. 40 The A-frames and fish tanks of ComCrop (Credit: Ching Sian Sia)

a tank measuring 1.6 × 0.8 × 0.8 m to rear tilapia fish. Apart from the 12 aquaponic units, there is also an area set aside for growing specialities in approximately 100 pots in order to educate the public on what plants can be grown in Singapore. The energy and water are sourced from the building as part of the deal (Fig. 40).

Production and Commercialisation Practices

The Comcrop Farm focuses on high-value crops to differentiate itself from the conventional local farms. In the A-frames, crops such as Italian basil, peppermint, wasabi greens, lettuce, mizuna and heirloom tomatoes are produced. In the pots and raised beds, specialities such as habanero pepper, ghost peppers, chilli, rosemary, Indian Glass Gem corn, beans and yellow pear tomatoes are grown.

The use of the vertical A-frames and the combination with fish production allows growing vegetables and fish at a much higher yield compared to conventional farming systems. The vegetables receive nutrients contained in the water with fish wastes circulated from the tilapia fish tanks, thus reducing the need to fertilise the crops with inorganic fertilisers as well as reducing the amount of fresh water needed to irrigate. As some crops require more nutrients than others, some fish tanks contain more tilapia fish than do other tanks. Also crops at the fruiting stage may require more nutrients. In addition to nutrients from fish wastes that are rich in nitrogen, micronutrients and potassium are also added to supplement plant growth. The produce is supplied to 30 bars and restaurants around Singapore.

Societal Impacts

Some ten volunteers assist in the farm work on weekdays; four of them are retired elderly women. On weekends an additional 15–20 volunteers come to work on the farm, since they like to do gardening for leisure. Comcrop is also in a trial period with the Movement for the Intellectually Disabled of Singapore (MINDS) involving ten intellectually disabled individuals in the seeding of the basil plants. Eventually, the company will hire some of these persons.

Workshops are conducted for schools that would like their students to learn more about how food is grown and harvested. The number of visitors varies on a monthly basis; 4–5 tours are conducted each month during busier periods. This is important for the many children who grew up in Singapore in a very urbanised environment and have limited knowledge of where their food comes from. Comcrop also plays a role in educating the public about the importance of urban food security and the possibilities for sustainable urban food production through utilising building rooftops for agriculture.

Economic Sustainability

The Comcrop farm was established with a private investment of SGD 300,000 (€198,000) with no subsidies from the government. A Special Employment Credit Scheme that encourages employers to employ disabled individuals is paying ComCrop up to 16% of the monthly salary for each disabled individual working with ComCrop. As of July 2016, Comcrop is making a profit only two years after having set up its rooftop farm on Scape Mall. It is now planning future investments to further enhance production efficiency.

Lessons Learned

An enclosed greenhouse would have been ideal to keep out pests, as birds feed on crops produced and pest infestations wipe out crops.

Automation processes will be introduced to monitor the water needs of crops and to regulate irrigation, as crops are exposed to a lot of heat and sunlight on the rooftop and tend to dry out quickly. With limited availability of labour in Singapore, there needs to be more focus on automation to improve productivity of the farm.

Marketing of produce from the farm needs to be improved to generate more interest within the community in urban farming. Comcrop holds many educational tours to create awareness about environmental sustainability, sustainable food systems, supporting the local food movement and food-safety issues. These tours educate people on problems with modern-day food systems and potential solutions. In turn, it may also influence building developers and architects in the way rooftops are being designed, so as to accommodate productive green roofs, thus encouraging more rooftop farms to be set up in the city.

Policy Relevance

Singapore is a highly urbanised city-state in Asia. With a land area of 720 km^2, it currently accommodates 5.54 million people and is targeted to reach 6.9 million people by 2030. This will mean a rise in demand for residential land to accommodate more people in an already high-density city-state. Despite an increase in land area through land reclamation since the 1960s, the area of farmland has been steadily declining. Self-sufficiency in terms of food has also reduced drastically because of rapid urbanisation and conversion of farmland to other land uses. Singapore currently produces only 10% of its own food, mainly fish, vegetables and poultry, in six Agrotechnology Parks. To secure a steady food supply, it heavily relies on food imports from other countries including Malaysia, Indonesia, Thailand, Vietnam, the Netherlands, India and China. Against this background, it is understandable that more and more initiatives are taken to start growing food on rooftops of buildings.

Although urban farming was initially not recognised by local authorities and despite a lack of policies to guide the establishment of urban farms in Singapore apart from land-use policies, there has been a steady increase in the number of urban farms on rooftops in the last years. This has led to an increase in awareness of the benefits of local food production.

Recently, a joint-ministerial taskforce in the Government was established that works with Comcrop and a number of other urban farm operators in Singapore to facilitate the setting up of urban farms in Singapore. Comcrop is expanding its operations on a rooftop in northern Singapore as part of an industrial transformation exercise in a food industrial park.

The Hague, The Netherlands – UrbanFarmers de Schilde: A Commercial Aquaponic Farm and the Largest Rooftop Farm in Europe

Shuang Liu, Henk de Zeeuw

Introduction

UrbanFarmers second farm (the first is in Basel, Switzerland) stands at 40m high on top of the building De Schilde, a former Philips telecommunications factory at a prime location close to the city centre of The Hague, the political capital of the Netherlands.

In autumn 2012, the Municipality of The Hague launched an Urban Agriculture Initiative to identify potential urban farming tenants for the De Schilde building's vacancies. February 2014 UrbanFarmers AG (UF), a Swiss-based technology company specialised in building and operating commercial food production units in

Fig. 41 The De Schilde building (Credit: Martijn Zegwaard)

cities, was awarded the rooftop and the 6th floor of the De Schilde building for their operations. In July 2015, UF secured full financing from two lead investors, and construction was initiated soon after, involving Dutch specialised firms Priva BV, Koppert Biological Systems and Rijk Zwaan. Production started in April 2016.

UF De Schilde is the first-mover of a series of new tenants re-purposing the building into a multi-story urban farming hotspot, following the Urban Agriculture Initiative by the Municipality of The Hague (Fig. 41).

Design

UF De Schilde includes on the rooftop a 1200 m^2 greenhouse for specialty vegetables and a 120 m^2 visitors' greenhouse. The 6th floor houses a recirculating aquaculture system (400 m^2) as well as a showroom and reception area dedicated for tours and events and other amenities.

UF's core production technique is aquaponics, whereby the fish-farming system discharges wastewater containing nutrients to the hydroponic unit as organic fertiliser and irrigation water for the plants and – after cleaning – is recirculated to the fish-farming unit.

UF De Schilde farm consists of two interconnected parts:

1. The **greenhouse:** The greenhouse envelop includes the hydroponic system with NGS (New Growing System), pumps and sensors for optimal water management and irrigation, and the interior electrical installations for shading, climate control and ventilation (Fig. 42).

2. The **recirculating aquaculture system:** This includes the fish tanks with drum filter, bio filter, UV and other disinfection equipment, the pumps and the aquaculture sub-controller (Fig. 43).

Fig. 42 The rooftop greenhouse (Credit: Martijn Zegwaard)

Fig. 43 The 6th floor with fish tanks (Credit: UrbanFarmers)

The key production processes, such as fish feeding, nutrient dosing and water circulation, are fully automated. The production processes are continuously monitored and recorded with help of the UF Controller: a LabView software protocol that connects all major farm sub-systems such as filters, pumps, timers, actors and sensors, enabling real-time performance monitoring and data analysis by the farm operator (with help of an operator dashboard) and cloud-based storage of data regarding harvests, delivery, water management, fertiliser addition and quality management.

Smart integration with the HVAC (Heating, Ventilation, and Air Conditioning) systems of the building allows for efficient use of energy and waste heat.

The rooftop greenhouse was extended to the end of the previously open balcony of the 6th floor to be able to produce more food. A small elevator was built connecting the rooftop and the 6th floor with the ground level to facilitate transport of inputs, produce and people from and to UF De Schilde.

Production and Commercialisation Practices

UF De Schilde expects to produce 50 t of rooftop vegetables (specialty varieties of heirloom tomatoes, chillies, herbs, salads and micro-greens) and 20 t of fresh fish (tilapia) each year. The production can serve about 900 families each week. Fish health is ensured through lower stocking densities and better water quality than in common aquaculture facilities, resulting in no need for antibiotics. No pesticides or herbicides are applied; only organic pest control is used.

UF De Schilde targets to sell its products to various local quality restaurants (20%) and directly to households (80%) through the "UF Fresh Weekly Basket" home-delivery scheme. Products will be delivered on the same day these are harvested and processed.

Economic Aspects

UF De Schilde required an initial investment of € 2.7 million to cover the development costs and the ramp-up costs, assuming a 6-month period after commissioning to reach cash break-even. The farm is expected to achieve an internal return on investment (IRR) of 9.1% with payback in 9.2 years (based on operating income as the most suitable cash proxy), assuming all equity financing.

(Expected) Societal Impacts and Policy Relevance

The aquaponic technology results in significant water savings, which are estimated to be up to 90% of the water use in systems that are not connected (actual savings depend on the selected crops and fish species). With the recirculation of fish waste into productive hydroponic growing, plants need less nitrogen-phosphorus-potassium (NPK) fertiliser, and no pesticides or herbicides are required in environmentally controlled hydroponics, resulting in a lower ecological footprint for food production. Additionally, the system has very little chemical inputs and therefore provides a "zero-residue" approach to farming (at par with or beyond most organic standards), highlighting health and food-safety benefits. Production is also closed-loop and therefore generates very low amounts of production wastes (we are still looking for a solution that enables to also re-use the harvested tomato plants on the farm, e.g. as packaging materials).

Toluca, Mexico – A Small-Scale Commercial Rooftop Greenhouse

Gloria Samperio Ruiz, Alfredo Rodriguez-Delfín

Introduction

The Mexican Hydroponics Association (MHA), a non-profit organisation based in the city of Toluca, aims at promoting hydroponics among small and medium-sized enterprises in Mexico through training courses, seminars and conferences. In order to enable practical training in hydroponics, the president of the MHA established a greenhouse on the roof of her house, of which the first floor is used for training workshops in hydroponics.

Design

The polycarbonate greenhouse measures 13 × 6 m. A computer and sensors regulate the heating and cooling, ventilation, irrigation and moist management in the greenhouse.

The greenhouse has three tanks of 1100 l fed with water from the public service and three small tanks to recover the recycled nutrient solution. The nutrient solution is monitored with electrical conductivity and pH metres. The drip-irrigation system is programmed with a timer.

Three types of hydroponic systems are applied in the greenhouse:

1. a floating-root system consisting of three units of 1.4 × 1.4 m with a production of 170 plants every 2 months;
2. a nutrient film technique (NFT) system with two units, each with ten PVC pipes 10 cm in diameter and 6 m long; each growing channel has 30 cavities, which allows a production of 600 heads of lettuce every 70 days from sowing to harvest;
3. polyethylene black bags with perlite as growing medium and drip irrigation (Fig. 44).

Fig. 44 Ms Gloria Samperio in her rooftop hydroponics training garden (Credit: Gloria Samperio)

Production and Commercialisation

In the greenhouse, different vegetables are produced for sale (mainly lettuce of the butterhead type) in the NFT and floating-root systems and peppers and tomatoes in the polyethylene bags with perlite and drip irrigation.

The NFT production consists of three stages: (1) seedling, (2) first transplanting and (3) final transplanting. The first transplanting is done in a container built of wood and covered with polyethylene film; the watering is by sprinklers and the nutrient solution is recycled. When the lettuce plants have the ideal age for the final transplanting, they are transplanted into the NFT growing channels. The production (five harvests) reaches up to 3000 heads of lettuce per year with an annual sales value of USD 2700 (€ 2400).

The floating-root system (five harvests) produces 850 plants/year with an annual sales value of USD 807.50 (€ 717). Hot peppers and tomatoes are grown in the growing media with the drip-irrigation system: 12 yellow hot pepper perennial plants that produce 96 kg of peppers per year with an annual sales value of USD 480 (€ 426), and 65 tomato plants that produce 910 kg/year with an annual sales value of USD 910 (€ 808). The vegetables are sold to five hospitals in Mexico City. Also tulips and rainbow flowers and lemon cedar plants are produced in polyethylene bags with perlite with an annual sales value of USD 1166 (€ 1036). Also 117 cedar lemon trees are produced each year with an annual sales value of USD 1053 (€ 936) (Fig. 45).

Chemical fertilisers are applied. To avoid the use of pesticides, the greenhouse has yellow and blue traps smeared with entomological glue.

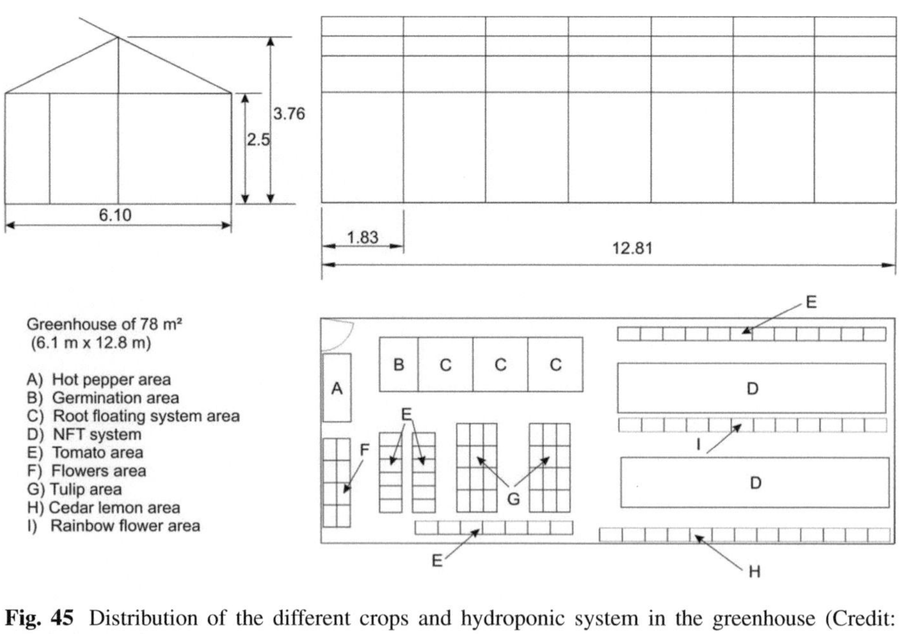

Fig. 45 Distribution of the different crops and hydroponic system in the greenhouse (Credit: Gloria Samperio)

Fig. 46 The rooftop greenhouse established by one of the trainees, Daniel Hernández Campusano, Colonia San Miguel, Toluca (Credit: Gloria Samperio)

Economic Sustainability

The sale of all crops allows an annual income of USD 7116 (€ 6322), while the annual expenditures come to USD 2600 (€) including the salary of an assistant for two days a week; seeds, electricity, fertilisers, contingency, handling losses and amortisation of the installation. This leaves an annual profit of USD 6022 (€ 5350).

Societal Impacts

The different growing systems demonstrated in the rooftop greenhouse and the hands-on training workshops organised there, the book *Hidroponia Basica* (*Basic Hydroponics*) that Gloria wrote based on her experiences, and the conferences organised by the MHA have contributed a lot to the spread of hydroponics in Mexico, including the simple greenhouses established by many poor families on the roofs of their houses to produce vegetables for self- consumption and sale of surpluses. But also some of the trainees have gradually built up commercial units (Fig. 46).

Policy Relevance

The experience shows that also small-scale rooftop greenhouses can generate a profit and be sustained over time as a commercial enterprise. Hands-on training by an experienced rooftop greenhouse farmer seems to be an important pre-condition for successful replication of popular hydroponics.

Toronto, Canada – Ryerson Urban Farm, Toronto: An Educational Rooftop Garden

Arlene Throness, June Komisar

Introduction

Ryerson Urban Farm began at Ryerson University's downtown Toronto campus on one small street-level plot of land. The project, begun by an informal student-led group, was meant to engage the university community, and was enabled by many volunteer hours and small grants from the University and the Faculty of Engineering and Architectural Science. Early on, a partnership with the Ontario-based Vineland Research Centre introduced exotic world crops (including Chinese long eggplant, okra and daikon radishes) to this small, highly visible garden. After this initial success, the group of students and one of the founders, by this time a recent graduate, obtained more in-ground plots, and finally they acquired an existing green roof on top of the university's four-story engineering building. With additional support from the University and the University Food Service, they established a small rooftop farm and hired a farm manager. Faculty and associate advisors from the Ryerson Centre for Food Security, and their partnership with Food Services, helped keep the project moving forward.

In the second season of the rooftop farm, the volunteers and farm manager extended the farm to almost the entire roof – over 900 m^2 of soil beds (Fig. 47).

Design and Construction of the Garden

This conversion from an extensive green roof of beds planted with daylilies to a farmed rooftop with planting beds in rows came with some challenges. On one hand, the roof composition, designed for direct planting, was perfect for a rooftop

Fig. 47 Views of Ryerson Urban Farm (Credit: Joe Nasr)

farm. Existing hose bibs that access the potable city water were already provided for the initial green roof, and this made a new drip-irrigation system easy to install. On the other hand, some site modifications were required. The addition of five cm of soil to the 15 cm soil bed of the initial green roof created deeper beds for crops. Furthermore, by shifting soil to make walking paths between the beds, they could maintain the same overall roof load and mound the rows about 25 cm high. Drip-irrigation hoses were then placed along the mounded rows and supplemented with hose watering by hand on hot sunny days.

Production and Commercialisation Practices

The rooftop garden is run by an experienced young urban farmer, Ms Arlene Throness, with help from a few interns that work over the summer months and a host of volunteers that take part in shifts and, in turn, learn by doing and receive weekly a bag of produce.

The farm had a yield of over 3600 kg of vegetables, herbs, greens and fruits in 2015. It has become a point of pride in the sustainability agenda of the university, and is seen as a good investment. With a yield of root vegetables, squash, eggplant, strawberries, leafy greens and more, the garden is thriving. By rotating crops each season, the soil nutrients are replenished. In addition, a large composting area tended directly on the roof contributes to the maintenance of rich soil. The farm also purchases additional organic compost and soil from nearby farms to replenish about 5 cm of the topsoil lost each year due to wind erosion, harvesting and compaction.

The University Food Service uses much of the produce for the school cafeteria, as this helped fund the project. Another part of the produce is sold to the public through a weekly farmers market on campus during the summer and through a weekly CSA (community supported agriculture) subscription that can be picked up at the farmers market.

Social Impacts

As indicated above, the rooftop garden has an important hands-on-learning function for students and other volunteers. Workshops for the community are organised to spread knowledge, from planting practices to seed saving. Just last year, over 600 visitors to the roof saw what was possible on an extensive green roof. At this point, engineers on campus are just beginning to measure the impact on storm-water run-off. However, the significance of a green roof to the heating and cooling loads will be harder to measure, since the engineering building always had a green roof with fairly deep soil beds.

Economic Sustainability

Faculty advisors helped with initial negotiations for funding the roof conversion but, in this case as in the first ground-level garden, it was mostly the work of students, former students and the farm manager, who sought the funds for the transformation of the green roof. The university invested in the construction of the higher railings.

Ryerson's Food Services was also very supportive, and – for some time – paid the farm manager's salary. Each year, the student group and farm manager have to apply for new funding from Ryerson's Food Services and the University administration. Additional income comes from selling crops at a weekly summer farmers market, while the community workshops also benefit the garden through small donations.

Even with considerable volunteer labour, the farm cannot sell enough to make the farm economically self-sustaining but will always have to rely on the university or other fundraising to maintain the farm. It is seen as a model for sustainable practices and a participatory learning opportunity, not a for-profit operation.

Challenges Encountered and Lessons Learned

Important challenges came from appropriating an existing green roof for a productive rooftop farm. While an existing green roof meant that the rooftop was already prepared for deep-bed growing, some needs of a productive green roof – from a tool shed, high parapets for safety, and elevator access, to a shaded rest area for farmers – are not easily added to the structure. Convenient access to the roof and human safety were never part of the initial roof design. To provide regular access for farm volunteers and visitors, the university had to invest in higher railings, and the farm manager was trained for roof safety because the design was still unsafe for the average visitor. The result is that volunteers can access the roof only when the farm manager is present, and the farm manager has to lead all roof tours.

Another challenge is the lack of a greenhouse. The last winter frost in Toronto arrives in late May, so local farmers and gardeners extend the short growing season by starting seedlings indoors. While Ryerson Urban Farm has always had access to a greenhouse for season extension, they have lost this facility and now have a challenge to find a new greenhouse for future seasons.

Policy Relevance

Toronto's green-roof bylaws encourage – and, for large buildings, mandate – the implementation of green roofs to minimise rainwater runoff. Ryerson Urban Farm has gone far beyond the minimum mandate of reducing rainwater runoff, and the ultimate challenge is to document the value of productive green roofs beyond mere economic ones, to make this a reproducible model.

Many universities across North America are introducing urban farms at various scales, often initiated by students and volunteers, and frequently on rooftops or paved plazas with issues of storm-water runoff. Universities own large buildings with vast rooftops and have students eager to find hands-on learning experiences. This combination is potent. Sharing funding models, growing and management strategies can help to scale up these initiatives.

References

GIZ, ICLEI (2014) Amman, Jordan Urban agriculture: finding multi-purpose Urban NEXUS solutions through collaborative action. Urban NEXUS Case Story 2014 No. 26, Amman, Jordan. Accessible at: www.iclei.org/publicationshttp://congresmtl.com/en/convention-center/sustainable-development/

http://construirtv.com/un-shopping-con-una-huerta-en-el-techo/

http://energaia.com/

http://fedsquare.com/shopvisit/pop-up-patch

http://gothamgreens.com/

http://greencitygrowers.com/fenway-farms/

http://hidroponia.org.mx

http://littleveggiepatchco.com.au/pages/pop-up-patch

http://recovergreenroofs.com/?projects=fenway-farms-boston-ma

https://ryersonurbanfarm.wordpress.com/

http://www.cbsnews.com/news/gotham-greens-viraj-puri-greenhouse-farm-on-roof-of-pullman-factory-chicago/

http://www.countryfarm-lifestyles.com/terrace-gardening.html

http://www.ecologieurbaine.net/en/activities-and-projects/projects/urban-agriculture/item/89-culti-vert-greening-roof-palais-des-congres-montreal

https://www.facebook.com/ComCrop

http://www.fao.org/ag/agp/greenercities/pdf/GGCLAC/FAO-Gaza-aquaponics.pdf

http://www.fao.org/fileadmin/templates/FCIT/PDF/Fact_sheet_on_aquaponics_Final.pdf

http://www.greenroofs.com/projects/rothenberg_rooftop_garden/rothenberg_rooftop_garden9.jpg

http://www.otrfoundation.org/Docs/Background%20Information%20%20Rothenberg%20Rooftop%20Garden.pdf

http://www.rooftoprepublic.com/

http://www.rothenbergrooftopgarden.com/

http://www.shoppingeldorado.com.br/card/telhado-verde

http://www.straitstimes.com/singapore/education/spectra-secondary-schools-programme-bears-fruit-and-veggies

http://www.thehindu.com/todays-paper/tp-national/tp-karnataka/govt-sowing-the-seeds-for-terrace-gardening/article5328587.ece

http://www.urbanfarmers.nl/nl/onze-farms/uf002-de-schilde/

https://www.youtube.com/watch?v=kD3gBEa3cGI&feature=youtu.beer

https://youtu.be/9H_aFpNu-8o

Laban P, Osman S (2013) Feasibility assessment of urban agriculture in Cairo, Report commissioned by PDP-GIZ. RUAF Foundation, Leusden

Orsini F, Gasperi D, Marchetti L et al (2014) Exploring the production capacity of rooftop gardens (RTGs) in urban agriculture: the potential impact on food and nutrition security, biodiversity and other ecosystem services in the city of Bologna. Food Secur 6:781–792. doi:10.1007/s12571-014-0389-6

Pirog R, Van Pelt T, Enshayan K, Cook E (2012) Food, fuel, and freeways: an Iowa perspective on how far food travels, fuel usage, and greenhouse gas emissions, Leopold Center for Sustainable Agriculture. Iowa State University, Iowa

Sanyé-Mengual E, Llorach-Massana P, Sanjuan-Delmás D et al (2014) The ICTA-ICP Rooftop Greenhouse Lab (RTG-Lab): closing metabolic flows (energy, water, CO_2) through integrated rooftop greenhouses. In: Roggema R, Keefer G (eds) "Finding spaces product spaces", 6th AESOP sustainable food planning conference. VHL University of Applied Sciences, Velp, pp 692–701

Sanyé-Mengual E, Oliver-Solà J, Montero JI, Rieradevall J (2015a) An environmental and economic life cycle assessment of rooftop greenhouse (RTG) implementation in Barcelona, Spain: assessing new forms of urban agriculture from the greenhouse structure to the final product level. Int J Life Cycle Assess. doi:10.1007/s11367-014-0836-9

Sanyé-Mengual E, Orsini F, Oliver-Solà J et al (2015b) Techniques and crops for efficient rooftop gardens in Bologna, Italy. Agron Sustain Dev 35:1477–1488. doi:10.1007/s13593-015-0331-0

Tohmé Tawk S, Moussa Z, Marroush Abi Saiid D, Abi Saiid M, Hamadeh S (2011) Redefining a sustainable role for urban agriculture in the Middle East and North Africa. Watch letter of CIHEAM #18: Urban and Periurban Agriculture in the Mediterranean Area. Accessible at: www.ciheam.org

Part VI
Conclusions

G. Gianquinto, F. Orsini, and M. Dubbeling

Conclusions

Marielle Dubbeling, Francesco Orsini, and Giorgio Gianquinto

Abstract This book has provided a panorama of rooftop agriculture. Examples are given where rooftop agriculture is practised for community building and health, commercial production, ecology and landscape enhancement or knowledge production and sharing. Case studies, from a variety of contexts and cities, describe rooftop home gardens, community rooftop gardens, therapeutic rooftop gardens, rooftop gardens serving a restaurant, hotel or shop, research oriented and educational rooftop gardens, amongst others. Applied technologies include soil based or hydroponic forms of growing; open-air or greenhouse type of production; hobby or highly technified production systems.

Rooftop agriculture can complement other forms of urban agriculture because of its unique use of built-up space. Although the potential of rooftop agriculture to contribute to urban sustainability (including climate change adaptation) is recognised, its scale of implementation is still limited, both in terms of production area and intensity of individual rooftop gardens as well as in terms of total rooftop production and area at city level. There is a need to address legal and regulatory issues, technical and infrastructural requirements and adapt cultivation practices to specific growing conditions and safety and sustainability requirements. Innovative practices addressed in this book show pathways to further development of rooftop agriculture.

Benefits of Rooftop Agriculture

Rooftop agriculture is practised in various forms and contexts. It is an unique form of urban agriculture because of its use of built-up rooftop space, thus allowing optimising use of – otherwise limited – vacant spaces in cities. Low-rise residential

The original version of this chapter was revised.
An erratum to this chapter can be found at https://doi.org/10.1007/978-3-319-57720-3_21

M. Dubbeling
RUAF Foundation, Leusden, The Netherlands
e-mail: m.dubbeling@ruaf.org

F. Orsini • G. Gianquinto (✉)
Department of Agricultural Sciences, University of Bologna, Bologna, Italy
e-mail: f.orsini@unibo.it; giorgio.gianquinto@unibo.it

F. Orsini et al. (eds.), *Rooftop Urban Agriculture*, Urban Agriculture,
https://doi.org/10.1007/978-3-319-57720-3_20

buildings often have sloped roofs that make them difficult to use for rooftop agriculture. However, apartment buildings, offices, and industrial buildings often have flat rooftop space and sufficient structural capacity to allow for rooftop farming. Temporal use of rooftop space, may also be appropriate in many residential building in the Global South, where new floors are added depending on family extensions or increased income.

Rooftop agriculture is practised for a variety of – and often multiple – objectives, including social and health benefits, commercial food growing, educational or therapeutic purposes, landscaping, environmental management and climate change adaptation. Rooftop agriculture may be one strategy to increase biodiversity, create green spaces, reduce the urban heat island effect and storm water runoff from buildings. Rooftop agriculture can enhance resource recycling and efficiency by connecting water, waste and energy flows with(in) the building or the community. Rooftop agriculture may provide growing spaces in otherwise densely built-up areas, allowing for reduction of food deserts, contributing to community food security and participation, providing educational opportunities and offering new forms of employment. Regulating the practice, improving design and production, as well as coordination and agreement between building owners and inhabitants/users as well as rooftop agriculture practitioners will be key to optimise the benefits and address potential conflicts as well as user rights.

Building a Facilitating Legal and Regulatory Framework

Rooftop agriculture is set by building type and related aspects such as tenancy, rooftop access and building structure. It requires infrastructure support and assessment (of load bearing capacity, seismic resistance, safety aspects and specific growing conditions) as well as related adaptations in local planning, regulatory frameworks and building codes. In many cases, rooftop agriculture is not (yet) formally recognised as a potential use of rooftop space. Growing legislation for green rooftop use may provide opportunities for further agricultural use. The recent inclusion of urban agriculture in BREEAM or LEED certification systems also provides new incentives for rooftop agriculture.

Next to specific legislation and regulations, rooftop agriculture is also influenced by other municipal planning instruments and measures such as building restrictions (height or access), social green or climate resilience planning amongst others. Incentives or levies for storm water management may for example promote rooftop agriculture to the extent it can help reduce storm water runoff for buildings. Insulation policies may similarly be a policy measure to promote agricultural coverage of rooftops. Zoning regulations may however limit building heights or maximum floor areas. Safety guidelines may limit user access to the rooftop. Load restrictions may limit soil depth required for growing or use of specific infrastructure (like water harvesting barrels). Multi-stakeholder dialogue between city officials, rooftop owners and rooftop users can be an important strategy to identify opportunities and find solutions to restrictions or constraints.

Improving Rooftop Design and Production

The development of improved technologies for green roofs has also spurred development of rooftop agriculture. Design of rooftop farms will be determined by set aims (commercial food growing versus recreational use for example), characteristics of building and growing space (load bearing capacity, size, presence of other rooftop uses, location, slope, availability of water and energy on the rooftop), economic factors (potential for investment) and the existing legal and regulatory framework. Innovations include growing in shallow soil depth or using non-soil based growing systems, use of light-weight and porous substrates, use of hardy plant species (that can grow in circumstance of reduced water availability, high temperatures and wind) and use of specific production techniques like the use of floor-raised planters, simplified or high-yield hydroponic growing techniques or aquaponics. Rooftop access (for both users as well as for transport of inputs – soil, plants, water – and products), rooftop waterproofing, railing and protection from air conditioners or electricity units also need to be considered.

The integration of rooftop agriculture in new buildings will greatly enhance inclusion of design and infrastructure requirements from the planning stage. Structural adaptations can be made allowing for good soil depth and even the production of shrubs or trees. Greenhouses and water tanks can be similarly integrated into building design. Roof retrofitting or upgrading in existing buildings also offer new opportunities for rooftop agriculture. Thermal insulation, storm-water or social benefits, or rooftop lifetime extension by protecting the rooftop from direct solar radiation, may offset costs in a 5–10 year time period, depending on financial investments made. Further development of production technologies, such as for example water management (both irrigation and drainage), sustainable pest and nutrient management; cultivar selection will be needed to reduce costs and enhance social, economic and environmental sustainability.

Guaranteeing Product Safety and Sustainability

As for other forms of urban agriculture, ensuring product quality and safety is essential. Product safety is specifically relevant for animal-based production systems (specifically chickens, pigeons) with regards to zoonosis risks and animal waste management. Air pollution may impact open-air based growing systems, especially when these are located close to heavy traffic locations or downstream of industrial areas. Use of organic waste and wastewater as well as certain recyclable materials (like old car tires; plastic containers) may be sources of biological or chemical contamination. Legislation and regulation as well as production management are key to minimise and avoid potential health and environmental risks. Innovations developed for other forms of urban agriculture can also be applied to rooftop agriculture to a large extent.

The specific rooftop growing conditions (high variability in temperatures, potential strong winds; water scarcity or excess) however make specific demands on pest and disease management. Sustainable water management is another area that requires further research. Use of potable water, and of electricity for pumping, should be minimised. Use of applied production techniques, as well as training are key.

Building on Existing Experiences and Innovations

Both practitioners as well as cities are innovating and testing different types of rooftop agriculture. Toronto (Canada), Kathmandu (Nepal) and Copenhagen (Denmark) are some of the cities having advanced in revision or formulation of new bylaws and regulations favouring rooftop agriculture. Vienna (Austria) and Melbourne (Australia) are promoting rooftop agriculture as part of green infrastructure and corridors.

Restaurants, malls and hotels are installing rooftop farms for own food provisioning or for offering their clients a diversified food and shopping experience. Private companies, mainly in the USA, Asia and increasingly in Europe, support rooftop agriculture as part of their social or environmental sustainability concerns or for seeking specific market niches, often applying new and innovative technologies and business models. NGOs, community and international organisations have promoted forms of rooftop gardening benefiting lower-income populations. Examples are provided from places as diverse as Bengularu (India), Gaza (PoT) and Brazil. Research and experimental rooftop gardens are developed in Barcelona (Spain), Bangkok (Thailand), the USA and several other places. Rooftop gardens are also integrated in social housing, settlement upgrading and community projects such as in Bologna (Italy) and Cairo (Egypt). New production systems such as aquaponics have been set up in Chicago (USA) and The Hague (The Netherlands), while other production technologies are tested and training in their use is provided in for example Toluca (Mexico), Lima (Peru) or Hong Kong.

Although economic, legal, technological and design aspects of rooftop agriculture will need to be further developed and strengthened, and adapted to specific contexts, these – and many other – experiences documented provide a clear roadmap for future rooftop agriculture development.

Erratum to: Rooftop Urban Agriculture

Francesco Orsini, Marielle Dubbeling, Henk de Zeeuw, and Giorgio Gianquinto

Erratum to: F. Orsini et al. (eds.), *Rooftop Urban Agriculture*, Urban Agriculture, https://doi.org/10.1007/978-3-319-57720-3

This book was inadvertently published with incorrect City name for two Editors. This has now been corrected in FM and Chapters 1, 2, 19 and 20 of this book as follows:

Marielle Dubbeling
RUAF Foundation
Leusden, The Netherlands

Henk de Zeeuw
RUAF Foundation
Leusden, The Netherlands

The updated original online version for this chapter can be found at
https://doi.org/10.1007/978-3-319-57720-3_1
https://doi.org/10.1007/978-3-319-57720-3_2
https://doi.org/10.1007/978-3-319-57720-3_19
https://doi.org/10.1007/978-3-319-57720-3_20
https://doi.org/10.1007/978-3-319-57720-3

F. Orsini et al. (eds.), *Rooftop Urban Agriculture*, Urban Agriculture,
https://doi.org/10.1007/978-3-319-57720-3_21

Index

F. Orsini et al. (eds.), *Rooftop Urban Agriculture*, Urban Agriculture,
https://doi.org/10.1007/978-3-319-57720-3

MIX
Papier aus verantwortungsvollen Quellen
Paper from responsible sources
FSC® C105338

If you have any concerns about our products,
you can contact us on
ProductSafety@springernature.com

In case Publisher is established outside the EU,
the EU authorized representative is:
Springer Nature Customer Service Center GmbH
Europaplatz 3, 69115 Heidelberg, Germany

Printed by Libri Plureos GmbH
in Hamburg, Germany